Springer Proceedings in Complexity

Springer Proceedings in Complexity publishes proceedings from scholarly meetings on all topics relating to the interdisciplinary studies of complex systems science. Springer welcomes book ideas from authors.

The series is indexed in Scopus.

Proposals must include the following:

- name, place and date of the scientific meeting
- a link to the committees (local organization, international advisors etc.)
- scientific description of the meeting
- list of invited/plenary speakers
- an estimate of the planned proceedings book parameters (number of pages/articles, requested number of bulk copies, submission deadline)

Submit your proposals to: Hisako.Niko@springer.com

Martin Gilje Jaatun · Cyril Onwubiko ·
Pierangelo Rosati · Aunshul Rege · Hanan Hindy ·
Arnau Erola · Xavier Bellekens
Editors

Proceedings of the International Conference on Cybersecurity, Situational Awareness and Social Media

Cyber Science 2024; 27–28 June; Edinburgh Napier University, Scotland

Springer

Editors
Martin Gilje Jaatun
SINTEF Digital
Trondheim, Norway

Cyril Onwubiko
Research Series Limited
London, UK

Pierangelo Rosati
J. E. Cairnes School of Business and Economics
University of Galway
Galway, Ireland

Aunshul Rege
Department of Criminal Justice
Temple University
Philadelphia, PA, USA

Hanan Hindy
Computer Science Department
Ain Shams University
Cairo, Egypt

Arnau Erola
Department of Computer Science
University of Oxford
Oxford, UK

Xavier Bellekens
Lupovis
Glasgow, UK

ISSN 2213-8684 ISSN 2213-8692 (electronic)
Springer Proceedings in Complexity
ISBN 978-981-96-0403-6 ISBN 978-981-96-0401-2 (eBook)
https://doi.org/10.1007/978-981-96-0401-2

This Springer imprint is published by the registered company Springer Nature Singapore Pte Ltd.
The registered company address is: 152 Beach Road, #21-01/04 Gateway East, Singapore 189721, Singapore

Cyber Science 2024 Committee

Committee Chairs

Arnau Erola, Department of Computer Science, University of Oxford, Oxford, UK
Aunshul Rege, Temple University, Pennsylvania, USA
Cyril Onwubiko, Centre for Multidisciplinary Research, Innovation and Collaboration, UK
Hanan Yousry Hindy, Computer Science Department, Faculty of computer and Information Sciences, Ain Shams University, Cairo, Egypt
Martin Gilje Jaatun, University of Stavanger, Norway
Pierangelo Rosati, University of Galway, Galway, Ireland
Xavier Bellekens, Lupovis, Glasgow, Scotland

Publicity Chair

Sean McKeown, Edinburgh Napier University, Scotland, UK

Publication Chair

Enoch Agyepong, Airbus Defence and Space, Newport, Wales, UK

Organising Committee

Bill Buchanan, Edinburgh Napier University, Scotland, UK
Petra Leimich, Edinburgh Napier University, Scotland, UK

Sean McKeown, Edinburgh Napier University, Scotland, UK
Sana Ullah Jan, Edinburgh Napier University, Scotland, UK
Belinda Onyeashie, Edinburgh Napier University, Scotland, UK
Matthew Thomson, Edinburgh Napier University, Scotland, UK

Programme Committee

Maria Bada, Queen Mary University London
David Brosset, Naval Academy Research Institute
F. Özgur Catak, University of Stavanger
Shamal Faily, Defence Science and Technology Laboratory
Hanan Hindy, Ain Shams University
Filip Holik, University of Glasgow
Martin Gilje Jaatun, University of Stavanger
Nektaria Kaloudi, SINTEF
Phil Legg, University of the West of England
Andrew Lenaghan, University of Oxford
Per Håkon Meland, SINTEF ICT
Cyril Onwubiko, Intelligence and Security Assurance, E-Security
Eckhard Pfluegel, School of CSM, Faculty of SEC, Kingston University
Aunshul Rege, Temple University, Department of Criminal Justice
Pierangelo Rosati, Irish Centre for Cloud Computing and Commerce
Lynsay Shepherd, Abertay University
Enoch Agyepong, Airbus, UK
Florian Skopik, AIT Austrian Institute of Technology
Eliana Stavrou, Open University of Cyprus
Zisis Tsiatsikas, University of the Aegean
Steffen Wendzel, Worms University of Applied Sciences
Tim Williams, BCS

Sponsors and Partners

UNIVERSITY OF OXFORD

DEPARTMENT OF COMPUTER SCIENCE

CYBER SECURITY OXFORD

Preface

Cyber Science is the flagship conference of the Centre for Multidisciplinary Research, Innovation and Collaboration (C-MRiC), a multidisciplinary conference focusing on pioneering research and innovation in Cyber Situational Awareness, Social Media, Cyber Security and Cyber Incident Response. Cyber Science aims to encourage participation and promotion of collaborative scientific, industrial and academic inter-workings among individual researchers, practitioners, members of existing associations, academia, standardisation bodies, and government departments and agencies. The purpose is to build bridges between academia and industry, and to encourage interplay of different cultures. Cyber Science invites researchers and industry practitioners to submit papers that encompass principles, analysis, design, methods and applications. It is an annual conference with the aim that it will be held in the future at various cities in different countries.

Cyber Science as a multidisciplinary event is maturing as a mainstream and notable conference, first, for its quality, second, for its uniqueness, and finally, for its structure, contribution, and originality; something existing mainstream conferences do not normally possess. A testament to the significant interest Cyber Science has thus so far gained.

In 2015, the first Cyber Science conference was held on June 8–9, 2015, at the Hotel Russell in Central London, UK. The conference was opened by the IEEE UK and Ireland Computer Society Chair, Professor Frank Wang, and featured 4 keynote speakers from academia, government and industry.

The next editions of the conference were held as follows:

- June 13–14, 2016, London, UK
- June 19–20, 2017, London, UK
- June 11–12, 2018, Glasgow, Scotland
- June 3–4, 2019, Oxford UK
- June 15–19, 2020, Dublin, Ireland (virtual)
- June 14–18, 2021, Dublin, Ireland (virtual)
- June 20–21, 2022, Cardiff, Wales

In 2023, the ninth episode of the Cyber Science conference, in partnership with Aalborg University Copenhagen was held at Aalborg University Copenhagen as a hybrid event (in-person & online) between July 3–4, 2023.

In 2024, the tenth episode of the Cyber Science conference, in partnership with Edinburgh Napier University was held as an in-person event between June 27–28, 2024.

This Cyber Science 2024 conference proceedings is a compilation of peer-reviewed and accepted papers submitted to the conference. The conference invited three keynote speakers who spoke at the event, namely:

- Dr. Lynsay Shepherd
- Dr. R. V. Gundur
- Dr. Uchenna Daniel Ani

Further, there was a panel discussion led by Prof. Rajendra K. Raj, Dr. Sara Vecchini and Dr. Cyril Onwubiko.

I would like to thank the organisers – Edinburgh Napier University, and special thanks to their academics, who helped with organising the event and performing several tasks including moderating and chairing sessions. Special thanks to Dr. Sean McKeown.

I am exceedingly grateful to the Programme Committee Members and the other Conference Reviewers for graciously contributing their time, assuring a scholarly fair and rigorous review process. I would also like to thank Springer for accepting to publish Cyber Science 2024 in their Proceedings on Complexity book series.

I express my profound gratitude to Dr. Hanan Hindy, Prof. Aunshul Rege, Dr. Pierangelo Rosati, Dr. Arnau Erola, and Dr. Xavier Bellekens for helping in many ways with the conference. Particular thanks to Dr. Cyril Onwubiko for his relentless efforts on maintaining the strategic vision of the conference from the beginning.

Finally, I would like to thank the authors of papers and the delegates present at the event; there would be no conference without you!

Trondheim, Norway

Martin Gilje Jaatun
Siv. Ing, Dr.Philos, Cyber Science 2024
Conference Chair

Keynote and Panel Speakers

Dr. Lynsay Shepherd, Reader at Abertay University, Scotland, UK

Dr. Lynsay Shepherd is a Reader in Cybersecurity and Human-Computer Interaction at Abertay University and works within the School of Design and Informatics. Lynsay holds a Ph.D. in Usable Security (a combination of HCI and security research), an M.Sc. in Internet Computing, and a B.Sc. (Hons) in Computing.

Lynsay focuses on applied research in the human aspects of cybersecurity, investigating how people interact with emerging technologies, and their behaviour in the context of secure interactions.

Lynsay is particularly interested in the use of eye-tracking measures, serious games, embodied agents, extended reality (XR), safety and accessibility in the metaverse, and romance fraud detection and prevention techniques using machine learning.

Dr. R. V. Gundur, Senior Lecturer at the University of the West of Scotland, Scotland, UK

Dr. R. V. Gundur is a Senior Lecturer in Criminology at Flinders University in Australia. He holds a Ph.D. in Criminology from Cardiff University, where he was an ESRC scholar; an M.Sc. in Criminology Research Methods from the University of Oxford; an MA in International Relations from The Australian National University, where he was a Hedley Bull Scholar; and a BA in Spanish and Latin American Studies from Tulane University.

His areas of expertise include illicit enterprise, gangs, and cybercrime. His work has appeared in reports published by the European Union and the City of London Corporation, and he has been published in several academic journals including Urban Affairs Review; Deviant Behavior; Crime, Law and Social Change; Trends in Organized Crime; Global Crime, and International Criminal Justice Review. He is the author of Trying to Make It: The Enterprises, Gangs, and People of the American Drug Trade, on Cornell University Press.

Dr. Uchenna Daniel Ani, Lecturer at Keele University, UK

Dr. Ani obtained his first degree in Computer Science and his Masters Degree in Computer Security and Forensics from the University of Bedfordshire, UK, and took up the role of Lecturer in Computer Science at Federal University Lokoja, Nigeria. Dr. Ani later went on to complete a Ph.D. in Industrial Control System Cybersecurity at Cranfield University, UK.

Prior to joining Keele University, Dr. Ani was a Senior Research Fellow in Cybersecurity at the PETRAS National Centre of Excellence for IoT Systems Cybersecurity at the Department of Science Technology Engineering and Public Policy (STEaPP), University College London (UCL). He was involved in leading and guiding the direction of some of the IoT-enabled critical infrastructure security projects in the PETRAS Centre Consortium, as well as supporting teaching in Cybersecurity Risk Assessment and Governance Module for Masters Students. His work is deeply ingrained in multidisciplinary research –exploring and understanding how technical cybersecurity solutions interact with social elements including human, standards, and policy attributes to create safe, secure, practical, and future-proof cyber security and resilient outcomes.

Prof. Rajendra K. Raj, Professor of Computer Science at Rochester Institute of Technology, USA

Professor Raj earned a B.Tech. in Electrical Engineering, IIT Madras, India; MS in Computer Science, University of Tennessee, Knoxville; Ph.D. in Computer Science, University of Washington, Seattle. He co-created the Computer Science 2023 curriculum, a joint task with IEEE CS, ACM and AAAI. He was also a member of the committee for Computer Science 2013 curriculum development efforts. He's an award winner of various foundations including the National Science Foundation to develop a hands-on data science course for non-computing majors.

You can reach him at https://www.rit.edu/directory/rkrics-rajendra-raj

Dr. Sara Vecchini, Security Operations and Data Science Leader for 2T Security, Oakham, UK

Dr. Sara Vecchini is a highly accomplished security operations and data science expert with a strong background in cyber and fraud data analytics. She holds a Ph.D. in Physics from the University of Modena, Italy. She combines her rigorous scientific training with cutting-edge data science techniques and is interested in the threats and benefits of emerging technologies, LLMs and AI.

Dr. Vecchini has held significant roles in organisations such as the Department for Work and Pensions (DWP) and Pearson. She is the Security Operations and Data Science Leader for 2T Security and has been instrumental in developing a specialised SOC for CNI. She is a passionate advocate for women in technology, committed to inspiring future generations of female tech leaders.

Dr. Cyril Onwubiko, CISO, Research Series Limited, London, UK

Dr. Cyril Onwubiko is the Chief Information Security Officer (CISO), Research Series Limited, London, UK. Also, he is the Vice President, Professional and Educational Activities, a Board of Governor (BoG) and a Distinguished Speaker (DVP) of the IEEE Computer Society. And a Member of the European Public Policy Committee for ICT and Artificial Intelligence. He was a Senior Director, Enterprise Security Architecture at Pearson, where he was responsible for directing enterprise security architecture. Prior to Pearson, he had worked in the Financial, Telecommunication, Government & Public Services Sectors.

Cyril is a leading scholar in Cyber Situational Awareness (Cyber SA), Cyber Security, and Cyber Science. He is knowledgeable in Information Assurance, HMG Security Policy Framework and Risk Assessment & Management, Cyber-Threat Analysis, Graph Theory, Intrusion Detection Systems and Data Fusion. And interested in the mathematical analysis of security and the application of mathematics in solving real-life problems. His current focus includes Artificial Intelligence, Machine Learning for Cyber Security, and Blockchain Technologies. Further details about him can be found on https://c-mric.org/cyril.

Contents

Artificial Intelligence Applications

RingGuard: A Privacy Protected Peer-to-Peer Federated Learning Framework ... 3
Narasimhan Kovalai, Ramsundar Tanikella, Rojalini Tripathy, and Padmalochan Bera

Improving Credit Card Fraud Detection with Combined Feature Extraction and Class Balance Optimisation Techniques ... 27
Boluwatife Ajibade-Ajibosin, Uchenna Daniel Ani, Theocharis Kyriacou, and Mark Turner

Attention Based Image Steganography Using CNNs in Integration with Quantization ... 45
Gatram Sravan Kumar, Kamalakanta Sethi, Piyush Joshi, and Rakesh Kumar Sanodiya

Advanced Cryptography in Emerging Applications

vSPACE: Voting in a Scalable, Privacy-Aware and Confidential Election ... 61
Se Elnour, William J. Buchanan, Paul Keating, Mwrwan Abubakar, and Sirag Elnour

Design of Pairing Free Attribute-Based Encryption for Smart Grid Applications ... 77
Anjali Kumari, Sreevallabh Karanam, Swejan Annabathini, and Kamalakanta Sethi

Critical Infrastructures in the Cloud ... 97
Martin Gilje Jaatun and Geir Kjetil Hanssen

Security Operations Centre (SOC) and Advanced Malware Detection Techniques

An Exemplar Incident Response Plan for Security Operations Centre Analysts 109
Enoch Agyepong and Cyril Onwubiko

Malware Detection Using Machine Learning Algorithms 129
Nureni Ayofe Azeez, Ogechukwu Juliet Nzeribe, Charles Van der Vyver, and Ademola Philip Abidoye

A Case Study of a Ransomware and Social Engineering Competition ... 149
Aunshul Rege and Rachel Bleiman

Exploring DTrace as an Incident Response Tool for Unix Systems 169
Joe Duin, Sean McKeown, and Mwrwan Abubakar

Utilizing YARA: An Effective Method for Phishing Attack Response ... 195
Ferenc Leitold

Threat Intelligence and Emerging Techniques for Fraud Detection and Prevention

TIPS: Threat Sharing Information Platform for Enhanced Security 209
Lakshmi Rama Kiran Pasumarthy, Hisham Ali, William J. Buchanan, Jawad Ahmad, Audun Josang, Vasileios Mavroeidis, and Mouad Lemoudden

TRIST: Towards a Container-Based ICS Testbed for Cyber Threat Simulation and Anomaly Detection 225
Carol Lo, Jack Christie, Thu Yein Win, Zeinab Rezaeifar, Zaheer Khan, and Phil Legg

Radio Frequency (RF) Cyber Threats and a Novel RF Cyber Kill Chain 241
Carolyn J. Swinney and John C. Woods

Cybersecurity Governance, Risk Management and Assurance

Perception of Cyber Risk for Assisted Living Technologies in the Norwegian Healthcare Sector—Key Stakeholders' Perceptions 265
Alvhild Skjelvik and Bian Yang

Workshop Insights: Navigating Cybersecurity Regulations for Device Manufacturers and Healthcare Operators 287
Andrea Skytterholm, Lars Halvdan Flå, and Martin Gilje Jaatun

Digital Evidence: A Legal Practitioner's Perspective 303
Ibeabuchi Egbu, Jacques Ophoff, and Annelize McKay

Critical Infrastructures, Systems and Applications

Fuzzing the ARM Cortex-M: A Survey 321
Silje Marie Sørlien, Åse Marie Solnør, Karin Bernsmed,
and Martin Gilje Jaatun

Towards a Secure Manufacturing Framework for Single Malt Whisky in Industry 4.0 349
Jacob Connell, Dimitrios Kasimatis, Pavlos Papadopoulos,
William J. Buchanan, Panagiotis Sarigiannidis, and Nikolaos Pitropakis

Implementation of Zones and Conduits in Industrial Control and Automation Systems 367
Lars Halvdan Flå, Mary Ann Lundteigen, Fredrik Gratte,
and Martin Gilje Jaatun

Author Index 383

Contributors

Ademola Philip Abidoye Kennesaw State University, Atlanta, USA

Mwrwan Abubakar Blockpass ID Lab, Edinburgh Napier University, Edinburgh, UK

Enoch Agyepong Airbus UK, Newport, UK

Jawad Ahmad Blockpass ID Lab, Edinburgh Napier University, Edinburgh, UK

Boluwatife Ajibade-Ajibosin School of Computer Science and Mathematics, Keele University, Keele, UK

Hisham Ali Blockpass ID Lab, Edinburgh Napier University, Edinburgh, UK

Uchenna Daniel Ani School of Computer Science and Mathematics, Keele University, Keele, UK

Swejan Annabathini Indian Institute of Information Technology, Sricity, India

Nureni Ayofe Azeez Department of Computer Sciences, Faculty of Science, University of Lagos, Lagos, Nigeria

Padmalochan Bera School of Electrical Science, Indian Institute of Technology Bhubaneswar, Bhubaneswar, Odisha, India

Karin Bernsmed IIK, NTNU, Trondheim, Norway;
SINTEF Digital, Trondheim, Norway

Rachel Bleiman Temple University, Philadelphia, PA, USA

William J. Buchanan Blockpass ID Lab, Edinburgh Napier University, Edinburgh, UK

Jack Christie University of the West of England, Bristol, UK

Jacob Connell Edinburgh Napier University, Edinburgh, UK

Joe Duin Edinburgh Napier University, Edinburgh, Scotland

Ibeabuchi Egbu Abertay University, Dundee, UK

Se Elnour Blockpass ID Lab, Edinburgh Napier University, Edinburgh, UK

Sirag Elnour TheSDEs, Austin, TX, USA

Lars Halvdan Flå SINTEF Digital, Trondheim, Norway

Fredrik Gratte Atina, Athens, Greece

Geir Kjetil Hanssen SINTEF Digital, Trondheim, Norway

Martin Gilje Jaatun SINTEF Digital, Trondheim, Norway

Audun Josang University of Oslo, Oslo, Norway

Piyush Joshi Indian Institute of Information Technology, Sricity, India

Sreevallabh Karanam Indian Institute of Information Technology, Sricity, India

Dimitrios Kasimatis Edinburgh Napier University, Edinburgh, UK

Paul Keating Democracy Counts, Warrington, Cheshire, UK

Zaheer Khan University of the West of England, Bristol, UK

Narasimhan Kovalai School of Electrical Science, Indian Institute of Technology Bhubaneswar, Bhubaneswar, Odisha, India

Anjali Kumari Indian Institute of Information Technology, Sricity, India

Theocharis Kyriacou School of Computer Science and Mathematics, Keele University, Keele, UK

Phil Legg University of the West of England, Bristol, UK

Ferenc Leitold Óbuda University, Budapest, Hungary

Mouad Lemoudden Blockpass ID Lab, Edinburgh Napier University, Edinburgh, UK

Carol Lo University of the West of England, Bristol, UK

Mary Ann Lundteigen Department of Engineering Cybernetics, Norwegian University of Science and Technology, Trondheim, Norway

Vasileios Mavroeidis University of Oslo, Oslo, Norway

Annelize McKay Abertay University, Dundee, UK;
University of Pretoria, Pretoria, South Africa

Sean McKeown Edinburgh Napier University, Edinburgh, Scotland

Ogechukwu Juliet Nzeribe Department of Computer Sciences, Faculty of Science, University of Lagos, Lagos, Nigeria

Cyril Onwubiko Cyber Security Intelligence, E-Security Group, Research Series, London, UK

Jacques Ophoff Abertay University, Dundee, UK;
University of Cape Town, Cape Town, South Africa

Pavlos Papadopoulos Edinburgh Napier University, Edinburgh, UK

Lakshmi Rama Kiran Pasumarthy Blockpass ID Lab, Edinburgh Napier University, Edinburgh, UK

Nikolaos Pitropakis Edinburgh Napier University, Edinburgh, UK

Aunshul Rege Temple University, Philadelphia, PA, USA

Zeinab Rezaeifar University of the West of England, Bristol, UK

Rakesh Kumar Sanodiya Indian Institute of Information Technology, Sricity, India

Panagiotis Sarigiannidis University of Western Macedonia, Kozani, Greece

Kamalakanta Sethi Indian Institute of Information Technology, Sricity, India

Alvhild Skjelvik Norwegian University of Science and Technology, Gjøvik, Norway

Andrea Skytterholm SINTEF Digital, Trondheim, Norway

Åse Marie Solnør IIK, NTNU, Trondheim, Norway

Silje Marie Sørlien IIK, NTNU, Trondheim, Norway

Gatram Sravan Kumar Indian Institute of Information Technology, Sricity, India

Carolyn J. Swinney Computer Science and Electronic Engineering, University of Essex, Colchester, UK

Ramsundar Tanikella School of Electrical Science, Indian Institute of Technology Bhubaneswar, Bhubaneswar, Odisha, India

Rojalini Tripathy School of Electrical Science, Indian Institute of Technology Bhubaneswar, Bhubaneswar, Odisha, India

Mark Turner School of Computer Science and Mathematics, Keele University, Keele, UK

Charles Van der Vyver School of Computer Science and Information Systems, Faculty of Natural and Agricultural Sciences, Vaal Triangle Campus, North-West University, Vanderbijlpark, South Africa

Thu Yein Win University of Sunderland, Sunderland, UK

John C. Woods Computer Science and Electronic Engineering, University of Essex, Colchester, UK

Bian Yang Norwegian University of Science and Technology, Gjøvik, Norway

Artificial Intelligence Applications

RingGuard: A Privacy Protected Peer-to-Peer Federated Learning Framework

Narasimhan Kovalai, Ramsundar Tanikella, Rojalini Tripathy, and Padmalochan Bera

Abstract Collaborative learning-based training has become popular for developing accurate ML models. However, data owners face constraints regarding privacy and regulatory compliance, limiting their ability to share data. To address this, Federated Learning (FL) has emerged as a solution, enabling collaborative model training without the need to share raw data. Instead, only model parameters are exchanged among participating parties. However, FL faces many challenges, particularly its reliance on a central server, managing data heterogeneity and security challenges of parameter sharing. In this paper, we propose a peer-to-peer federated learning framework that leverages secure multiparty computation (SMC) to overcome these challenges. Our framework eliminates the dependency on a central server, ensuring enhanced security in parameter sharing. We implement our proposed framework for both Independent and Identically Distributed (IID) and non Independent and Identically Distributed (non-IID) data distributions, using FedProx for non-IID data and FedAvg for IID data. We evaluated the performance of our framework using MNIST and FEMNIST datasets, incorporating varying numbers of clients, and conducted a comprehensive analysis. Experimental results demonstrate that our framework performs better in terms of test accuracy and smooth learning than existing centralized FL frameworks.

N. Kovalai (✉) · R. Tanikella · R. Tripathy · P. Bera
School of Electrical Science, Indian Institute of Technology Bhubaneswar, Bhubaneswar 752050, Odisha, India
e-mail: 20CS01075@iitbbs.ac.in

R. Tanikella
e-mail: 20CS01071@iitbbs.ac.in

R. Tripathy
e-mail: a22ee09003@iitbbs.ac.in

P. Bera
e-mail: plb@iitbbs.ac.in

M. G. Jaatun et al. (eds.), *Proceedings of the International Conference on Cybersecurity, Situational Awareness and Social Media*, Springer Proceedings in Complexity,
https://doi.org/10.1007/978-981-96-0401-2_1

1 Introduction

Machine Learning (ML) uses data from various sources to design accurate models through collaborative and distributed training. The practical challenges associated with the distributed machine learning approach include communication overhead, high training time, and violations of security and user data privacy. User data privacy and security have emerged as critical focal points within contemporary research, a trend underscored by legislative measures such as the General Data Protection Regulation (GDPR) in the European Union [1] and the Health Insurance Portability and Accountability Act (HIPAA) [2]. To overcome this, McMahan et al. proposed a novel distributed machine learning technique named Federated Learning (FL) [3]. The fundamental idea of federated learning revolves around training a unified statistical model using data dispersed across numerous remote devices, ranging from a few to potentially millions. The overarching objective is to develop this model while adhering to the principle that data generated by individual devices remains localized, with only periodic transmission of incremental updates to a central server. The objective of FL is to develop an unbiased model distributively, without disclosing the data to participating clients. FL operates on a client-server architecture, where clients represent data owners, such as organizations, IoT devices, individual business owners, etc. Initially, the server distributes initial model parameters to all participating clients simultaneously. After receiving these parameters, each client trains its model locally with its dataset. Subsequently, each client shares its updated local model parameters with the server. The server then aggregates these parameters and redistributes the updated parameters back to the clients. This iterative process continues until convergence or for a specified number of rounds. Figure 1 illustrates the workflow of FL.

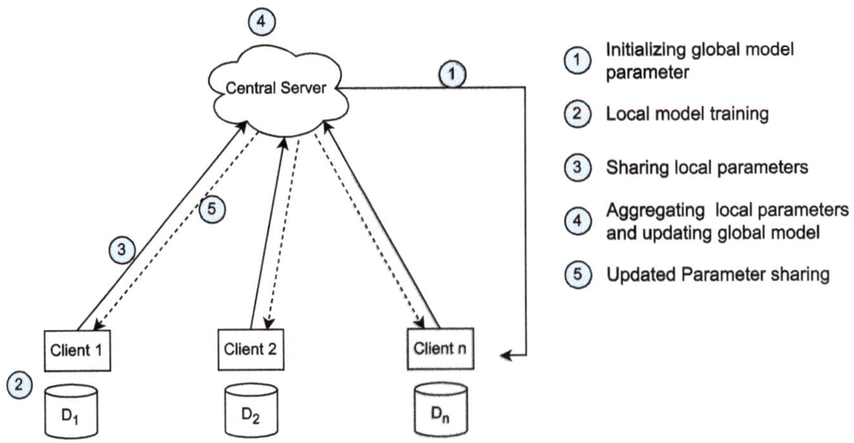

Fig. 1 A generic workflow of federated learning

However, FL faces some major challenges. In FL, clients are distributed across various environments owned by different data owners, resulting in non-identical datasets in real life scenarios. While individual clients aim to optimize their local models, this optimization may not align with the aggregation of the global model [4]. Managing data heterogeneity poses a significant challenge in FL. Additionally, recent research has shown that sharing model parameters in FL is not fully secure [5]. The central server in FL plays a crucial role as it aggregates all the local updates. To maintain model security, the central server needs to be a trusted party. However, in some studies, it is observed that attackers target the server through methods such as Generative Adversarial Model Inversion (GAMIN) [6], membership inference attacks [7], and reverse engineering. These emerging attacks pose a threat as they enable the service provider to leak information about the actual training data of individual clients. To address the shortcomings of FL, in this paper, we propose a fully decentralized FL framework by eliminating the dependency on a central server. We have used Secure Multiparty Computation (SMC) for sharing model parameters between clients. Our framework uses a peer-to-peer architecture, ensuring complete security and effectively managing data heterogeneity. The proposed framework has been tested for both non-IID and IID distribution scenarios. For non-IID data distribution, the Fedprox [8] algorithm is used, while the Fedavg algorithm is used for IID data. We summarize the key contributions of our research as follows:

Peer-to-Peer Architecture: Our framework operates on a fully decentralized system, eliminating the need for a trusted third-party server to aggregate model parameters. In our framework, clients are distributed over a circular network, and we utilize the secure sum protocol from SMC to aggregate trained local model parameters. We use a ring architecture in particular, but since each node can be a client or server in any communication round, we consider our architecture as a special case of a P2P network.

Handling Unbalanced Data: Addressing the significant variance in data quantity among clients, we employ batch-wise model training and leverage the FedProx algorithm.

Model Security: Our framework ensures model security by utilizing secure sum, which divides parameters into secret shares, ensuring that individual shares do not reveal any information about the original secret. The model and intermediate model parameters are fully secure.

Accuracy: We conducted experiments on MNIST and FEMNIST datasets, considering both IID and non-IID distributions, varying the number of rounds and clients. Our findings suggest that our model exhibits promising performance when compared to existing state-of-the-art research in terms of accuracy.

Organisation: The rest of the paper is structured as follows: Sect. 2 reviews related work in the field of FL. Section 3 introduces the foundational components. Section 4 presents our proposed peer-to-peer FL framework. Section 5 conducts a comprehensive analysis and comparison with existing models through experiments. Section 6 outlines future directions. Finally, Sect. 7 concludes the paper. Section 8 contains the appendix to refer to some terms commonly used in tables and text throughout our paper.

2 Literature Review

Federated Learning provides a promising approach to distributed model training. Recently, researchers are improving FL in various aspects: enhancing the privacy of parameter sharing [9–12], addressing data heterogeneity [8, 13, 14], and to make it peer-to-peer instead of client server architecture [15–17].

Truex et al. [9] introduced a combined SMC and differential privacy (DP) method for secure parameter transfer. Their approach leads to enhanced model accuracy and stronger resistance against inference attacks, particularly compared to models using DP. However, the complexity of this hybrid model increases due to high computational costs and an increased number of iterations. Li et al. [10] presented a variant of FL called chained secure multiparty computation, integrating single masking technique and chained communication among clients. In this method, the server shares a mask to a chosen client, initiating the learning process. This reduces dependency on the central server, which lowers computational costs. However, in this framework, all clients are interconnected in a chained network, and the final updates are retrieved from the last client. Consequently, if an attacker targets the last client, the entire secret could be retrieved. But the randomness of choosing a client for aggregation reduces the chances of the attacker guessing the current aggregator. Bonawitz et al. [11] proposed a framework focused on preserving privacy by securely aggregating local model parameters. However, their approach employs homomorphic encryption to execute algebraic operations on encrypted data without decryption, leading to high computational cost and increased communication rounds throughout the learning process. Xu et al. [12] introduced HybridAlpha, a FL approach designed to preserve privacy, utilizing both a DP and functional encryption scheme derived from SMC. Their framework shows faster convergence compared to other existing approaches using security techniques. However, it requires a third-party involvement for key setup prior to training, making the process computationally complex.

In [13], Pang et al. proposed an approach for FL aggregation to address data heterogeneity. The method employs a self-organizing process to determine the weights of participating clients, utilizing Reinforcement Learning (RL). This technique transforms the FL central server into an RL agent, actively gathering feedback from clients to optimize weight allocation. In [14], Zhao et al. showed that the decrease in accuracy observed in FL can be attributed to differences in model weights. They proposed a method to enhance the training of non-IID data by utilizing Earth Mover's Distance (EMD) to share a portion of the global data among clients. While this strategy leads to improved accuracy, obtaining such a dataset is not feasible, and it introduces additional complexity to the model. In [8], Li et al. proposed a framework to handle statistical heterogeneity while also safeguarding privacy. In [15], Wing et al. proposed a peer-to-peer FL framework using n-out-of-n secret sharing scheme from SMC, providing security for parameter sharing. The n-out-of-n secret sharing scheme is a naive approach in SMC, and there is potential for further enhancement. To address these concerns, we propose a FL framework based on a peer-to-peer architecture, utilizing the k-secure sum technique from SMC. Our framework ensures security in the parameter sharing process and enables smoother learning curves. Federated learning involves training local clients with their local datasets and using an aggregate mechanism to update the global model. This however suffers from communication bottleneck of client server parameters updates and a single point of failure as highlighted in [18]. [18, 19] aimed to increase network utilization performance for parameter updates to get around the bottleneck issues. In [20], the authors proposed FedAvg: an aggregation mechanism to average client parameters to update the global parameters to reduce parameter updates but still the network suffers from congestion at the server side. Recent research on Decentralized (peer to peer) federated learning shows a lot of promise in reducing communication overhead and costs over centralized federated learning. In [21], authors propose a model segment level gossip approach for full node to node bandwidth utilization and better convergence but communication overhead costs still remain. In [22], the authors compare gossip learning as an alternative to centralized federated learning offering the advantages of no single point of failure and having a peer to peer architecture to significantly reduce communication overheads overall, though data privacy and security concerns are not addressed. Authors in [23] propose another peer to peer approach to tackle data heterogeneity problem of clients during personalized training. The authors propose a dense model for all clients to start the training with and progressively prune the model during training to create sparse networks. Pruning depends on the gradient information across the loss during training. This reduces the communication costs of parameter transmission as the training progresses.

All these approaches still suffer from data privacy and security leakage. In this paper we propose another gossip based protocol for parameter transmission with added security (using SMC based technique of k-secure sum and its optimization for parameter aggregation and using FHE and further security). One of the similar gossip mechanisms to our frameworks can be found in [24].

3 Background

3.1 FedAvg Algorithm

Federated learning has emerged as a revolutionary approach to train machine learning models across decentralized devices while preserving data privacy. Among the various federated learning algorithms, Federated Averaging (FedAvg) stands as a prominent method due to its simplicity and effectiveness. Federated learning addresses the challenges associated with centralized model training by enabling collaborative learning across distributed devices. FedAvg, proposed by McMahan et al. [25], represents a foundational algorithm in the federated learning paradigm. It takes into account the collective intelligence of decentralized devices while mitigating privacy concerns, finding applications in healthcare, finance, and edge computing.

At its core, FedAvg operates on the principle of aggregating model updates from decentralized devices to train a global model stored at the server. The algorithm iteratively alternates between local model training parallely on individual devices and global model aggregation on a central server. Each device computes a model update using its local data, and these updates are averaged to update the global model at the server. The weights are the proportion of data points present in clients w.r.t to total data points across clients. Not all the clients, though, participate in each round for training, due to connectivity and power constraints [25]. The FedAvg algorithm is depicted in Algorithm 1. Note that K represents the client set, R the total communication rounds and D_k represents the training set of the kth client.

Algorithm 1 FedAvg Algorithm (proposed by [25])

Require: $K, w_0, r = 0, R$
Ensure: Final Model Parameter w
while $r \leq R$ **do**
    Aggregator A selects a subset of clients k_r;
    Aggregator A broadcasts W_0 to k_i, where $0 < i < K_r$;
    $k_i \in K$ in parallel trains local model and updates $w_r^{k_i}$;
    k_i sends $w_r^{k_i}$ to A;
    A aggregates $w_r^{k_i}$ and updates w_r;
    $w_{r+1} \leftarrow \frac{\sum_{i=1}^{K_r} |D_k| w_r^{k_i}}{\sum_{i=1}^{K_r} |D_k|}$;
    $r \leftarrow r + 1$;
end while

3.2 *FedProx Algorithm*

The FedProx algorithm, given by Li et al. [26] is a generalised version of the FedAvg algorithm proposed by McMahan et al. [25]. It incorporates a proximal term to the client's local optimisation problem in addition to the Mean Square Error (MSE) loss term (denoted by $F_k(w)$ in Algorithm 2), which forces it to update its weights in such a way that the squared difference of local weights and global weights is also minimised. This proximal term effectively limit the impact of variable local updates. It eliminates the need to set each client's epochs according to data heterogenity and allows for variable amounts of local work [26]. Algorithm 2 gives a formal procedure of the FedProx algorithm. In our proposed ring P2P federated framework, the global model weights is the weight matrix of the model which is present at the starter node after a complete circular communication round.

Algorithm 2 FedProx (proposed by Li et al. [26])

Input: $K, T, \mu, \gamma, w_0, N, p_k, k = 1, \ldots, N$
for $t = 0, \ldots, T-1$ **do**
    Server selects a subset S_t of K devices at random (each device k is chosen with probability p_k)
    Server sends w_t to all chosen devices
    for each chosen device $k \in S_t$ **do**
        $w_{t+1}^k \leftarrow \arg\min_w h_k(w; w_t) = F_k(w) + \frac{\mu}{2}\|w - w_t\|^2$
        where w_{t+1}^k is a γ_t-inexact minimizer
    end for
    for each device $k \in S_t$ **do**
        Device k sends w_{t+1}^k back to the server
    end for
    Server aggregates the w's as $w_{t+1} = \frac{1}{|S_t|}\sum_{k \in S_t} w_{t+1}^k$
end for

3.3 K-Secure Sum

In SMC, multiple parties (denoted as P_i) with private inputs x_i, operating in a distributed computing environment, aim to collaboratively compute an agreed function $f(x_1, x_2, x_3, \ldots, x_n) = (y_1, y_2, y_3, \ldots, y_n)$ based on their respective private inputs. Following the computation, each party P_i should receive its individual output y_i without divulging any additional information. SMC protocols are designed to facilitate this computation securely, enabling mutually distrusting participants to jointly compute a function over their inputs while ensuring output correctness and maintaining input privacy, even in the presence of dishonest behavior [27]. In K-secure sum algorithm, each party contributes its share to the sum without revealing its private input directly [28]. Inputs are divided into segments, and computation is performed on these segments in a distributed manner [29]. Each step ensures that no party can gain unauthorized access to sensitive information (as inputs of each party is divided into k segments and then segments are redistributed). Figure 2 illustrates steps involved in K-Secure Sum algorithm.

As depicted in Algorithm 3, parties $P_0, P_1, P_2, \ldots, P_{n-1}$ are involved in a peer-to-peer (P2P) fashion employing federated learning. Aggregation of each party's weights is achieved via k-secure sum computations. k is defined as an integer representing the number of segments in each data block. n represents the number of parties. w_i denotes the weights matrix for each client. For every entry w_i in the weights matrix of all clients, the algorithm iterates through each party P_i. Each data block D_i belonging to an entry of weight matrix w_i is broken into random segments

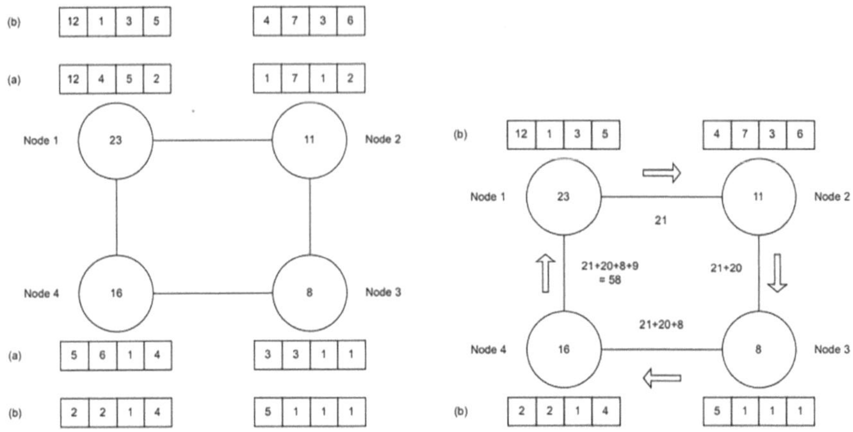

(a) Every node with their secrets:
(i) refers to the secrets and (ii) refers to inputs after redistribution

(b) Computation of the final sum of secrets

Fig. 2 K-Secure Sum in steps

Algorithm 3 K-Secure Sum Algorithm

Assume parties $P_0, P_1, P_2, \ldots, P_{n-1}$ involved in a P2P ring.
Let k be an integer representing the number of segments in each data block, n be the number of parties and w_i represent the weights matrix for each client.
for every entry wi in the weights matrix of all clients **do**
  for $i = 0$ to $n - 1$ **do**
    Let D_i be the data block belonging to an entry of weight matrix w_i of party P_i.
    Break D_i into random segments $D_{i0}, D_{i1}, D_{i2}, \ldots, D_{i(k-1)}$ such that $D_i = \sum_{j=0}^{k-1} D_{ij}$.
  end for
end for
Let $rc = k$ and $S_{ij} = 0$ (where S_{ij} is the partial sum).
while $rc \neq 0$ **do**
  for $j = 0$ to $k - 1$ **do**
    for $i = 0$ to $n - 1$ **do**
      P_i sends $S_{ij} = D_{ij} + S_{ij}$ to $P_{(i+1) \bmod n}$.
    end for
    $rc = rc - 1$.
  end for
end while
P_0 announces S_{ij}.

$D_{i0}, D_{i1}, D_{i2}, \ldots, D_{i(k-1)}$ such that D_i is the sum of these segments. Weight aggregation in each round involves communication of the weights between parties which involves the computation of partial weight sums S_{ij} for each entry of the weights matrices, where j iterates through segments and i iterates through parties. Each party

P_i sends its partial sum S_{ij} to the next party in the P2P network. After k steps, party P_0 announces the final result S_{ij}.

Through the secure sum algorithm, we outline a process for performing federated learning among multiple parties in a peer-to-peer manner, ensuring privacy through secure sum computations. It demonstrates how parties can collaborate to train a model without sharing sensitive information about each party's weights.

4 Our Proposed P2P Framework

In this paper, we present a federated learning framework based on a ring architecture, aimed to facilitate sequential knowledge accumulation among nodes within the network. Additionally, our framework affords clients the autonomy to opt into the current training round, thereby fostering a flexible and inclusive learning environment. The sequential communication mechanism is similar to the rumour-mongering gossip protocol used in P2P architectures [30], and the nodes forget their old weights/parameters at the end of each communication round. We now describe with formal notations, the above proposed framework. Consistent with the classical FL notion, nodes only transfer learned model parameters instead of the local dataset between them.

Assuming a network comprising n data owners denoted as $(P_1, P_2, \ldots, P_n)$, each entrusted with a distinct dataset D_i, we initialize the system by arranging all data owners in a ring structure as depicted in Fig. 3. Herein, each data owner P_i is responsible for managing dataset D_i. The initiation of training rounds commences with the selection of a random client. For instance, if P_1 is chosen as the initiator, a round is defined as the sequential traversal from P_1 to P_n, with each node being visited at

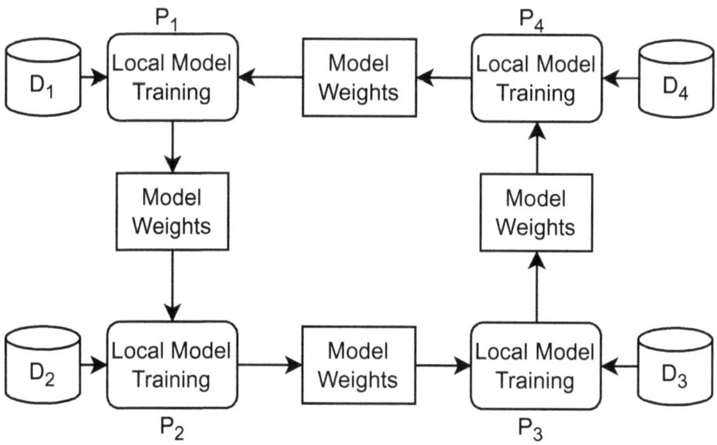

Fig. 3 Diagram of proposed P2P federated learning framework

most once. During each round, the initiating client P_i undertakes the training of a local model M_i based on its corresponding dataset D_i and subsequently transmits the learned model parameters to the succeeding participating node, P_{i+k}, where k denotes the index gap in the ring between the current node and the next participating node. Subsequent to receiving the model parameters, P_{i+k} initializes its weights with the one sent by P_i and initiates the training process anew with its dataset. This iterative process persists until the model parameters circle back to the originating client as mentioned in the loop from line 11 to line 22 in Algorithm 4. This constitutes of one circular communication round of training.

Algorithm 4 A P2P FedProx based Aggregate Algorithm with Secure Aggregation

1: Initialise a peer-to-peer ring with n nodes. Let N be the node set containing all nodes.
2: Let M_i be the model architecture of $node_i$. We assign $M_1 = M_2 = M_3 = \ldots = M_i = M_{i+1} = \ldots = M_n = M$, where M is the globally agreed model architecture.
3: Assign the custom FedProx optimizer to be used in $M_i, \forall i \in [1, n]$.
4: Let there be R circular communication rounds. Let D_i be the local dataset of $node_i$.
5: **for** r in $[1, R]$ **do**
6: start_node = random.choice(N)
7: current_node = start_node
8: $i \leftarrow$ start_node.id
9: $w_{\text{new}_i} \leftarrow \text{train}(w_{\text{old}_i}, D_i)$
10: $prev\_weight \leftarrow w_{\text{new}_i}$
11: **while true do**
12: **if** `current_node.active==True` **then**
13: $w_{\text{old}_i} \leftarrow$ prev_weight
14: $w_{\text{new}_i} \leftarrow \text{train}(w_{\text{old}_i}, D_i)$
15: **end if**
16: prev_weight $\leftarrow w_{\text{new}_i}$
17: **if** `current_node` $\equiv$ `starter_node` **then**
18: **break**
19: **end if**
20: `current_node = current_node->next`
21: $i \leftarrow$ `current_node.id`
22: **end while**
23: Test Model using Secure Algorithm mentioned in Algorithm 5
24: **end for**

At the conclusion of each communication round, the originating client consolidates the weights reflecting the most accrued knowledge from the participating clients. While it may be tempting to utilize these weights for evaluating performance metrics on the test set of the nodes, caution must be exercised to prevent overfitting, particularly in scenarios involving a large number of active nodes during a communication round. This communication mechanism has been described formally in algorithm 4. To mitigate the risk of overfitting and ensure robustness in testing, we employ a cautious approach. Instead of solely relying on the weights of the starter client, we opt to aggregate weights across all participating nodes using the secure sum protocol. This strategy allows for a more comprehensive evaluation by leveraging a collective representation of model parameters. By averaging weights from multiple sources, we gain greater confidence in predicting outcomes on previously unseen test data. We assume that the starter node which is randomly selected at each round will have the final averaged weights (using secure sum protocol). Algorithm 5 shows this secure aspect of our framework and how it is deployed in the framework.

As demonstrated later in this paper, our proposed framework achieves performance parity with traditional central server architectures. In addition, it offers enhanced security during testing, as well as the advantage of operating without a centralized server susceptible to potential attacks. Several attacks that could occur in a server during test/inference time is mentioned in [31–33]. But since the starter client is random at each round in our framework, no single node acts as a server for a prolonged period of time.

4.1 Secure Aggregation Algorithm for Testing

Algorithm 5 Secure Aggregation Algorithm

1: Let M_i be the model architecture of $node_i$, M be the globally agreed model architecture and t be the total number of parameters/ weights in each client model.
2: Let $w_1^{(i)}, w_2^{(i)}, \ldots, w_t^{(i)}$ be the weights of M_i, $w_1^{(agg)}, w_2^{(agg)}, \ldots, w_t^{(agg)}$ be the weights of the global model.
3: **for** $i = 1, 2, \ldots, t$ **do**
4: $\quad w_i^{(agg)} \leftarrow$ K - Secure_Sum$(w_i^{(1)}, w_i^{(2)}, \ldots, w_i^{(n)})/n$ ▷ Using Algorithm 3
5: **end for**

5 Experimental Results and Discussion

5.1 Experimental Setup

Instead of employing physical devices connected to actual IP addresses, we opted for a simulation approach within a Python environment. This simulation replicates the behavior and interactions of the devices, providing a controlled and scalable environment for experimentation and analysis.

For the non-IID study, the simulation settings have been majorly inspired by the repository of the paper [8], including the FEMNIST dataset source. The MNIST dataset has been sourced from the .mat file present in the github repository of [34]. The list of clients are created using an object-oriented-programming approach and clients are accessed using a regex search on their unique identifier number. The index is incremented in a loop and the modulus operator ensures that a ring network is simulated. The approach is chosen so that the code design remains similar for a fair comparison with the original server-client architecture using FedProx in the paper [8].

For the IID study, the simulation for developing the peer-to-peer framework was done using the *NetworkX* python library (in which clients are modeled as nodes of the graph), and the simulation settings for FedAvg are used as mentioned in the repository of the paper [8]. Both the FEMNIST and MNIST datasets have been sourced from the repository of [34].

Henceforth in this paper, devices/clients correspond to pythonic nodes, created using the methods mentioned above. We have simulated our experiments for 40 communication rounds. *Number of rounds* is a hyperparameter that can be changed. Using *number of rounds* = 40, we observe that all the loss curves saturate, with changes in loss values in the scale of 10^{-3}. We also note that adjusting the hyperparameter *Number of rounds* would not alter the conclusions drawn in the paper.

MNIST : We investigate the task of image classification on the MNIST dataset introduced by LeCun et al. (1998) [35], focusing on 70,000 samples of handwritten digits 0 through 9. Our approach utilizes multinomial logistic regression, employing a model architecture consisting of a single dense layer with 10 output units, tailored for image classification tasks. The input layer encompasses 784 neurons, representing the flattened dimensions of the input images. This architecture is deliberately selected for its simplicity and computational efficiency, rendering it well-suited for fundamental image classification endeavors. The same architecture has been used for both the IID and non-IID settings.

To emulate a heterogeneous environment, in case of non-IID settings, we distribute the dataset across a varying number of devices, specifically (10, 20, 30, 40, 50), ensuring that each device contains samples corresponding to only two digits. Moreover, we adhere to a power law distribution for the allocation of samples per device, as proposed in prior literature [26].

FEMNIST : We study an image classification problem on the 62-class EMNIST dataset [36], which consists of 805,263 samples. The model architecture for the non-IID setting comprises of convolutional layers followed by dense layers, designed for image classification tasks. To simulate a heterogeneous setting for non-IID case, we distribute the data among (10, 20, 30, 40, 50) devices such that each device has only 5 classes. All sub-sampling schemes has been adopted from [26].

We describe the model architecture used for training in case of non-IID and IID settings in Tables 1 and 2 respectively.

We applied FedProx on both ring P2P framework and Server-Client architecture for comparison in non-IID settings, while FedAvg on both ring P2P framework and Server-Client architecture for comparison in IID settings.

Table 1 Model architecture for FEMNIST non-IID setting

Layer	Configuration	Output shape	Activation
Input	Reshape: (784,1) to (28, 28, 1)	(28, 28, 1)	–
Convolutional 1	32 filters, 5 × 5 kernel	(24, 24, 32)	ReLU
Max pooling 1	2 × 2 pool size, strides 2	(12, 12, 32)	–
Convolutional 2	64 filters, 3 × 3 kernel	(10, 10, 64)	ReLU
Max pooling 2	2 × 2 pool size, strides 2	(5, 5, 64)	–
Flatten	Flattens output	1600	–
Dense	128 units	128	ReLU
Output	62 units	62	–

Table 2 Model architecture for FEMNIST IID setting

Layer	Configuration	Output shape	Activation
Input	Reshape: (784,1) to (28, 28, 1)	(28, 28, 1)	–
Convolutional 1	32 filters, 3 × 3 kernel	(26, 26, 32)	ReLU
Max pooling 1	2 × 2 pool size, strides 2	(13, 13, 32)	–
Convolutional 2	64 filters, 3 × 3 kernel	(11, 11, 64)	ReLU
Max pooling 2	2 × 2 pool size, strides 2	(5, 5, 64)	–
Convolutional 3	64 filters, 3 × 3 kernel	(3, 3, 64)	ReLU
Flatten	Flattens output	576	–
Dense	128 units	128	ReLU
Dense	128 units	128	ReLU
Output	62 units	62	–

5.2 *Implementation Results of FedProx for Non-IID Datasets*

The experimental results for MNIST dataset is shown in Figs. 4, 5, while 6, 7 shows the results for the FEMNIST dataset. The non-IID data partitioning has been done as given in the repository of [26]. We observe that our proposed P2P framework gives better accuracy than the baseline server client architecture for the MNIST dataset, in

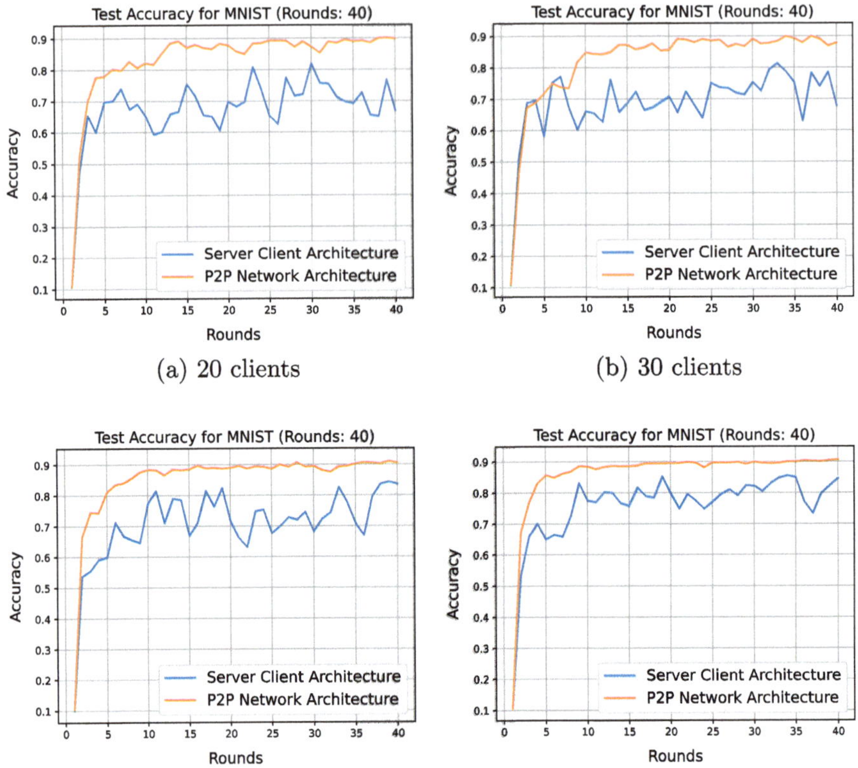

(a) 20 clients (b) 30 clients (c) 40 clients (d) 50 clients

Fig. 4 Accuracy versus rounds for non-IID MNIST dataset

terms of accuracy and offers the advantages of using a secure aggregation protocol as outlined in Algorithms 4 and 5. We get an accuracy advantage of 12.75 % on average as calculated from Table 4, which shows the comparison of terminal accuracy and loss for the two architectures.

In case of the FEMNIST dataset, constructed as given in [26], we observe that our framework is about par with the server client architecture, with a accuracy drop of a mere 0.9% on average. Our P2P framework, though, has very smooth test accuracy and loss curves compared to the baseline cases, which shows that our proposed framework (Algorithm 4), tries to nullify the non-IID nature of the dataset in the network. The choice of hyperparameters for these non-IID experiments can be found in Table 3, while Table 5 shows the comparison of terminal accuracy and loss for the two architectures.

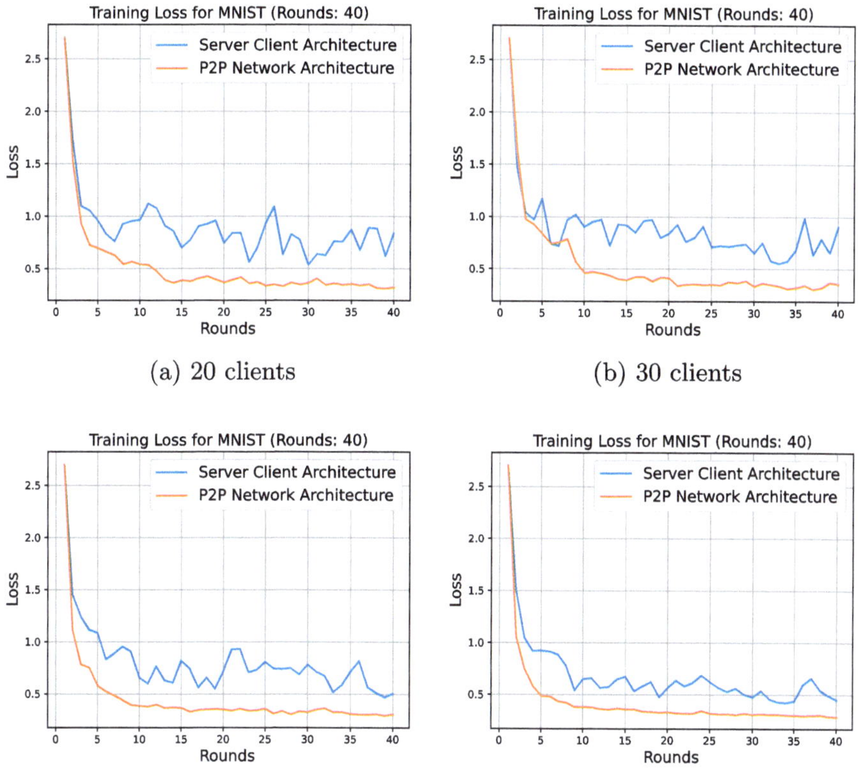

(a) 20 clients

(b) 30 clients

(c) 40 clients

(d) 50 clients

Fig. 5 Loss versus rounds for non-IID MNIST dataset

5.3 *Implementation Results of FedAvg for IID Datasets*

The experimental results have been shown in Figs. 8, 9, 10, 11. Hyperparameter choice has been mentioned in Table 8. We have used IID local dataset samples for each client. datasets. We observe that our proposed P2P framework performs nearly on par with the baseline server client architecture in terms of accuracy, with a 1.59% accuracy drop for FEMNIST and 0.712% accuracy drop for MNIST on average. These metrics have been calculated from the terminal accuracy and loss given in Tables 7 and 6. Unlike the case of non-IID settings, we see smooth learning curves for both the architectures for both the datasets (Fig. 12).

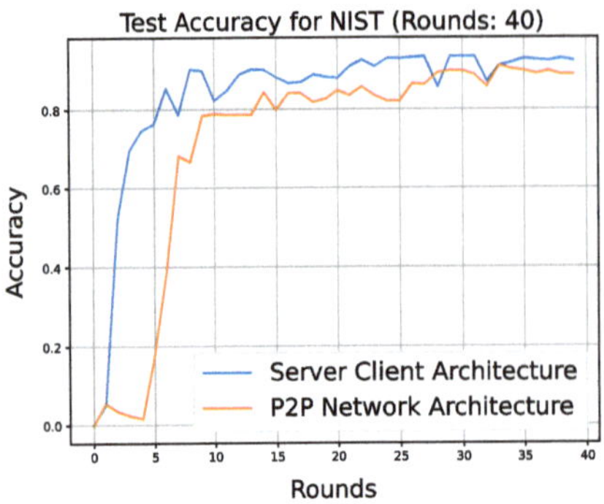

(a) 20 clients

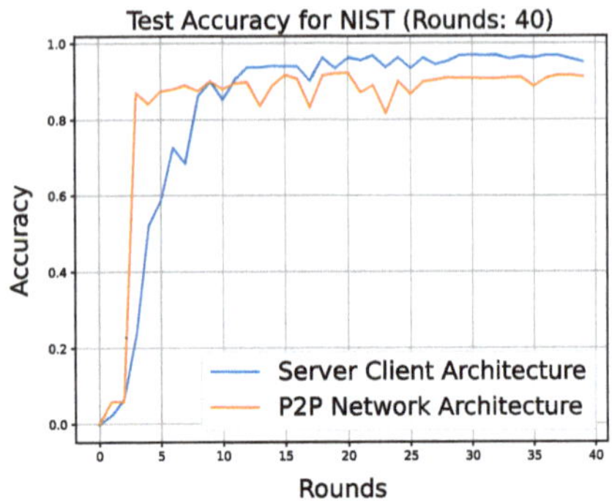

(b) 30 clients

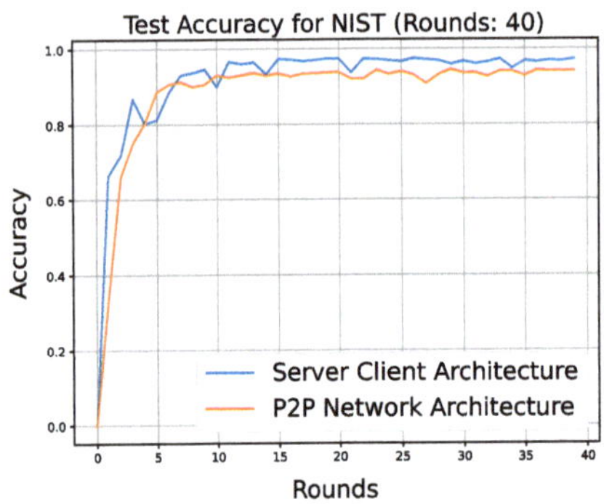

(c) 40 clients

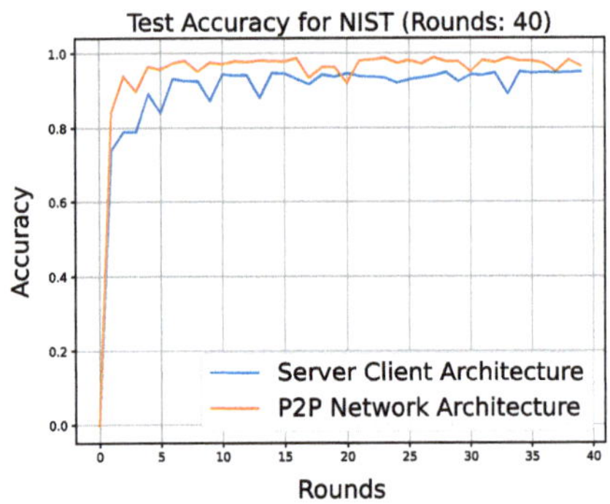

(d) 50 clients

Fig. 6 Accuracy versus rounds for non-IID FEMNIST dataset

6 Future Scope

Though this paper provides better results over current baseline architecture for non-IID datasets, we aspire to gain the same with more informative client selection in a circular round. Greater number of clients participating in each round leads to good accuracy gains in non-IID cases, but it is not always feasible in a real life setting due to limitation of bandwidth and energy as mentioned in [37–39]. Certain threshold-based algorithms like CMFL [40] can be integrated in our P2P network in order to reduce communication costs and choose unbiased clients for model updates. We also aim to make the secure aggregation process more efficient, by aggregating the

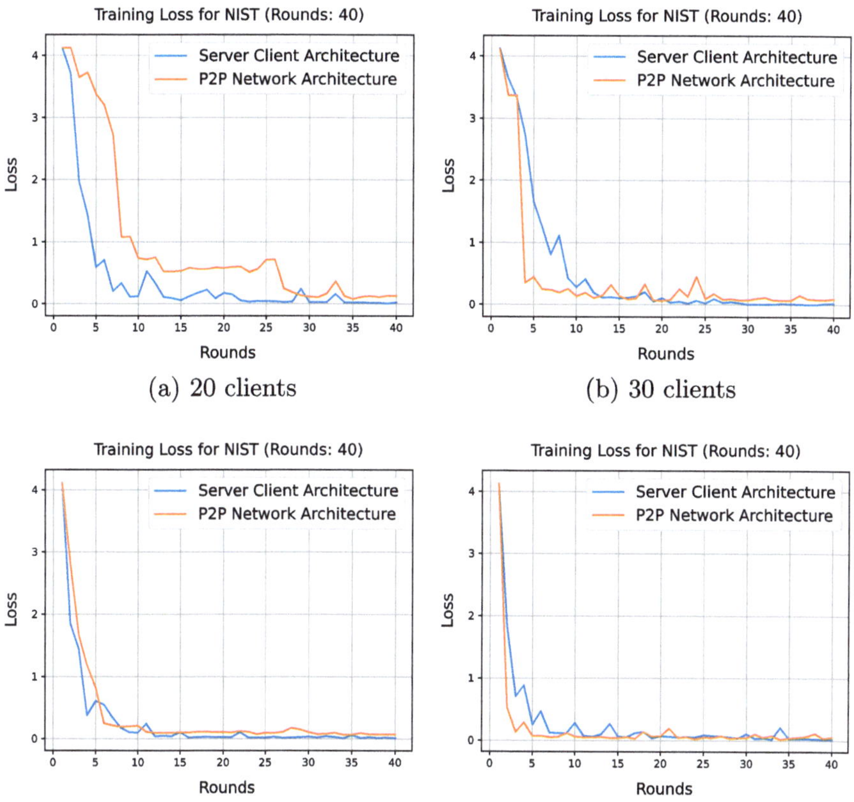

(a) 20 clients (b) 30 clients

(c) 40 clients (d) 50 clients

Fig. 7 Loss versus rounds for non-IID FEMNIST dataset

Table 3 Hyperparameter choice

Dataset	Number of rounds	Epochs/client/round	Learning rate	Method	μ
MNIST	40	1	0.001	FedProx	0.001
FEMNIST	40	20	0.001	FedProx	0.001

difference of weights (w.r.t previous round's weights) instead of the current weights themselves, to reduce the scale of data on which computations are done, which can potentially reduce inference time even further. This reduction in inference time (due to rescaling of data) is highlighted greatly when we used advanced security schemes, like FHE [41].

Table 4 Final accuracy and loss: non-IID MNIST dataset

Clients	Ring P2P architecture		Server-client architecture	
	Acc[1]	Loss[2]	Acc[1]	Loss[2]
10	89.00	0.348	72.16	0.764
20	89.92	0.325	73.04	0.777
30	88.04	0.362	79.65	0.542
40	90.66	0.308	76.41	0.602
50	90.78	0.292	83.40	0.446

Table 5 Final accuracy and loss: non-IID FEMNIST dataset

Clients	Ring architecture		Server-client architecture	
	Acc[1]	Loss[2]	Acc[1]	Loss[2]
10	91.46	0.029	87.1	0.059
20	88.79	0.137	92.31	0.028
30	91.33	0.097	95.16	0.023
40	94.08	0.076	97.32	0.076
50	96.71	0.058	95.02	0.058

Table 6 Final accuracy and loss: IID FEMNIST dataset

Clients	Server-client architecture		Ring P2P architecture	
	Acc[1]	Loss[2]	Acc[1]	Loss[2]
10	91.55	0.102	90.53	0.121
20	91.04	0.126	89.01	0.149
30	87.05	0.136	85.58	0.152
40	85.04	0.167	83.42	0.169
50	81.89	0.175	80.08	0.183

7 Conclusion

Through this paper, we described a P2P alternative for federated learning to eliminate a central server aggregation method. We also used the secure sum protocol (algorithm 5) for weight aggregation of the global model, with no trade-off in accuracy (as no information loss occurs in algorithm 5. We observe improvement in the said metrics over the baseline architecture in case of non-IID datasets (MNIST and FEMNIST), for a range of total clients in the P2P network. We expect that if we choose for more advanced privacy-preserving aggregation algorithm for testing (like Fully Homomorphic encryption) , there will be a slight decrease in accuracy, because of precision loss, as explained in [42], which should still be on par with current client server architecure metrics, with a very less tolerance rate.

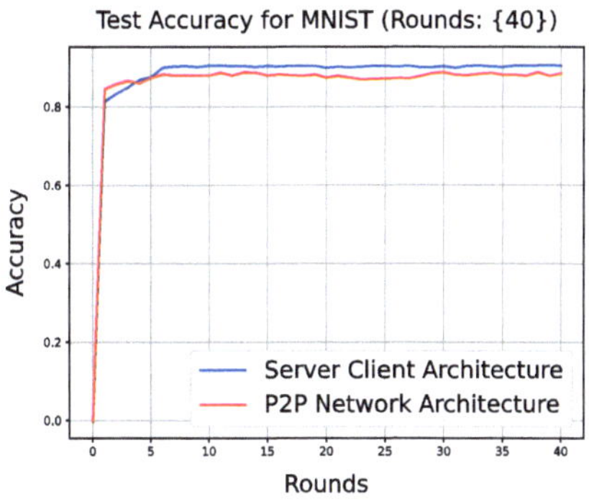

(a) 20 clients

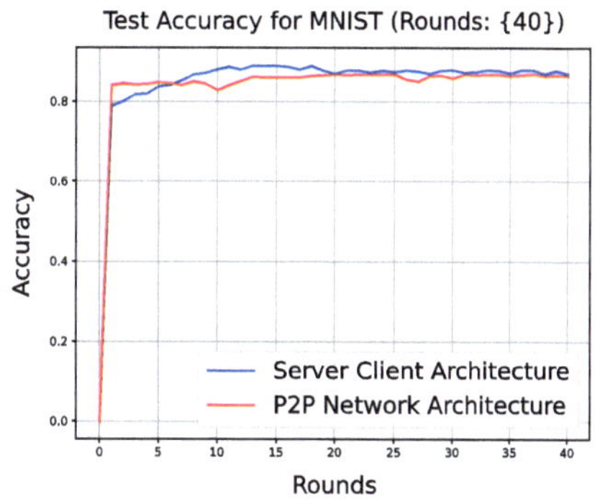

(b) 30 clients

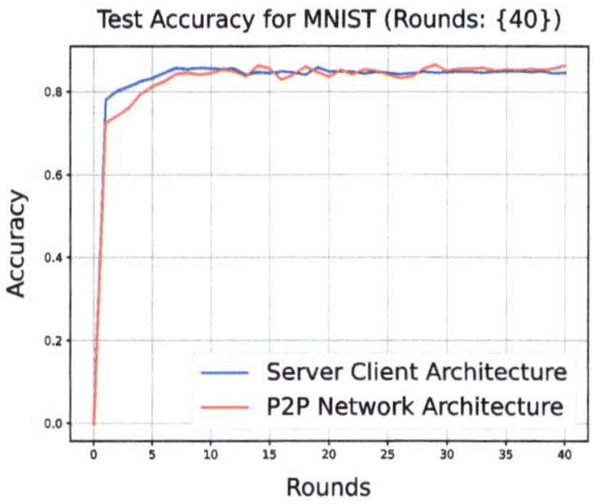

(c) 40 clients

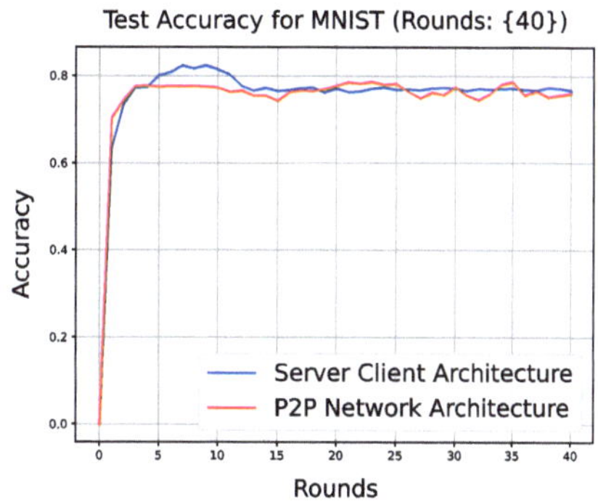

(d) 50 clients

Fig. 8 Accuracy versus rounds for IID MNIST dataset

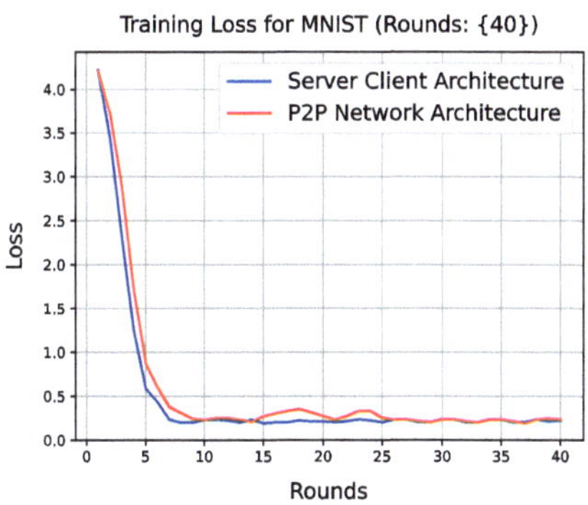

(a) 20 clients

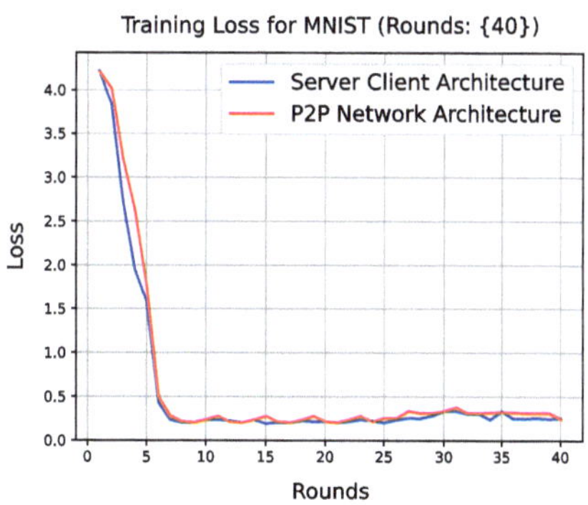

(b) 30 clients

Fig. 9 Loss versus rounds for IID MNIST dataset 20 and 30 clients

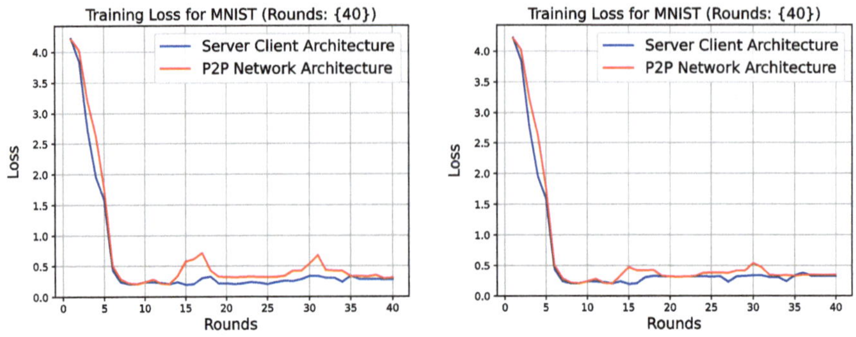

(a) 40 clients (b) 50 clients

Fig. 10 Loss versus rounds for IID MNIST dataset for 40 and 50 clients

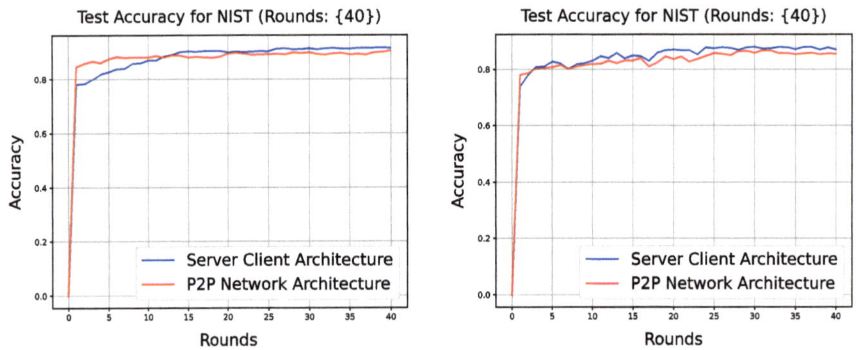

(a) 20 clients (b) 30 clients

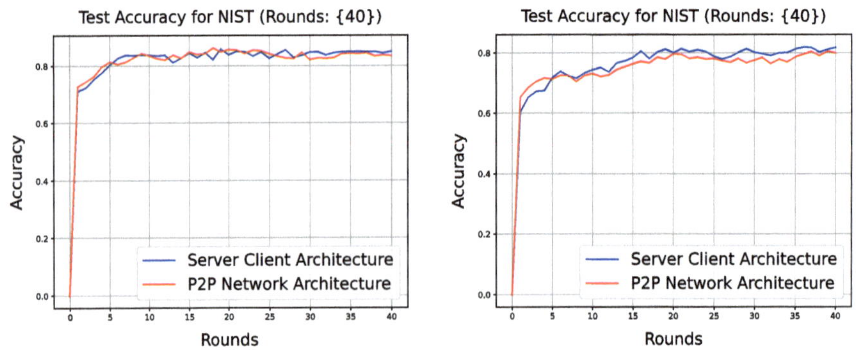

(c) 40 clients (d) 50 clients

Fig. 11 Accuracy versus rounds for IID FEMNIST dataset

Table 7 Final accuracy and loss: IID MNIST dataset

Clients	Server-client architecture		Ring P2P architecture	
	Acc[1]	Loss[2]	Acc[1]	Loss[2]
10	91.04	0.192	88.79	0.123
20	90.55	0.231	88.53	0.252
30	87.05	0.263	86.78	0.271
40	84.59	0.286	86.42	0.324
50	76.89	0.328	76.04	0.356

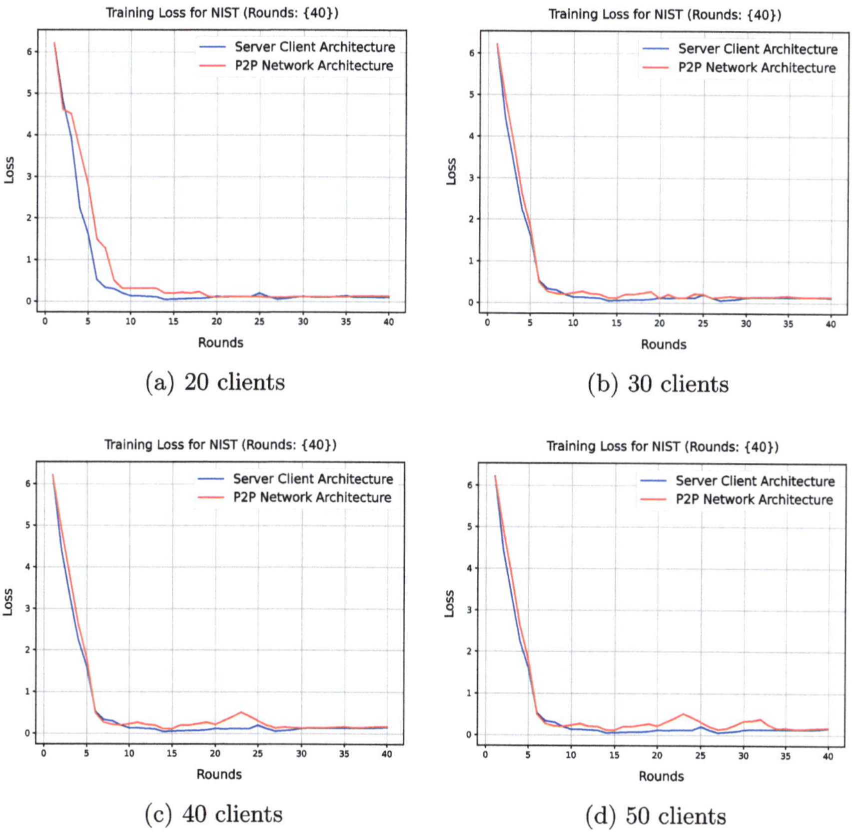

(a) 20 clients (b) 30 clients

(c) 40 clients (d) 50 clients

Fig. 12 Loss versus rounds for IID FEMNIST dataset

Table 8 Hyperparameter Choice

Dataset	Number of rounds	Epochs/client/round	Learning rate	Method
MNIST	40	1	0.001	FedAvg
FEMNIST	40	10	0.001	FedAvg

Table 9 Notation Table

Abbreviation	Description
IID	Identically and Independently Distributed
non-IID	Non-Identically and Independently Distributed
P2P	Peer-to-Peer
[1]Acc	Test accuracy at round 40
[2]Loss	Training loss at round 40
FHE	Fully Homomorphic Encryption
FL	Federated Learning
μ	Proximal hyperparameter in FedProx
Epochs/Client/Round	Number of epochs per client in each circular communication round.

8 Appendix

In table 9, we provide a list of abbreviations used throughout the paper. These abbreviations are commonly employed in the context of the research and provide concise representations of key concepts, methodologies, or terms discussed within the text.

References

1. Kotsios, A., Magnani, M., Vega, D., Rossi, L., Shklovski, I.: An analysis of the consequences of the general data protection regulation on social network research. ACM Trans. Soc. Comput. **2**(3), 1–22 (2019)
2. Act, A.: Health insurance portability and accountability act of 1996. Public Law **104**, 191 (1996)
3. Konečný, J., McMahan, H.B., Yu, F.X., Richtárik, P., Suresh, A.T., Bacon, D.: Federated learning: strategies for improving communication efficiency. arXiv preprint arXiv:1610.05492 (2016)
4. Mendieta, M., Yang, T., Wang, P., Lee, M., Ding, Z., Chen, C.: Local learning matters: rethinking data heterogeneity in federated learning. In: Proceedings of the IEEE/CVF Conference on Computer Vision and Pattern Recognition, pp. 8397–8406 (2022)
5. Zhu, L., Liu, Z., Han, S.: Deep leakage from gradients. Adv. Neural Inform. Process. Syst. **32** (2019)
6. Aïvodji, U., Gambs, S., Ther, T.: Gamin: an adversarial approach to black-box model inversion. arXiv preprint arXiv:1909.11835 (2019)
7. Shokri, R., Stronati, M., Song, C., Shmatikov, V.: Membership inference attacks against machine learning models. In: 2017 IEEE Symposium on Security and Privacy (SP), pp. 3–18. IEEE (2017)
8. Li, T., Sahu, A.K., Zaheer, M., Sanjabi, M., Talwalkar, A., Smith, V.: Federated optimization in heterogeneous networks. Proc. Mach. Learn. Syst. **2**, 429–450 (2020)
9. Truex, S., Baracaldo, N., Anwar, A., Steinke, T., Ludwig, H., Zhang, R., Zhou, Y.: A hybrid approach to privacy-preserving federated learning. In: Proceedings of the 12th ACM Workshop on Artificial Intelligence and Security, pp. 1–11 (2019)

10. Li, Y., Zhou, Y., Jolfaei, A., Yu, D., Xu, G., Zheng, X.: Privacy-preserving federated learning framework based on chained secure multiparty computing. IEEE Internet Things J. **8**(8), 6178–6186 (2020)
11. Young, T., Hazarika, D., Poria, S., Cambria, E.: Recent trends in deep learning based natural language processing. IEEE Comput. Intell. Mag. **13**(3), 55–75 (2018)
12. Xu, R., Baracaldo, N., Zhou, Y., Anwar, A., Ludwig, H.: Hybridalpha: an efficient approach for privacy-preserving federated learning. In: Proceedings of the 12th ACM Workshop on Artificial Intelligence and Security, pp. 13–23 (2019)
13. Pang, J., Huang, Y., Xie, Z., Han, Q., Cai, Z.: Realizing the heterogeneity: a self-organized federated learning framework for IoT. IEEE Internet Things J. **8**(5), 3088–3098 (2020)
14. Zhao, Y., Li, M., Lai, L., Suda, N., Civin, D., Chandra, V.: Federated learning with non-IID data. arXiv preprint arXiv:1806.00582 (2018)
15. Wink, T., Nochta, Z.: An approach for peer-to-peer federated learning. In: 2021 51st Annual IEEE/IFIP International Conference on Dependable Systems and Networks Workshops (DSN-W), pp. 150–157 (2021). IEEE
16. Roy, A.G., Siddiqui, S., Pölsterl, S., Navab, N., Wachinger, C.: Braintorrent: A peer-to-peer environment for decentralized federated learning. arXiv preprint arXiv:1905.06731 (2019)
17. Li, Z., Lu, J., Luo, S., Zhu, D., Shao, Y., Li, Y., Zhang, Z., Wang, Y., Wu, C.: Towards effective clustered federated learning: a peer-to-peer framework with adaptive neighbor matching. IEEE Trans. Big Data (2022)
18. Li, M., Andersen, D.G., Park, J.W., Smola, A.J., Ahmed, A., Josifovski, V., Long, J., Shekita, E.J., Su, B.-Y.: Scaling distributed machine learning with the parameter server. In: 11th USENIX Symposium on Operating Systems Design and Implementation (OSDI 14), pp. 583–598. USENIX Association, Broomfield, CO (2014). https://www.usenix.org/conference/osdi14/technical-sessions/presentation/li_mu
19. Bonawitz, K., Eichner, H., Grieskamp, W., Huba, D., Ingerman, A., Ivanov, V., Kiddon, C., Konečný, J., Mazzocchi, S., McMahan, B., et al.: Towards federated learning at scale: system design. Proc. Mach. Learn. Syst. **1**, 374–388 (2019)
20. McMahan, B., Moore, E., Ramage, D., Hampson, S., Arcas, B.A.: Communication-efficient learning of deep networks from decentralized data. In: Artificial Intelligence and Statistics, pp. 1273–1282. PMLR (2017)
21. Hu, C., Jiang, J., Wang, Z.: Decentralized federated learning: a segmented gossip approach (2019). arXiv preprint arXiv:1908.07782
22. Hegedűs, I., Danner, G., Jelasity, M.: Gossip learning as a decentralized alternative to federated learning. In: Distributed Applications and Interoperable Systems: 19th IFIP WG 6.1 International Conference, DAIS 2019, Held as Part of the 14th International Federated Conference on Distributed Computing Techniques, DisCoTec 2019, Kongens Lyngby, Denmark, June 17–21, 2019, Proceedings 19, pp. 74–90. Springer (2019)
23. Dai, R., Shen, L., He, F., Tian, X., Tao, D.: Dispfl: Towards communication-efficient personalized federated learning via decentralized sparse training. In: International Conference on Machine Learning, pp. 4587–4604. PMLR (2022)
24. Hegedűs, I., Danner, G., Jelasity, M.: Gossip learning as a decentralized alternative to federated learning. In: Pereira, J., Ricci, L. (eds.) Distributed Applications and Interoperable Systems, pp. 74–90. Springer, Cham (2019)
25. McMahan, H.B., Moore, E., Ramage, D., Arcas, B.A.: Federated learning of deep networks using model averaging (2016). CoRR arXiv:1602.05629
26. Sahu, A.K., Li, T., Sanjabi, M., Zaheer, M., Talwalkar, A., Smith, V.: On the convergence of federated optimization in heterogeneous networks (2018). CoRR arXiv:1812.06127
27. Goldreich, O.: Secure multi-party computation. Manuscript. Preliminary version **78**(110), 1–108 (1998)
28. Sheikh, R., Kumar, B., Mishra, D.K.: A distributed k-secure sum protocol for secure multi-party computations (2010). arXiv preprint arXiv:1003.4071
29. Sheikh, R., Kumar, B., Mishra, D.K.: Changing neighbors k secure sum protocol for secure multi party computation (2010). arXiv preprint arXiv:1002.2409

30. Naik, D., Grace, P., Naik, N., Jenkins, P., Mishra, D., Prajapat, S.: An introduction to gossip protocol based learning in peer-to-peer federated learning. In: 2023 IEEE International Conference on ICT in Business Industry and Government (ICTBIG), pp. 1–8 (2023). https://doi.org/10.1109/ICTBIG59752.2023.10456324
31. Lyu, L., Yu, H., Ma, X., Chen, C., Sun, L., Zhao, J., Yang, Q., Yu, P.S.: Privacy and robustness in federated learning: attacks and defenses. IEEE Trans. Neural Netw. Learn. Syst. 1–21 (2022). https://doi.org/10.1109/TNNLS.2022.3216981
32. Nasr, M., Shokri, R., Houmansadr, A.: Comprehensive privacy analysis of deep learning: Passive and active white-box inference attacks against centralized and federated learning. In: 2019 IEEE Symposium on Security and Privacy (SP), pp. 739–753 (2019). https://doi.org/10.1109/SP.2019.00065
33. Shokri, R., Stronati, M., Song, C., Shmatikov, V.: Membership inference attacks against machine learning models. In: 2017 IEEE Symposium on Security and Privacy (SP), pp. 3–18 (2017). https://doi.org/10.1109/SP.2017.41
34. ML-LAB: FedProx: An Implementation of the FedProx Algorithm for Federated Learning. GitHub (2018)
35. Lecun, Y.: The MNIST database of handwritten digits. http://yann.lecun.com/exdb/mnist/
36. Cohen, G., Afshar, S., Tapson, J., Schaik, A.: EMNIST: an extension of MNIST to handwritten letters (2017). CoRR arXiv:1702.05373
37. Fu, L., Zhang, H., Gao, G., Zhang, M., Liu, X.: Client selection in federated learning: principles, challenges, and opportunities. IEEE Internet Things J. **10**(24), 21811–21819 (2023). https://doi.org/10.1109/JIOT.2023.3299573
38. Xu, J., Wang, H.: Client selection and bandwidth allocation in wireless federated learning networks: a long-term perspective. IEEE Trans. Wireless Commun. **20**(2), 1188–1200 (2021). https://doi.org/10.1109/TWC.2020.3031503
39. Amiri, M.M., Gündüz, D., Kulkarni, S.R., Poor, H.V.: Update aware device scheduling for federated learning at the wireless edge (2020). CoRR arXiv:2001.10402
40. WANG, L., WANG, W., LI, B.: Cmfl: Mitigating communication overhead for federated learning. In: 2019 IEEE 39th International Conference on Distributed Computing Systems (ICDCS), pp. 954–964 (2019). https://doi.org/10.1109/ICDCS.2019.00099
41. Benaissa, A., Retiat, B., Cebere, B., Belfedhal, A.E.: TenSEAL: A Library for Encrypted Tensor Operations Using Homomorphic Encryption (2021)
42. Costache, A., Curtis, B.R., Hales, E., Murphy, S., Ogilvie, T., Player, R.: On the precision loss in approximate homomorphic encryption. In: Carlet, C., Mandal, K., Rijmen, V. (eds.) Selected Areas in Cryptography—SAC 2023, pp. 325–345. Springer, Cham (2024)

Improving Credit Card Fraud Detection with Combined Feature Extraction and Class Balance Optimisation Techniques

Boluwatife Ajibade-Ajibosin, Uchenna Daniel Ani, Theocharis Kyriacou, and Mark Turner

Abstract Credit card fraud is a persistent and evolving challenge that poses significant financial harm to cardholders, financial institutions, national, and global economies. The use of Machine Learning (ML) methods has heavily enhanced the detection of credit card fraud, offering improvements to other traditional credit card fraud detection approaches such as human manual checks and rule-based methods. However, there are limitations that impact the performance and efficiency of the credit card detection using ML. This work addresses some of the current challenges associated with employing ML for credit card fraud by developing a robust credit card fraud detection model. The proposed model employs Synthetic Minority Oversampling Technique (SMOTE) to address the class imbalance issue typically akin to credit card fraud datasets. A Recursive Feature Elimination with Cross-Validation (RFECV) scheme was utilized as a feature selection technique to select the optimal subset of features that can improve the accuracy of the credit card fraud detection model. Following rigorous evaluation, the proposed credit card fraud detection system demonstrates exceptional performance above other known existing systems across critical metrics including accuracy, recall, precision, F1 score, and ROC-AUC. Thus, is considered a better solution for more accurate detection of credit card fraud instances.

B. Ajibade-Ajibosin · U. D. Ani (✉) · T. Kyriacou · M. Turner
School of Computer Science and Mathematics, Keele University, Keele, UK
e-mail: u.d.ani@keele.ac.uk

B. Ajibade-Ajibosin
e-mail: x9a03@students.keele.ac.uk

T. Kyriacou
e-mail: t.kyriacou@keele.ac.uk

M. Turner
e-mail: m.turner@keele.ac.uk

M. G. Jaatun et al. (eds.), *Proceedings of the International Conference on Cybersecurity, Situational Awareness and Social Media*, Springer Proceedings in Complexity,
https://doi.org/10.1007/978-981-96-0401-2_2

1 Introduction

Credit cards are used for online banking, online shopping, and everyday transactions. With the increased use of credit card, there has been an emergence of credit card fraud [1], malicious activities that aim and tend to exploit the technology and associated processes and service for illegitimate gains. Credit card fraud has been defined as the use of credit card details to make transactions by unauthorized persons, resulting from a range of illegal or criminal activities such as the theft of credit cards, phishing for credit card details, and utilizing fake cards to mimic legitimate user behaviour amongst others [2].

One of the most critical consequences of credit card fraud is the economic impact on cardholders and issuing financial institutions. Fraudulent action often results in the depletion of funds from affected customers' accounts, and unauthorized credit transactions leading to the loss of funds. Financial institutions are also impacted as they spend millions of funds reimbursing defrauded customers and spending more to cover the costs of investigating these fraudulent activities. For example, credit card frauds were the most prevalent incidences in a portfolio of about 1579 data breaches resulting in 179 million data points [3]. Card fraud cases continues to increase globally, with an estimated 30 billion U.S. dollars in losses for merchants and card acquirers in 2022 [4]. It is forecast that global card frauds for 2024 would reach 35.8 billion dollars, which equates to 6.5 cents per $100 spent on cards [5].

Beyond the financial impact, credit card fraud can negatively affect the trust and reputation of financial institutions and the associated electronic payment services. Thus, the need for an effective credit card fraud detection system to help mitigate the risks and damaging impacts associated with credit card frauds, thus helping to preserve the security of the credit card electronic transaction systems and services.

Credit card fraud not only affects individuals and financial institutions, but also poses a threat to the normal functioning of other critical infrastructure sectors such as e-commerce, retail, hospitality, and healthcare. For example, due to convenience and accessibility, e-commerce platforms have become an increasingly preferred means of shopping as opposed to traditional physical shops [6]. However, a prevailing challenge is the same enabling efficiencies means that online retailers and e-commerce platforms are increasingly reachable targets of credit card fraudsters, often exacerbated by the potential lack or limitation in security capabilities enabled with the system which can often lead to data breach of customers card details [7]. Another pathway for credit card fraud occurrences in e-commerce setting is the use of stolen card credentials to make purchases, posing as a legitimate customer [8]. Also, the healthcare sector and the hospitality industry, especially hotels, restaurants, and online travel agencies are predominantly at risk of credit card fraud due to the high volume of credit card transactions they process. Thus, mitigating the risks associated with credit card fraud in various sectors and preserving the security and privacy of the cardholders requires an effective and reliable solution approach, especially from a detection point of view. This is because, it is only when the fraud actions and/or activities are well-detected can appropriate countermeasure be employed to curb the menace caused.

Detecting credit card fraud entails the continuous monitoring of transactions and user behaviours with the aim of recognizing irregular transactional patterns, anomalies, or suspicious activities. However, the dynamics and unpredictable tactics employed by credit card fraudsters pose a formidable challenge for financial institutions to efficiently and timely identify and prevent unauthorized credit card transactions [9], and the associated financial and reputational impacts that can occur. In detecting credit card fraud cases, several methods that employ manual checks by humans and rule-based systems in isolation or combination have been used with varied results. However, a common limitation of these methods has been the limited effectiveness in detection due to the inability to correctly identify and adapt to the complex and the evolving credit card fraud patterns as well as the extensive reliance on manual methods for defining fraud detection rules, which pruned to delays and human-error. Machine learning (ML) techniques are more recently being adopted to address the above challenges with the potentials for improved outcomes. However, despite the effectiveness of ML in supporting the detection of credit card fraud, there are still challenges that need to be addressed to optimise the outcomes. One of main issues with credit card fraud detection using ML approach is class imbalance [10], a problem that stems from the inherent nature of credit card transactions, where legitimate transactions often significantly outnumber illegitimate transaction. As a result, the dataset used for training and evaluation is heavily imbalanced, with most examples falling into the legitimate class. This tends to affect the quality of training, accuracy level of detection, and associated outcomes, often characterised by a range of false positives and false negatives.

This paper aims to address the class imbalance issue associated with credit card fraud detection, as well as optimising the feature selection approach to improve the classification model's capability, all towards achieving a more effective credit card fraud detection system. To achieve the above aim, this paper focuses on comparing the performance of specific algorithms–Random Forest, Logistic Regression and Decision Trees which have been noted to yield favourable outcomes from prior literature. Then, a comparative analysis of three sampling techniques namely, Random Undersampling, Random Oversampling and Synthetic Minority Over-Sampling Technique (SMOTE) is explored to help address the class imbalance problem present in the chosen European Cardholders dataset used in this study. Also, the feature selection approach named Recursive Feature Elimination with Cross-Validation (RFECV) is explored for selecting the most optimal features in the dataset to help solve the issue of high dimensionality due to the large number of features in the dataset.

This paper contributes novelty by presenting a new ML model that improves the accuracy of detecting credit card fraud, and minimising associated losses and impacts. This is demonstrated through the combination of multiple techniques to address existing challenges to attaining more accurate detection of credit card fraud instances. Employing Synthetic Minority Over-sampling Technique (SMOTE) to address the class imbalance issues typical to credit card fraud datasets, and applying Recursive Feature Elimination with Cross-Validation (RFECV) scheme for optimising the selection of features for an improved detection accuracy of the proposed model, all serve to highlight a novel approach that yields better outcome that existing techniques, and can inform the development of improved credit fraud detection systems.

The remainder of this paper is organised as follows. Section 2 provides a brief overview of the ML classifiers explored in the research. Section 3 provides a review of related works. Section 4 provides an overview of the research methodology comprising dataset description, feature engineering, class imbalance resolution, and credit card fraud detection methodology framework. Section 5 presents the results and discussion. Section 6 presents the conclusion.

2 Related Work

Credit card fraud incidents presents a persistent threat globally that continues to evolve in form. Advancements in digital technology and adoptions appear to be supporting the evolution and ever-changing nature of credit card fraud tactics and patterns [11].

Credit card fraud practices started out with skimming in the early 1970s where the magnetic strip data from legitimate cards are electronically copied illegally and/ or secretly using skimming devices on Automated Teller Machines (ATMs) or Point of Sale terminals (POS) [12]. The practice then progressed into the counterfeiting of cards by fraudsters as the technology and skills for replicating card details of unsuspecting user on to blank cards became more accessible and easier to implement [13]. With the emergence of e-commerce in the 1990s, new credit card fraud methods emerged which involved the use of fake merchant websites and credit card number generators became common [14]. This progressed into high impact bulk data breaches that saw the compromise and stealing of highly sensitive credit information using malicious tools. For example, in 2009, a huge data breach in the U.S.A. involving the use of malicious software to compromise a payment processor's systems resulted in the compromise of about 130 million cards, which further impacted a minimum of 670 banks [13]. the TJX data breach in 2017 resulted in about 94 million credit card records being exposed [15]. Although, the use of Europay, MasterCard and Visa (EMVs) technology have emerged with improved encryption [12] to mitigate the menace of growing credit card fraud occurrences (van den [16]), the situation has not been completely addressed and cases continues to emerge. Thus, today, credit card fraud instances can take various forms ranging from 'Card not present fraud–(CNP fraud)' [17], 'Counterfeit Card Fraud–(CCF)' [20, 21], 'Lost or stolen Card Fraud' [18], 'Internet-related Frauds methods such as phishing-linked [19], fraudulent websites, site cloning/page jacking, and credit card generators [20], 'Bulk data breaches' [21], and 'Identify theft' [22]. All these bear the singular end goal of illegally making away with funds on unsuspecting victims.

In response to these range of credit card fraud forms emerging, over the years, several approaches–tools and techniques–starting out as traditional methods involving rule-based systems for identifying suspicious card transaction activities which relied on human experts [23] or the newer machine learning (ML) techniques [24], various solution techniques continue to be explored towards detecting credit card fraud instances in their varied forms in order to help mitigate their damaging impacts on individuals, businesses, and financial institutions.

It is crucial to highlight that credit card fraud detection using ML techniques is essentially a classification problem, hence, ML algorithms such as Naïve Bayes (NB), Logistic regression (LR), Decision Tree (DT), Support Vector Machine (SVM), K-Nearest Neighbour (KNN) and Neural Networks can be used to classify transactions into either a 'Fraud' or 'Not Fraud' class. Ensemble ML algorithms such as Random Forest (RF) and Extreme Gradient Boosting (XGBoost) can also be used for developing robust credit card fraud detection systems. Given the ability of outlined ML algorithms to uncover and learn patterns from large datasets to make appropriate predictions [24], they are well-suited for the task of detecting credit card fraud transactions as shown in some of the outlined works that follow.

In [1], the performances of five ML algorithms in the detection of credit card fraud were evaluated using a heavily imbalanced European credit card dataset obtained from Kaggle. The algorithms used were LR, RF, NB, DT and SVM. The researchers experimented with various training and testing sizes and the results concluded that RF with 80% training set and 20% testing set yielded the best performance of 99.95% accuracy, 0.74 recall, and 0.97 precision values. Similarly, [19] evaluated the efficiency of RF for credit card fraud detection, and obtained a high performance of 99.9% accuracy, 1.0 recall, and 0.99 precision values.

Authors in [25] proposed a strategy for obtaining a high recall score in credit card fraud detection using KNN, LDA and LR algorithms. An integrated algorithm created from the combination of KNN, LDA and LR was proposed and tested on four different datasets. The proposed integrated methodology was noted to have outperformed the other individual models with recall scores of 1.00, 0.97, 1.00 and 0.93 from the four datasets.

Tanouz et al. [10] introduced a credit card fraud detection system characterised by high accuracy and efficient prediction within a shorter time frame. The paper details the comparison of multiple classification algorithms to determine the algorithm with the best performance. To address the class imbalance issue present in the heavily imbalanced dataset obtained from Kaggle, undersampling technique was applied. The ML algorithms used in this study were DT, RF, LR and NB. The performance of each algorithm was then evaluated using accuracy, recall, precision, F1 score and ROC-AUC metrics. The RF algorithm appeared to have performed best among all the other ML models with an accuracy of 96.77%, 1.00 precision, 0.91 recall, 0.9535 F1 score and 0.9555 ROC-AUC score.

Padhi et al. [26] introduces a novel feature selection method called Rock Hyrax Swarm Optimization (RHSO) which is inspired by rock hyrax swarm behaviour, for improving credit card fraud detection. RHSO was used to identify an optimal subset of features from the European credit card dataset, which were then utilised to train four machine learning models: DT, SVM, KNN, and NB. Also, the paper compares RHSO with other feature selection techniques like Differential Evolution (DE), Genetic Algorithm (GA), Particle Swarm Optimization (PSO), and Ant Colony

Optimization (ACO). The findings showed that RHSO surpassed these methods in performance, achieving average accuracies of 99.58% with DT, 99.11% with SVM, 99.61% with KNN, and 98.85% with NB classifiers.

Authors in [27] proposed a credit card fraud detection system using Genetic Algorithm (GA). To handle the class imbalance in the European dataset used in this study, the SMOTE sampling technique was used. GA with RF as a fitness method was implemented as a feature selection technique to solve the high dimensionality issue. The authors detail that RF was used due to its ability to perform well with missing values and imbalanced data. DT, RF, LR, NB and ANN were employed in the development of the credit card fraud detection system. Similar to the research by [10] the performance metrics used were accuracy, recall, precision, F1 score and ROC-AUC. The results showed that GA contributed to better performance of the models in which RF produced the best results with accuracy ranging from 99.1–99.98% on the 5 optimal feature vectors selected by the GA. The AUC of RF was also the best at 0.96.

Madkaikar et al. [28] conducted a comparative analysis of LR, NB, RF, KNN, GB, SVM and neural network models for the detection of credit card fraud. The dataset used in the study was imbalanced with 51,149 legal transactions and 3,312 fraudulent transactions. To address the class imbalance, undersampling technique was used. The authors employed seven feature selection methods to remove unnecessary features in the dataset. The feature selection techniques included SelectKBest, feature importance, extra trees classifier, Pearson's correlation, mutual information, step forward selection and Recursive Feature Elimination (RFE). The features that were common to all feature selection methods were used to train the machine learning models. The results of the comparative analysis revealed that GB yielded the best accuracy score of 95.9% among the other models.

The study by [29] proposed a credit card fraud detection framework that utilises undersampling to overcome the class imbalance issue and feature selection techniques to ensure selection of only the relevant features. The feature selection techniques employed were SelectKBest, RFE and embedded method using RF as a classifier. It was concluded that employing feature selection techniques positively impacted the models resulting in higher accuracy of 93% as opposed to the absence of feature selection technique which resulted in the accuracy value of 90%.

Bearing all of the works and outcomes reviewed, it is clear that certain key challenges exist with credit card detections using ML techniques. One of these challenges is the class imbalance, also highlighted in [10]. This stems from the inherent nature of credit card transactions as often comprising of largely legitimate transactions significantly outnumbering the illegitimate ones. Thus, the dataset used for training and evaluation becomes heavily imbalanced, with most examples falling into the legitimate class. The class imbalance has been shown to have a detrimental effect on the effectiveness of SVM and other ML models [30]. With an imbalanced dataset, models tend to exhibit bias toward the majority class. In the credit card fraud detection context, this means that a model may become less sensitive to detecting fraudulent transactions, consequently, resulting to some fraudulent transactions going undetected, which is a significant concern. A key objective is achieving a credit card fraud detection model with resolve class imbalance and significantly enhanced sensitivity to fraudulent transactions, ultimately enhancing the system's overall performance. Oversampling technique [31], Synthetic Minority Oversampling Technique (SMOTE) [32], and Random Walk-Oversampling (RWO) [33] are approaches that have shown promising results in addressing the class imbalance issue.

Another issue that can affect the quality of outcome for credit card fraud detection is related to feature engineering. This describes where in a credit card transaction dataset containing rows of transactions is also comprised of a column data each representing a crucial feature of the transaction, and tends to constitute continuous, categorical, and some binary forms. Most ML algorithms cannot perform with categorical features hence, the need to transform the categorical features to numerical forms. Also, these datasets may contain many features that can negatively impact on the performance of the ML algorithm. To resolve this, dimensionality reductions techniques such as Principal Component Analysis (PCA) [34], t-distributed Stochastic Neighbour Embedding (t-SNE) [35], or Linear Discriminant Analysis (LDA) [25]. Feature selection techniques such as Genetic Algorithm (GA) [36] and Particle Swarm Optimization (PSO) [37] have been explored to help with selecting and maintaining the essential information that optimises the features for use.

Concept drift is another challenge with credit card fraud detection with ML, where unforeseen statistical changes occur in the data being analysed [3]. This can happen from the changing fraud patterns due to evolving tactics and smartness of fraudsters. This can result in new fraud patterns emerging which can shift the statistical distribution of illegitimate transactions in the dataset, thus, leading to reduced detection accuracy as the model's training data will not match with real-time data fed to the model. Similarly, evolving patterns from legitimate transactions can introduce a drift capable of negatively impacting a models detection accuracy.

Using ensemble models, implementing real-time monitoring of model's performance, regular retraining of models with updated data, or implementing drift detection algorithms to identify and measure concept drift can help towards addressing the impact of concept drift credit card fraud detection systems.

In this study, we focus on building on existing efforts towards addressing class imbalance and feature engineering issues towards enhancing the accuracy of detection for credit card fraud transactions. Addressing the class imbalance in credit card datasets and employing appropriate feature engineering techniques can lead to improved data used to train the ML model, thus, improving the quality of outcomes.

3 Classifiers

3.1 *Random Forest (RF)*

Random Forest is an ensemble learning algorithm which consists of multiple random decision trees. When constructing each individual tree, it employs bagging and feature randomness to produce an uncorrelated forest of trees with predictions that are more accurate than those of a single tree. Each tree in the random forest gives a class prediction then majority voting is used to pick the models prediction [38].

3.2 *Logistic Regression (LR)*

Logistic Regression is a supervised learning algorithm that is best used for binary classification. The output of the LR model is a probability score that represents the likelihood of each sample belonging to the positive class which in this project is the "Fraud" class. This score is typically a value between 0 and 1, with a decision boundary, often set at 0.5 by default to distinguish the two classes [39]. LR is particularly well-suited for binary classification problems such as the credit card fraud detection use case where each sample is either classified as 'Fraud' or 'Not Fraud'.

3.3 *Decision Tree (DT)*

Decision Tree is a tree-like structured supervised learning method that uses a set of rules to make decisions [38]. The idea is to build yes/no questions using the dataset features and keep splitting the dataset until all the data points for each class are isolated.

4 Research Methodology

In this section, the methodology engaged in carrying out the research is described.

4.1 *Process Flow Description*

As shown in Fig. 1, Following on from defining research scope, and literature review of credit card fraud detection forms and their progressive evolutionary instances and applicable approaches for detection, a review of related ML-driven detection

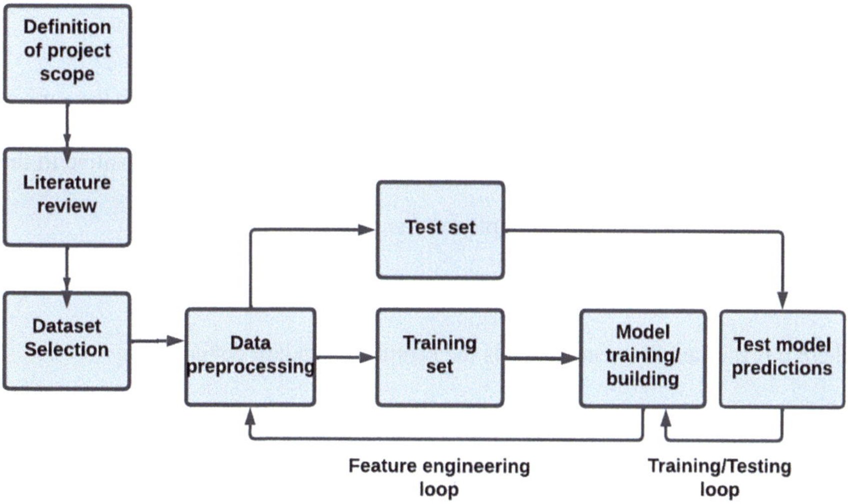

Fig. 1 Research methodology

techniques proposed by prior research was conducted. Specific focus was on the performance metrics and associated results obtained by prior research. Also, we highlighted some of the common challenges akin to the use of ML techniques for credit card fraud detection and potential resolution approaches found in literature. Based on these, we define the focus and gaps our research seeks to address, hence highlighting the contribution made by our work. This is the followed by dataset selection process, data pre-processing, a looped back or iterative training and testing for model building and optimisation to yield improved detection outcome.

4.2 Dataset Selection and Description

The dataset used in this research study obtained from Kaggle. The dataset contains recorded transactions carried out in September 2013 by European cardholders within a span of two days. Due to confidentiality issues, PCA was performed on the dataset leading to anonymised feature names; V1- V28. *Time, Amount* and *Class* are the only features in the dataset that were not transformed with PCA. The dataset is considered to be heavily imbalanced with 284,315 instances of the *'Not Fraud'* class accounting for 99.828% and only 492 *'Fraud'* instances which makes up only 0.172% of the total transactions.

4.3 Data Pre-Processing

Standardisation, which is a process of scaling numerical variables so that it has a mean of zero and a standard deviation of 1 was performed on the dataset. It was important to standardize the dataset as some machine learning algorithms are sensitive to the scale of the data.

The formula to determine a sample (x) standard score (z), is.

$$z = (x - u)/s \tag{1}$$

where u is the sample mean, and s is the samples' standard deviation.

4.4 Handling Class Imbalance

Addressing class imbalance is crucial, as data imbalance can negatively impact the performance of machine learning models, resulting in bias towards the majority class which in this case is the*'Not Fraud'* class. Such bias can make it more difficult for the model to accurately identify the more important but infrequent *'Fraud'* transactions which is the main goal of credit card fraud detection systems.

Recognising the importance of handling class imbalance, three different sampling techniques namely Random undersampling, Random Oversampling and Synthetic Minority Oversampling Technique (SMOTE) are evaluated. To provide an unbiased evaluation of the efficiency of each technique in mitigating class imbalance in the credit card fraud dataset, a systematic comparative study of sampling techniques was carried out.

4.5 Feature Selection

Due to the large number of features in the dataset, it was imperative to perform feature selection to obtain a subset of optimal features that may improve the performance of the machine learning models. For this, we adopted a wrapper-based feature selection method known as Recursive Feature Elimination with Cross-Validation (RFECV) [40].

4.5.1 Recursive Feature Elimination with Cross-Validation (RFECV)

Recursive Feature Elimination with Cross-Validation (RFECV) is a feature selection technique that selects the optimal number of features based on cross-validation performance of the model. RFECV combines the efficiency of Recursive Feature

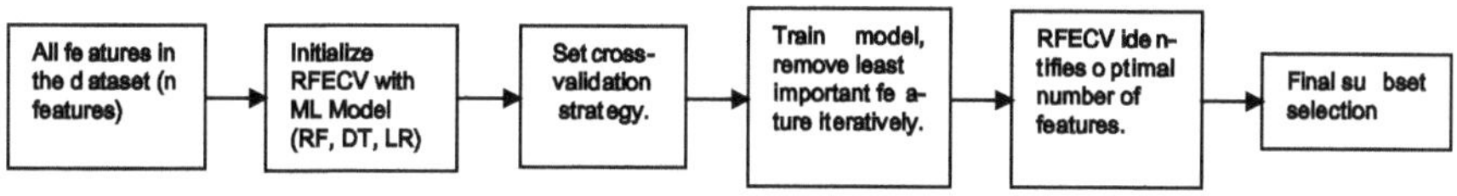

Fig. 2 Flowchart of the RFECV process

Elimination (RFE) with the reliability of cross-validation to determine the most relevant set of features for the model.

It starts with all features and removes the least important one at each step. After each step, it calculates the cross-validation score for the remaining set of features. The process continues until all features are explored. Finally, it selects the number of features that yielded the best performance during cross-validation. By incorporating cross-validation, RFECV ensures that the feature selection process is reliable and generalizable to unseen data.

Figure 2 presents a stepwise representation of how RFECV works to achieve optimised feature selection. It starts with initialising every feature in the dataset and moves on to RFECVs initialisation with ML algorithm (RF, DT, and LR) as an estimator, setting cross-validation and scoring metric selection, model training with feature elimination, optimal feature counting, and final subset selection based on the assessment of model performance using key performance metrics such as accuracy, precision, recall, F1, and ROC-AUC scores. Appropriate methods such as the scikit-learn library in python were used to support importing and implementing the RFECV scheme.

The proposed methodology is illustrated in Fig. 3. Data pre-processing techniques such as standardisation is performed on the dataset to ensure the numerical variables are within a specific desirable range. The dataset is then split into test and train splits. Data sampling techniques: Random undersampling, Random Oversampling and Synthetic Minority Oversampling Technique (SMOTE) are applied to the training data to address the class imbalance. The feature selection approach RFECV was then applied and the DT, LR and RF models were trained with the optimal subset of features obtained.

The choice of machine learning algorithms used in this study was based on a variety of factors, including their track records from literature in predictive modelling, and most importantly their capacity to rank features. These models are all well-suited to the iterative feature elimination process of RFECV since they can all provide importance scores to features, either through coefficients (in LR) or feature importance (as in DT and RF).

4.6 Performance Metrics

Given the complexity and importance of the task, the ML models' efficiency in detecting credit card fraud could not be effectively assessed using one-dimensional

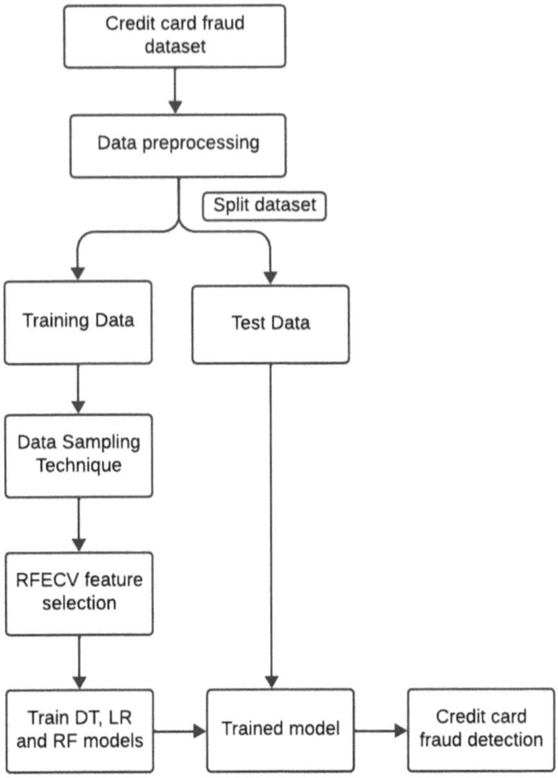

Fig. 3 The optimised credit card fraud detection process methodology detailing each step

metric. As a result, a combination of metrics was used to offer a thorough assessment of each model's performance. These include accuracy, recall, precision, and F1 and ROC-AUC scores which are further described below.

i. True Positives (TP): samples that the model properly classifies as positive and are truly positive.
ii. True Negatives (TN): samples that the model properly classifies as negative but are negative.
iii. False Positive (FP): samples that the model mistakenly classifies as positive even when they are negative.
iv. False Negative (FN): samples that are positive but incorrectly classified as negative by the model.

$$Accuracy = \frac{TP + TN}{TP + TN + FP + FN} \tag{2}$$

$$Recall = \frac{TP}{TP + FN} \tag{3}$$

$$Precision = \frac{TP}{TP + FP} \quad (4)$$

$$F1score = 2 * \frac{precision * recall}{precision + recall} \quad (5)$$

5 Results and Discussion

To choose a baseline model to compare the effects of the three sampling techniques, RF, DT and LR models were trained on the imbalanced dataset. As shown in Table 1, RF obtained the highest accuracy, precision, and F1 Scores of 99.96, 0.97 and 0.86 respectively with a high recall of 0.77. Hence, RF was chosen as the baseline model to compare the effects of oversampling, under-sampling, and SMOTE sampling techniques.

While all techniques showed high accuracy as shown in Table 2, oversampling and SMOTE improved the accuracy of the RF model while a slight decrease in accuracy was recorded when under-sampling was employed. Under-sampling showed a remarkably high recall of 0.94, but this was at the cost of very low precision of 0.06.

The results indicated that SMOTE significantly improved the RF model's performance in terms of recall, F1-score, and ROC-AUC without extreme trade-offs in accuracy and precision values. The high recall and precision made it a great choice for credit card fraud detection where both false positives and false negatives can carry significant consequences in terms of inconvenience, customer experience, and potential loss due to missed cases of fraud. This result aligns with the views from prior works [41, 42] where SMOTE measured as the optimal sampling technique also for credit card fraud detection use cases.

Table 1 Performance metrics of Initial model development

Model	Accuracy	Precision	Recall	F1 score	ROC-AUC
RF	99.96	0.97	0.77	0.86	0.96
DT	99.95	0.90	0.80	0.85	0.94
LR	99.91	0.86	0.58	0.70	0.97

Table 2 Comparative analysis of sampling methods

	Accuracy	Precision	Recall	F1 score	ROC-AUC
Initial RF model	99.96	0.97	0.77	0.86	0.96
Undersampled RF	97.62	0.06	0.94	0.12	0.98
Oversampled RF	99.96	0.99	0.78	0.87	0.95
SMOTE RF	99.96	0.91	0.85	0.88	0.99

The RFECV was implemented with RF, DT and LR models as the estimators to conduct a comparative analysis in order to determine the best performing machine learning model for developing a more accurate and robust credit card fraud detection system. The cross-validation strategy employed accuracy as the scoring metric in identifying the optimal features.

Figure 4 shows the visualisation of the performance metrics of the three different models: (i) RFECV with a Random Forest (RF) model, (ii) a Decision Tree (DT) model, and (iii) a Logistic Regression (LR) model. The results indicate that the RFECV RF model bears high scores across all metrics, indicating a well-performing and robust model for credit card fraud detection. The RFECV DT model has a high accuracy but significantly lower precision and F1 scores compared to the RF model. Its recall is relatively high, which is good for fraud detection as it indicates a higher true positive rate. The RFECV LR model has the least precision and F1 scores, but its recall is the highest among the three models. This suggests that the model is best at identifying most actual fraudulent transactions despite a high number of false positives (as indicated by the low precision).

After considering the trade-offs between the accuracy, precision, recall and F1 metrics across the models, RFECV with RF model was determined as the best performing model. The high performance of this model for each of the metrics suggest that the model performs well in correctly identifying fraudulent credit card transactions leading to less false alarms and capturing most of the fraudulent transactions to ensure less chances of missed credit card fraud cases. The F1 score shows the model's ability to correctly balance the predicting of fraudulent credit card transactions and identifying most of the fraudulent credit card cases. This result aligns with the results from previous works [1, 10], and [19],where the RF model yielded

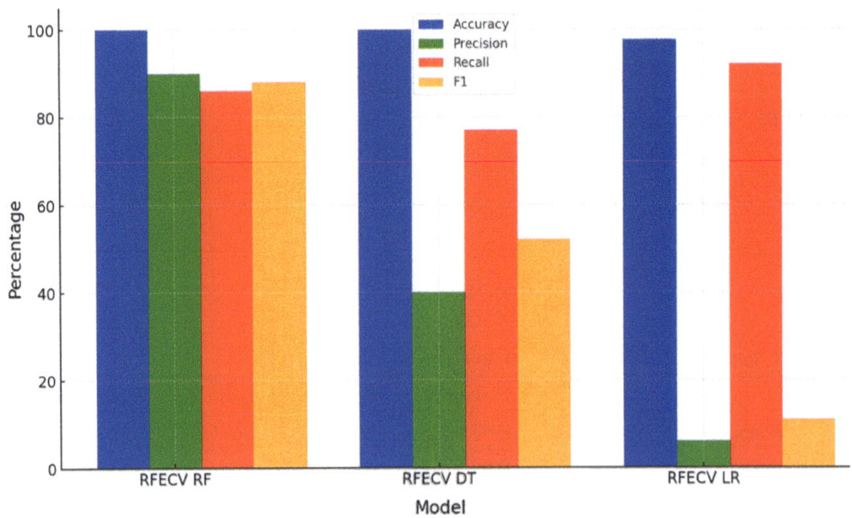

Fig. 4 Comparison of performance metrics across models

the best performance in each case with the highest accuracy values among other ML models explored. Also, the proposed RFECV RF model for credit card fraud detection surpasses the KNN model by [43] in terms of recall and F1 scores. This indicates that the RFECV model implemented in this research study bears better ability of correctly identifying fraudulent transactions as well as a better balance between precision and recall. In the implementation of feature selection approach, the proposed RFECV RF model obtained higher accuracy scores than the RHSO approach proposed by Padhi et al. [44]. Also, it achieved higher ROC-AUC score compared to the GA RF model proposed by [27]. This suggests that the RFECV RF model proposed for credit card fraud detection yields a better performance comparatively to prior related works, and thus, is an improved solution for achieving a more accurate detection of credit card fraud instances.

Despite the success in developing a more robust RFECV RF model for credit card fraud detection, certain issues caused some limitation in this project. An important limitation was the lack of knowledge behind the features in the dataset due to the PCA transformation. Understanding the features pre-data processing could lead to improved domain knowledge on credit card transactions, and the creation of new features that can potentially improve the predictive power of the models. However, this was not possible with this dataset. Another limitation associated with the PCA transformed dataset was due to the nature of PCA components. The components/features generated during PCA transformation is the amalgamation of all features in the dataset which can lead to spreading the importance across the dataset such that each feature contributes to the model. As the RFECV is a feature selection approach, the aim is to select optimal subset of features, thereby reducing the number of features to train the model on. However, due to the potential distribution of importance across the dataset, the RFECV identified most of the features as optimal therefore not reducing the number of features. Using raw data not transformed with PCA for training a RFECV RF model can yield better results associated with selected most optimal features within a dataset.

6 Conclusion

This project focused on addressing the pressing issue of credit card fraud, which poses significant risks to individual cardholders, financial institutions, and the economy. The limitations of existing credit card fraud detection systems were identified and measures to mitigate these limitations were explored in this study. By implementing SMOTE sampling technique, the class imbalance issue associated with credit card fraud detection was handled. The application of feature selection approach; RFECV ensured that the optimal subset of features was employed in the dataset. Approaches implemented along with proper data pre-processing steps ensured this project met the primary aim of creating a robust and efficient solution to detect fraudulent transaction. The effectiveness of the credit card fraud detection system developed in this project is reflected in its high performance across various evaluation criteria which includes accuracy, recall, precision, F1 score and ROC-AUC, indicating the models ability to overcome challenges associated with imbalanced dataset and feature engineering complexities.

References

1. Vejalla, I., Battula, S.P., Kalluri, K., Kalluri, H.K.: Credit card fraud detection using machine learning techniques. In: 2023 2nd International Conference on Paradigm Shifts in Communications Embedded Systems, Machine Learning and Signal Processing (PCEMS), pp. 1–4. IEEE (2023)
2. Maniraj, S.P., Saini, A., Ahmed, S., Sarkar, S.: Credit card fraud detection using machine learning and data science. Int J Eng Res **8**(9), 110–115 (2019)
3. Dornadula, V.N., Geetha, S.: Credit card fraud detection using machine learning algorithms. Procedia Comput Sci **165**, 631–641 (2019)
4. Raynor de Best.: Total value of losses due to card fraud, either credit card fraud or debit 497 card fraud, worldwide from 2014 to 2022. In: Card fraud – credit cards and debit cards 498 combined – worldwide 2014–2022. Crime & Law Enforcement, Online Statistics, 24 September, STATISTA (2024). Available at: https://www.statista.com/statistics/1394119/global-card-fraudlosses/#statisticContainer
5. Merchant Savvy, Payment fraud statistics, trends, and forecasts 2024. Merchant Savvy (2024). https://www.merchantsavvy.co.uk/payment-fraud-statistics/
6. Taher, G.: E-commerce: advantages and limitations. Int J Acad Res Ac-Count Financ Manag Sci **11**(1), 153–165 (2021)
7. Jamra, R.K., Anggorojati, B., Sensuse, D.I., Suryono, R.R.: Systematic review of issues and solutions for security in e-commerce. In: 2020 International Conference on Electrical Engineering and Informatics (ICELTICs), pp. 1–5. IEEE (2020)
8. Chaudhary, K., Mallick, B.: Credit card fraud: bang in e-commerce. Editorial Board, p. 935
9. Cherif, A., Badhib, A., Ammar, H., Alshehri, S., Kalkatawi, M., Imine, A.: Credit card fraud detection in the era of disruptive technologies: a systematic review. J. King Saud Univ.-Comput. In-Form. Sci. (2022)
10. Tanouz, D., Subramanian, R.R., Eswar, D., Reddy, G.P., Kumar, A.R., Praneeth, C.V.: Credit card fraud detection using machine learning. In: 2021 5th International Conference on Intelligent Computing and Control Systems (ICICCS), pp. 967–972. IEEE (2021)
11. Benchaji, I., Douzi, S., El Ouahidi, B., Jaafari, J.: Enhanced credit card fraud detection based on attention mechanism and LSTM deep model. J Big Data **8**, 1–21 (2021)
12. Gold, S.: The evolution of payment card fraud. Comput Fraud Secur **2014**(3), 12–17 (2014)
13. Sullivan, R.J.: The US adoption of computer-chip payment cards: implications for payment fraud. Economic Review-Federal Reserve Bank of Kansas City, p.59 (2013)
14. Bhatla, T.P., Prabhu, V., Dua, A.: Understanding credit card frauds. Cards business review **1**(6), 1–15 (2003)
15. Sullivan, R.J.: May. Issues for Industry and Public Policy, The Changing Nature of US Card Payment Fraud (2010). In WEIS
16. van den Breekel, J., Ortiz-Yepes, D.A., Poll, E., de Ruiter, J.: EMV in a nutshell. Technical Report (2016)
17. European Central Bank: Report on card fraud in 2020 and 2021. Eur CentL Bank (2023). https://doi.org/10.2866/607986
18. Ross, D.E.: Credit card fraud Prevention & Detection Strategies (2023). https://www.britannica.com/topic/credit-card-fraud
19. Agarwal, S., Usha, J.: Detection of fraud card and data breaches in credit card transactions (2022)

20. Almudaires, F., Almaiah, M.: Data an overview of cybersecurity threats on credit card companies and credit card risk mitigation. In: 2021 International Conference on Information Technology (ICIT), pp. 732–738. IEEE (2021)
21. Statista, UK number of breached data points in Q1 2020-Q2 2023 (2023). https://www.statista.com/statistics/1386806/uk-number-of-leaked-records/. Accessed 10 Nov 2023
22. Consumer Sentinel Network Data Book 2022, 2023. Federal Trade Commission. https://www.ftc.gov/reports/consumer-sentinel-network-data-book-2022. Accessed 07 Nov 2023
23. Ryman-Tubb, N.F., d'Avila Garcez, A.: SOAR – sparse oracle-based adaptive rule extraction: knowledge extraction from large–scale datasets to detect credit card fraud. In: International Joint Conference on Neural Networks (IJCNN), Barcelona, Spain, pp. 1–9 (2010). https://doi.org/10.1109/IJCNN.2010.5596631
24. Mohsen, O.R., Nassreddine, G., Massoud, M.: Credit Card Fraud Detector Based on Machine Learning Techniques. J Comput. Sci Technol Stud. **5**(2), 16–30 (2023)
25. Chung, J., Lee, K.: Credit card fraud detection: an improved strategy for high recall using KNN, LDA, and linear regression. Sensors **23**(18), 7788 (2023)
26. Padhi, B.K., Chakravarty, S., Naik, B., Pattanayak, R.M., Das, H.: RHSOFS: feature selection using the rock hyrax swarm optimization algorithm for credit card fraud detection system. Sensors **22**(23), 9321 (2022)
27. Ileberi, E., Sun, Y., Wang, Z.: A machine learning based credit card fraud detection using the GA algorithm for feature selection. J. Big Data **9**(1), 1–17 (2022)
28. Madkaikar, K., Nagvekar, M., Parab, P., Raikar, R., Patil, S.: Credit card fraud detection system. Int. J. Recent. Technol. Eng. (IJRTE) **10**, 158–162 (2021)
29. Mniai, A., Tarik, M., Jebari, K.: A novel framework for credit card fraud detection. IEEE Access (2023)
30. Wu, G., Chang, E.Y.: Class-boundary alignment for imbalanced dataset learning. In: ICML 2003 Workshop on Learning from Imbalanced Data Sets II, Washington, DC, pp. 49–56. (2003)
31. Zheng, M., Wang, F., Hu, X., Miao, Y., Cao, H., Tang, M.: A method for analyzing the performance impact of imbalanced binary data on machine learning models. Axioms **11**(11), 607 (2022)
32. Chawla, N.V., Bowyer, K.W., Hall, L.O., Kegelmeyer, W.P.: SMOTE: synthetic minority oversampling technique. J. Artif. Intell. Res. **16**, 321–357 (2002)
33. Zhang, H., Li, M.: RWO-Sampling: a random walk over-sampling approach to imbalanced data classification. Inf. Fusion **20**, 99–116 (2014)
34. Trisanto, D., Rismawati, N., Muhamad Femy, M., Felix Indra, K.: Effectiveness undersampling method and feature reduction in credit card fraud detection. Int. J. Intell. Eng. Syst. **13**(2), 173–181 (2020)
35. Deb, K., Ghosal, S., Bose, D.: A comparative study on credit card fraud detection (2021)
36. Saheed, Y.K., Hambali, M.A., Arowolo, M.O., Olasupo, Y.A.: Application of GA feature selection on Naive Bayes, random forest and SVM for credit card fraud detection. In: 2020 International Conference on Decision Aid Sciences and Application (DASA), pp. 1091–1097. IEEE (2020)
37. Shukla, S., Rakesh, D.: Dynamic ensemble-based feature selection model for credit card fraud detection. In: 2020 IEEE 17th India Council International Conference (INDICON), pp. 1–6. IEEE (2020)
38. Ali, J., Khan, R., Ahmad, N., Maqsood, I.: Random forests and decision trees. Int J Comput Sci Issues (IJCSI) **9**(5), 272 (2012)
39. Hilbe, J.M.: Logistic regression. Int Encycl Stat Sci **1**, 15–32 (2011)
40. Akhtar, F., Li, J., Pei, Y., Xu, Y., Rajput, A., Wang, Q.: Optimal features subset selection for large for gestational age classification using gridsearch based recursive feature elimination with cross-validation scheme. In Frontier Computing: Theory, Technologies and Applications (FC 2019) 8 pp. 63–71. Springer Singapore (2020)
41. Varmedja, D., Karanovic, M., Sladojevic, S., Arsenovic, M., Anderla, A.: Credit card fraud detection-machine learning methods. In: 2019 18th International Symposium INFOTEH-JAHORINA (INFOTEH), pp. 1–5. IEEE (2019)

42. Puh, M., Brkić, L.: Detecting credit card fraud using selected machine learning algorithms. In: 2019 42nd International Convention on Information and Communication Technology, Electronics and Microelectronics (MIPRO), pp. 1250–1255. IEEE (2019)
43. Vengatesan, K., Kumar, A., Yuvraj, S., Kumar, V., Sabnis, S.: Credit card fraud detection using data analytic techniques. Adv. Math.: Sci. J. **9**(3), 1185–1196 (2020)
44. Padhi, B.K., Chakravarty, S., Biswal, B.N.: Anonymized credit card transaction using machine learning techniques. In: Mohanty, M., Das, S. (eds.) Advances in Intelligent Computing and Communication. Lecture Notes in Networks and Systems, vol. 109 (2020). Springer, Singapore. https://doi.org/10.1007/978-981-15-2774-6_49
45. De Best, Total value of losses due to card fraud, either credit card fraud or debit card fraud worldwide from 2014–2022. Statista (2014). https://www.statista.com/statistics/1394119/global-card-fraud-losses/#:~:text=Card%20fraud%20losses%20across%20the,from%20the%20United%20States%20alone
46. Statista, UK debit & credit cards: Counterfeit losses on plastic cards 2002–2020 (2022). https://www.statista.com/statistics/286254/united-kingdom-uk-counterfeit-losses-on-plastic-cards/. Accessed 08 Nov 2023
47. Dekker, N.: Staggering payment fraud statistics you need to know (2023). https://eftsure.com/statistics/payment-fraud-statistics/#source-wrapper. Accessed 16 Nov 2023

Attention Based Image Steganography Using CNNs in Integration with Quantization

Gatram Sravan Kumar, Kamalakanta Sethi, Piyush Joshi, and Rakesh Kumar Sanodiya

Abstract Steganography is a security mechanism used to conceal sensitive data in various forms of digital media. This is useful for transmitting sensitive information without disclosing its existence. Our paper introduces an advanced steganographic method based on deep learning, which combines Convolutional Neural Networks (CNNs) with Quantization and Convolutional Block Attention Module (CBAM). Our steganography method addresses several issues with traditional approaches, such as low embedding capacity and steganographic image distortions. The integration of CNNs in our proposed approach allows for effective feature extraction, while Quantization refines and enhances the extracted features which helps to preserve the image quality after embedding. In addition, the CBAM attention mechanism captures useful features while filtering out noise and irrelevant information. This helps us to reduce the feature distortions and improve the feature representations. We have implemented and tested the effectiveness of our steganographic approach with various evaluation metrics. The experimental results show that our model outperforms the state-of-the-art approaches.

1 Introduction

Image Steganography is a process of embedding secret information into digital images and the output generated as a result of embedding is known as a steganographic image. The embedding is performed in such a way that the changes are not

G. Sravan Kumar (✉) · K. Sethi · P. Joshi · R. K. Sanodiya
Indian Institute of Information Technology, Sricity, India
e-mail: sravankumar.g21@iiits.in

K. Sethi
e-mail: kamalakanta.s@iiits.in

P. Joshi
e-mail: piyush.j@iiits.in

R. K. Sanodiya
e-mail: rakesh.s@iiits.in

M. G. Jaatun et al. (eds.), *Proceedings of the International Conference on Cybersecurity, Situational Awareness and Social Media*, Springer Proceedings in Complexity,
https://doi.org/10.1007/978-981-96-0401-2_3

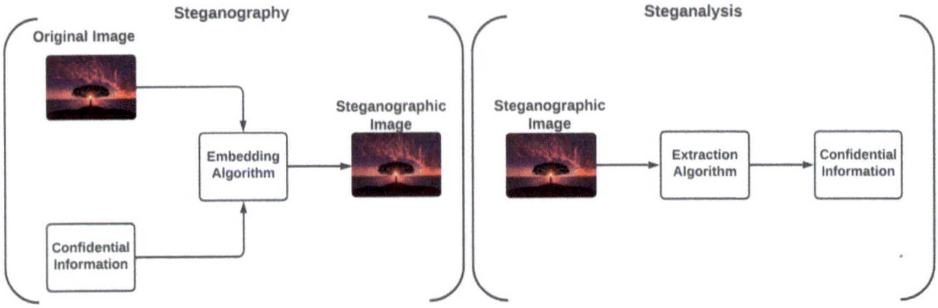

Fig. 1 General working process of steganography

perceptible to the human eye. The steganographic image is transmitted over a network in such a way that the attacker believes that it is just a normal image but the secret information is encoded inside the pixels of the image. This is useful for sending secret information without revealing its existence. Figure 1 illustrates the working of an image steganographic approach. Researchers in the literature used several image steganographic techniques, which are described below.

Least Significant Bit (LSB) steganography is a process where secret information is embedded by altering the least significant bits of pixel values. LSB is simple and efficient but it is vulnerable to statistical analysis [1]. Another popular approach involves Discrete Cosine Transform (DCT) steganography, where secret data is embedded in the frequency domain by modifying the DCT coefficients of the image. Despite its effectiveness in hiding information, DCT-based steganography can introduce image distortions which decrease the visual quality of the steganographic image [3]. In Discrete Wavelet Transform (DWT) steganography, we divide the image into frequency subbands and embed data in selected coefficients. DWT offers improved robustness and security compared to LSB and DCT but the embedding capacity of DWT is less [5].

To address the limitations of traditional approaches mentioned above, deep learning techniques such as Convolutional Neural Networks (CNNs) and Generative Adversarial Networks (GANs) have evolved in the field of image steganography [12]. CNNs enable automatic feature extraction and learning of complex patterns, GANs help in the generation of realistic-looking steganographic images.

The major challenges in the field of image steganography includes the data embedding capacity and preserving the steganographic image quality. To address these limitations and enhance the efficiency of image steganography, we propose a novel steganographic approach that combines CNNs, Quantization, and Convolution Block Attention Module (CBAM). CNNs help to extract features efficiently, while the integration of Quantization and CBAM attention mechanisms improves the extraction of refined essential features. This helps in preserving the image quality of the steganographic image.

The remainder of the paper is structured as follows. In Sect. 2, we discussed the state-of-the-art steganographic approaches. Our proposed DL-based steganographic approach is discussed in Sect. 3. The detailed description of the dataset and its pre-processing is discussed in Sect. 4. The experimental results are shown in Sect. 5. We conclude the paper with future work in Sect. 6.

2 Related Work

In this section, we will discuss some of the existing works that use traditional approaches like LSB, DCT, DWT, and deep learning techniques like CNNs and GANs for image steganography.

Bharatkumar et al. [1] proposed an enhanced LSB based image steganography approach that uses triple prime image encryption. In paper [2], authors presented an image steganography approach that uses LSB and AES-128 encryption. Although LSB steganography is simple and extensively used, it has two major drawbacks: low embedding capacity and vulnerability to statistical analysis. Therefore, researchers started investigating other approaches such as, DCT and DWT for image stenography. Discrete Cosine Transform (DCT) based steganography provides high embedding capacity due to its ability to spread information across the entire frequency domain. In paper [3], authors presented a DCT based steganographic approach that uses edge pixel embedding. In [4], Mohammed Baziyad et al. explores the significance of maintaining the phase information in DCT base steganographic approach. DCT-based techniques degrade the overall visual quality of the image and are susceptible to image manipulations. On the other hand, DWT-based steganography offers high capacity for hiding data due to its ability to divide images into multiple frequency components, allowing for efficient embedding of information in various frequency components. In paper [5] authors studied the importance of DWT-based steganography with Haar filters. Vyas et al. [6] explores the integration of cryptography and steganography by leveraging both DWT and DCT techniques. It achieves higher resistance against different attacks. In another paper [7], the authors presented a comparative study on DCT and DWT based steganography. They found that DCT generally outperforms DWT in terms of PSNR and payload capacity. DWT steganography may suffer from less resistance against different attacks since the modifications in the high-frequency subbands are easier to detect.

To increase the embedding capacity along with preserving the quality of the image, researchers explored different DL based techniques for image steganography. In paper [8], the authors presented a comparative study between the traditional and DL based image steganography techniques. Qiu et al. [9] explores the fusion of image steganography and deep learning techniques, proposing a system with higher accuracy than existing methods. The authors in [10], analyzed various DL network architectures, layers, and activation functions to enhance the accuracy of steganography. Wei et al. [11] presented a DL based image steganographic algorithm using convolutional neural networks, U-Net, and Swin Transformer. They evaluated the effectiveness of their model using various evaluation metrics. In the next paragraph, we will discuss DL based image steganographic approaches that use GANs.

In [12], Sultan et al. developed a DL based steganography model using DCGANs (Deep Convolutional GAN) that can hide two messages in a single cover image. It achieved high extraction rates and maintained good image quality. The authors in [13], introduced an attention GAN-based steganographic method that includes an encoder, decoder, and discriminator networks. Adversarial training is used to optimize network weights, resulting in high PSNR and low MSE.

Thenmozhi et al. [14] discussed the advantages of thinner network layers in CNNs for image steganography. In paper [15], authors present a steganography method by combining U-Net architecture and MobileNet CNN to embed images into other images. This gives better results in comparison with traditional approaches. Geetha Devi et al. [16] employed the MSS-SE approach, in steganography to protect secret information by sending a randomized mix of cover and confused images. Gurumurthy et al. [17] proposed a new steganography using DWT and CNNs. DWT is used for encoding and decoding whereas CNN is used for preprocessing and training. In paper [18], authors presented a novel image steganography approach by combining a customized CNN model with an Encoder-Decoder network. The CNN extracts important features from steganographic images, while the Encoder-Decoder network reconstructs the original image from the steganographic image.

In the paper [19] authors presented the importance of CBAM attention mechanism in CNN. The CBAM attention mechanism helps to capture useful features while filtering out noise and irrelevant information. Kwak et al. [20] discussed the performance improvement by using Quantization layers with CNN. Quantization helps to refine the extracted features. Although CBAM and Quantization helps to preserve the image quality by refining the extracted features, they are not explored yet in the field of image steganography. This provides the motivation of our current research. In the next section, we will discuss our proposed DL based steganographic approach.

3 Proposed Stegnography Approach

In this section, we will discuss the methodology used to perform image steganography. Our steganographic model primarily consists of three networks: the preparation network, the hiding network, and the reveal network. The preparation network is used to extract features from the secret image. The embedding network is used to embed the extracted features of the cover image with the secret image and generate a steganographic image. The reveal network is used to extract the secret image from the steganographic image. The overall architecture of the model is depicted in Fig. 2. The detailed description of each network is discussed below.

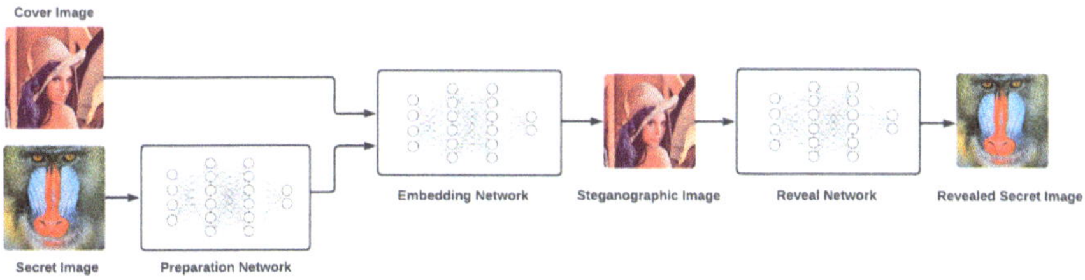

Fig. 2 General working architecture of our proposed approach

3.1 Preparation Network

The preparation network is used to extract features from the secret image and make it ready for embedding into the cover image. This network takes a secret image as the input and it is passed through a set of three convolutional layers (L1, L2, L3). Layer L1 consists of 50 filters with kernel size 3×3. Layer L2 consists of 10 filters with kernel size 4×4 where as layer L3 consists of 5 filters with kernel size 5×5. The stride used in each layer is 1. It is noted that the stride determines how many squares or pixels our filters skip when they move across the image, from left to right and from top to bottom. The activation function used for all three layers is Relu which is used to introduce nonlinearity to the network that helps in identifying complex patterns. The number of filters in each layer is reduced (L1–50 filters, L2–10 filters, and L3–5 filters) and the size of the kernel in each layer is increased (L1–3×3, L2–4×4, and L3–5×5). This helps the preparation network to learn global patterns and high-level representations of the secret image. The output of each layer are feature maps and the number of feature maps generated by each layer is equal to the number of filters used in that layer.

After each convolutional layer, a custom quantizer layer is applied to quantize the feature maps (generated from each convolutional layer). The custom quantizer layer uses 8-bit quantization which scales the floating-point values of the feature maps to the range [0, 255]. Subsequently, it rounds the scaled values to the nearest integer. After rounding, the values are scaled back to the original range to preserve the original magnitude of the features. Quantization reduces distortions in features by rounding the feature maps. This helps to refine and enhance the extracted features.

In the next step, the output feature maps from the three convolutional quantized layers are concatenated. This combines the extracted features from different filter sizes into a single feature map. In our model L1, L2, and L3 capture 50,10, and 5 features respectively. Therefore, the concatenated feature map contains a total of 65 (50+10+5) features. Concatenation allows the model to capture diverse features extracted from different convolutional layers of the network. By combining feature maps from multiple layers, the model gains the ability to learn complex patterns and improve overall performance. Further, the concatenated feature maps are iteratively passed through the same layers in the architecture as shown in Fig. 3. This helps in extracting more abstract and high-level features of the secret image.

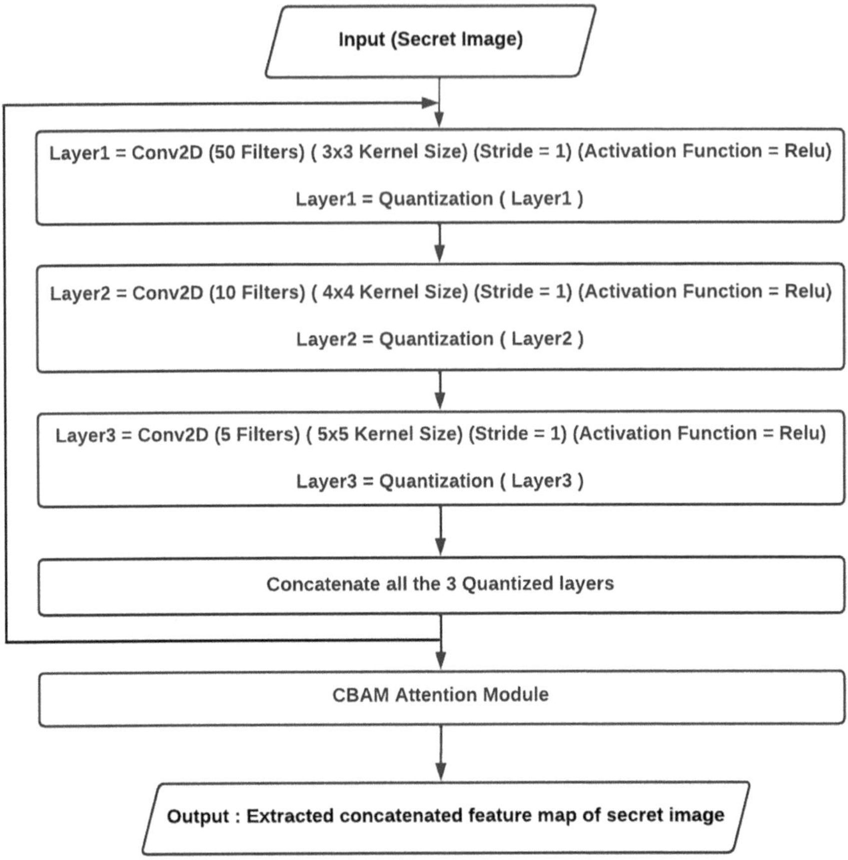

Fig. 3 Preparation network architecture

Finally, the CBAM attention mechanism [19] is applied on the concatenated feature map which is a combination of channel-wise attention mechanism and spatial-wise attention mechanism. It is a powerful tool used to improve CNN by enhancing its ability to extract features from images. Channel-wise attention focuses on inter-dependencies between different channels (Here channel indicates different types of features) of feature maps. It helps to focus on relevant features and suppress the irrelevant features. Spatial attention focuses on capturing spatial (positions of features) dependencies. It helps to prioritize relevant spatial regions in the feature maps. The overall CBAM attention mechanism helps to capture useful features while filtering out noise and irrelevant information. The general architecture of CBAM is illustrated in Fig. 4. The complete feature extraction of the secret image is done and is ready for embedding into the cover image. This completes the total working of the preparation network.

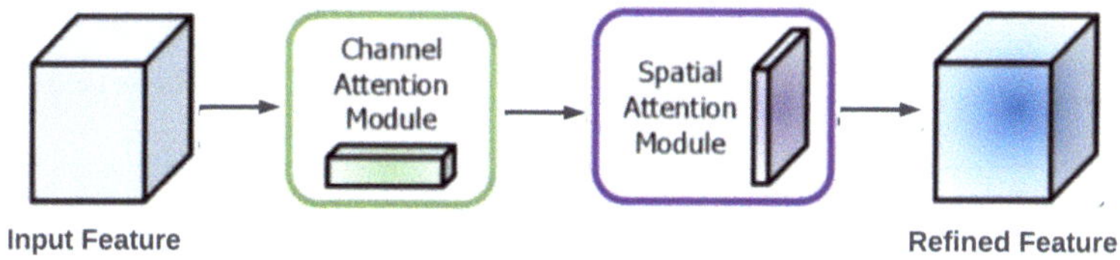

Fig. 4 General working architecture of CBAM

3.2 *Embedding Network*

The embedding network is used to hide the secret image into the cover image while maintaining the visual quality of the cover image. The output of the CBAM attention module (from the preparation network) is concatenated with the cover image. The purpose of concatenation is to combine the information extracted from the secret image with the cover image. This combined representation is then fed into the embedding network where the secret image is completely embedded into the cover image without much image distortion.

Following the concatenation, feature maps undergo a series of convolutional operations similar to the peripheral network as shown in Fig. 5. The activation function used is Relu which introduces nonlinearity to the network which helps in identifying complex patterns. These convolutional layers are responsible for extracting hierarchical features from the concatenated data.

After each convolutional layer, a custom quantizer layer is applied to quantize the feature maps which helps to reduce the distortions in the extracted features. All the extracted features from the three feature maps are concatenated. The concatenated feature map is iteratively passed through the layers in the architecture as shown in Fig. 5. Finally, the CBAM attention mechanism is used to refine the features. The final output is a steganographic image which contains the secret image inside the cover image.

3.3 *Reveal Network*

The Reveal network is responsible for reconstructing the secret image from the steganographic image. This network takes the steganographic image as input and Gaussian noise with a standard deviation of 0.01 is added to the steganographic image. This noise addition is intended to enhance the robustness of the decoding process. The architecture for extracting the features from steganographic image is similar to the preparation and embedding network. The whole process performed by the revealed network is shown in Fig. 6.

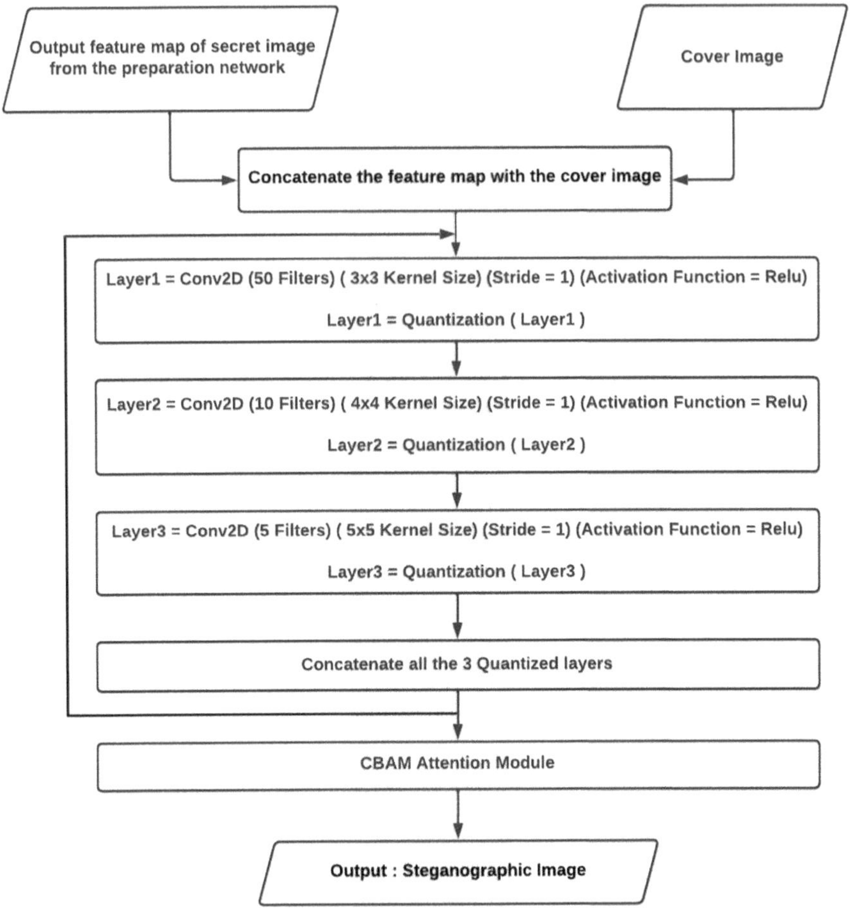

Fig. 5 Embedding network architecture

3.4 *Loss Functions*

The Loss functions used for training the preparation and embedding networks is given in Eq. 1. The loss function used for reveal network is shown in Eq. 2.

$$\text{Full\_loss} = \sum_{i,j}(c_{\text{true}}(i,j) - c_{\text{pred}}(i,j))^2 + \beta \cdot \sum_{i,j}(s_{\text{true}}(i,j) - s_{\text{pred}}(i,j))^2 \quad (1)$$

$$\text{Rev\_loss}(s_{\text{true}}, s_{\text{pred}}) = \beta \cdot \sum_{i,j}(s_{\text{true}}(i,j) - s_{\text{pred}}(i,j))^2 \quad (2)$$

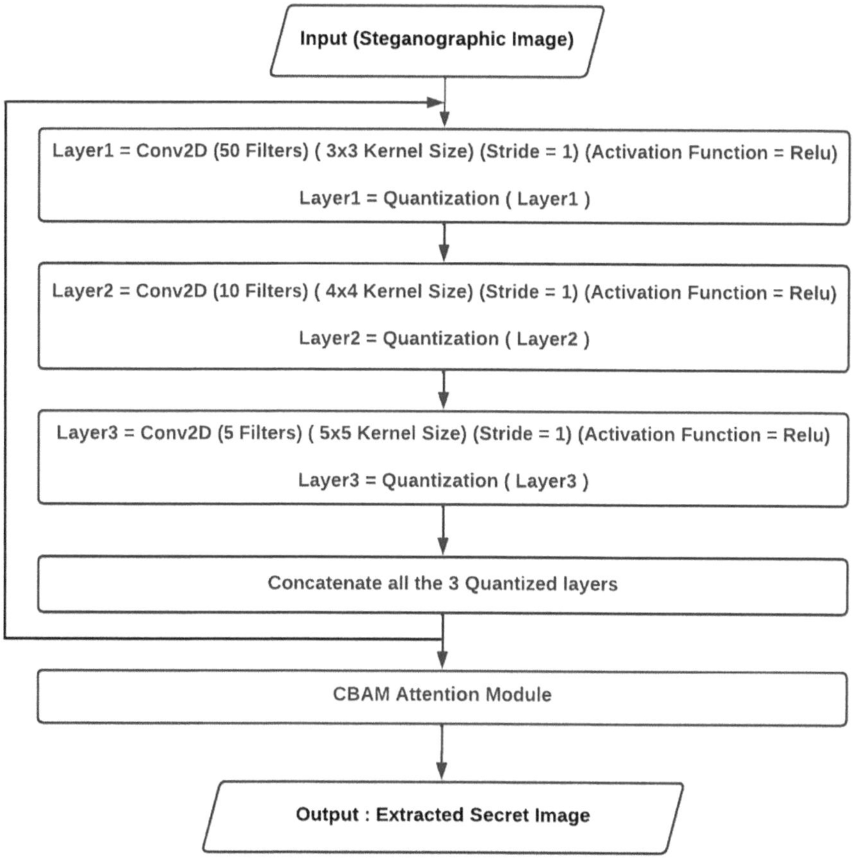

Fig. 6 Reveal network architecture

In Eqs. 1 and 2, s_{true} and s_{pred} represent the true and predicted secret images, c_{true} and c_{pred} represent the true and predicted cover images, and β is a hyperparameter representing the weight of the secret image loss term. The sums are taken over all pixel locations (i, j) in the cover and secret images. These loss functions help to achieve better performance in image steganography by balancing the trade-off between embedding and image quality.

4 Dataset Description and Preprocessing

Our proposed steganographic approach is evaluated on the Tiny ImageNet dataset [14]. Tiny Imagenet is a widely used dataset in the field of steganography. It contains a diverse collection of images categorized into 200 classes. A total of 1000 images are chosen at random from various classes for training, and they are split evenly into 500 images each for secret and cover images. The order of the images in the dataset is randomized to introduce variability and prevent overfitting. This randomization ensures that the model encounters diverse examples during training, thereby improving its ability to generalize on unseen data.

All the selected images are resized to dimensions of 64×64 pixels to ensure uniformity in input size for the steganographic model. Each image has three color channels (red, green, and blue), which represent the color information. The chosen images in the dataset are converted into a multi-dimensional array for efficient processing and computation. Before model training, the pixel values in each image array are normalized between 0 and 1. This normalization improves model convergence and numerical stability during training.

5 Results and Discussions

The performance of our proposed steganographic approach is evaluated on the Tiny Imagenet dataset. We selected 200 images randomly from various classes for testing, and they are split evenly into 100 images each for secret and cover images.

The proposed steganography is implemented on the Google Colab platform using Python programming language. We employed PSNR (Peak Signal-to-Noise Ratio), MSE (Mean Square Error), and SSIM (Structural Similarity Index) [15] as evaluation metrics to assess the effectiveness of our proposed method. These metrics are commonly used in image steganography to measure the quality of steganographic images with respect to their original images. The PSNR computes the peak signal-to-noise ratio, in decibels, between the original image and the steganographic image which helps to identify the quality of the steganographic image. On the other hand, Mean Squared Error (MSE) calculates the average squared differences between corresponding pixels of the original and steganographic images. Structural Similarity Index (SSIM) is a perceptual metric that assesses the similarity between the cover image and steganographic image based on luminance, contrast, and structure. The formulas used to calculate PSNR, MSE, and SSIM are given in the below equations.

$$\text{MSE} = \frac{1}{N} \sum_{i=1}^{N} (\text{original}_i - \text{modified}_i)^2 \tag{3}$$

$$\text{PSNR} = 20 \cdot \log_{10} \left(\frac{255}{\sqrt{\text{MSE}}} \right) \tag{4}$$

In Eq. (3) $original_i$ and $modified_i$ are the pixel values at position i in the original and modified images, respectively and N is the total number of pixels.

Table 1 Comparison of proposed approach with state-of-the-art approaches

Sl. no.	Methodology	PSNR (db)	MSE	SSIM
1	LSB	53.32	0.23	0.936
2	DCT	56.13	0.16	0.942
3	DWT	51.74	0.27	0.921
4	GAN (U-NET)	40.66	0.31	0.944
5	CNN	73.41	0.006	0.957
6	Proposed method	77.52	0.0043	**0.973**

Table 2 Comparison of different activation functions for our proposed approach

Sl. no.	Activation function	PSNR (db)	MSE	SSIM
1	Selu	42.51	10.16	−0.096
2	Tanh	55.29	0.57	0.671
3	Sigmoid	62.32	0.25	0.936
4	ReLU	77.52	0.0043	0.981

$$\text{SSIM}(x, y) = \frac{(2\mu_x\mu_y + C_1)(2\sigma_{xy} + C_2)}{(\mu_x^2 + \mu_y^2 + C_1)(\sigma_x^2 + \sigma_y^2 + C_2)} \tag{5}$$

In Eq. (5) μ_x and μ_y are the means of the cover and steganographic images, respectively. σ_x^2 and σ_y^2 are the variances of the cover and steganographic images, respectively. σ_{xy} is the covariance between the cover and steganographic images. The C_1 and C_2 are small constants to stabilize the division with a weak denominator.

Table 1 presents a comparison of PSNR, MSE, and SSIM values for various steganographic methods, including traditional steganographic approaches (LSB, DCT, DWT), and deep learning techniques like GAN (U-NET), CNN, and the proposed method. LSB, DCT, DWT steganography techniques are tested on Lena and Baboon images. Deep learning techniques are tested on 100 cover images and 100 secret images from the Tiny Imagenet dataset. The average PSNR, MSE, and SSIM values are calculated with the produced steganographic images and are shown in Table 1. Among these methods, our proposed approach achieved the highest PSNR value of 77.52 dB, indicating the high quality of the steganographic image with respect to the cover image. In Addition to this, it achieves the lowest MSE value of 0.0043, signifying minimal distortion between the cover and steganographic images. Moreover, the proposed method exhibits the highest SSIM value of 0.973, indicating the closest structural similarity between the cover and steganographic images.

The comparison of the various activation functions that can be used in our steganography model are shown in Table 2. Table 2 demonstrates that the ReLU activation function performs better than all other activation functions. ReLU activation function is used to introduce nonlinearity to the network that helps in identifying

Fig. 7 Results of our proposed methodology

complex patterns which helps our steganography model for better embedding and improves image quality better than other activation functions. These results show the effectiveness of our approach in achieving high-quality embedding of the secret image into the cover image. Figure 7 represents cover images, secret images, and their respective steganographic images produced using our proposed method. We can observe that the produced steganographic image is similar to the cover image. The minimal distortions that are produced after embedding are not perceptible to the human eye. This shows the effectiveness of our approach in comparison with the state-of-the-art approaches.

6 Conclusion and Future Work

In this paper, we proposed a novel image steganography approach by leveraging the concept of CNNs with Quantization and CBAM attention mechanism. Our method gives high PSNR, low MSE, and high SSIM values in comparison to the traditional steganographic and other deep learning approaches. The utilization of CNNs allows for efficient feature extraction and integrating it with Quantization and CBAM attention mechanisms refines feature extraction. In future, we will explore the possibility of integrating other advanced techniques, such as adversarial training and reinforcement learning. This could lead to further improvements in the steganographic performance. In addition to this, we will work to integrate advanced cryptosystems such as ABE (Attribute based encryptions) with steganography algorithms. This could tighten the security of confidential information and provide fine-grained access control to authorized users.

References

1. Bharathkumar, B., Logeswar, S., Sudharsun, S., Karthikeyan, B.: An enhanced triple prime encryption approach for image encryption in LSB steganography. In: 2023 First International Conference on Advances in Electrical, Electronics and Computational Intelligence (ICAEECI), Tiruchengode, India, pp. 1–6 (2023).https://doi.org/10.1109/ICAEECI58247.2023.10370817
2. Dumre, R., Dave, A.: Exploring LSB steganography possibilities in RGB images. In: 2021 12th International Conference on Computing Communication and Networking Technologies (ICCCNT), Kharagpur, India, pp. 1–7 (2021),.https://doi.org/10.1109/ICCCNT51525.2021.9579588
3. Lahiri, S., Paul, P., Banerjee, S., Mitra, S., Mukhopadhyay, A., Gangopadhyaya, M.: Image steganography on coloured images using edge based data hiding in DCT domain. In: IEEE 7th Annual Information Technology, Electronics and Mobile Communication Conference (IEMCON), Vancouver, BC, Canada, pp 1–8 (2016). https://doi.org/10.1109/IEMCON.2016.7746270
4. Baziyad, M., Obaidat, M.S.: On the importance of the DCT phase for image steganography schemes. In: 2020 IEEE 5th International Conference on Computing Communication and Automation (ICCCA), Greater Noida, India, pp. 791–795 (2020). https://doi.org/10.1109/ICCCA49541.2020.9250849
5. Farahani, M.R.D., Parsayan, A.: A DWT based perfect secure and high capacity image steganography method. In: 2013 International Conference on Parallel and Distributed Computing, Applications and Technologies, Taipei, Taiwan, pp. 314–317 (2013). https://doi.org/10.1109/PDCAT.2013.56
6. vyas, A.O., Dudul, S.V.: Hybrid DWT-DCT image steganography for encrypted secret image. In: 2019 International Conference on Recent Advances in Energy-efficient Computing and Communication (ICRAECC), Nagercoil, India, pp. 1–7 (2019). https://doi.org/10.1109/ICRAECC43874.2019.8995111
7. Nururrahmah, A.T., Ahmad, T.: Analysis of image steganography using wavelet and cosine transforms. In: 2021 13th International Conference on Information and Communication Technology and System (ICTS), Surabaya, Indonesia, pp. 40–45 (2021). https://doi.org/10.1109/ICTS52701.2021.9609062
8. AbdelRazik, H.F., ELSayed, A.S., Ismail, S.S., Mahmoud, A.M.: Traditional and deep learning techniques in image steganography: recent advances. In: 2023 Eleventh International Confer-

ence on Intelligent Computing and Information Systems (ICICIS), Cairo, Egypt, pp. 38–44 (2023). DOIurlhttps://doi.org/10.1109/ICICIS58388.2023.10391084
9. Qiu, J.: Generative image steganography scheme based on deep learning. In: 2022 International Conference on Education, Network and Information Technology (ICENIT), Liverpool, United Kingdom, pp. 191–194 (2022). https://doi.org/10.1109/ICENIT57306.2022.00049
10. Fan, R., Wei, T., Wen, Y., Yi, C.: Research on image multi-feature steganography based on deep learning. In: 2023 2nd International Conference on 3D Immersion, Interaction and Multi-sensory Experiences (ICDIIME), Madrid, Spain, pp. 379–383 (2023). https://doi.org/10.1109/ICDIIME59043.2023.00079
11. Wei, B., Duan, X., Nam, H.: Image steganography with deep learning networks. In: 2022 13th International Conference on Information and Communication Technology Convergence (ICTC), Jeju Island, Republic of Korea, pp. 1371–1374 (2022). https://doi.org/10.1109/ICTC55196.2022.9952432
12. Sultan, B., Wani, M.A.: Multi-data image steganography using generative adversarial networks. In: 2022 9th International Conference on Computing for Sustainable Global Development (INDIACom), New Delhi, India, pp. 454–459 (2022). https://doi.org/10.23919/INDIACom54597.2022.9763273
13. Shilpa, N., Alassedi, Z., Almoussawi, Z.A., Shakir, A.M.: Image steganography: uncovering hiding image using deep learning for attention-generative adversarial network. In: 2023 3rd International Conference on Mobile Networks and Wireless Communications (ICMNWC), Tumkur, India, pp. 1–6 (2023). https://doi.org/10.1109/ICMNWC60182.2023.10435894
14. Thenmozhi, S.: CNN based image steganography techniques: a cutting edge, state of art review. In: Smart Technologies, Communication and Robotics (STCR). Sathyamangalam, India, pp. 1–5 (2022). https://doi.org/10.1109/STCR55312.2022.10009102
15. Jenynof, A., Ahmad, T.: Image to image steganography using U-Net architecture with MobileNet convolutional neural network. In: 2023 14th International Conference on Computing Communication and Networking Technologies (ICCCNT), Delhi, India, pp. 1–7 (2023). https://doi.org/10.1109/ICCCNT56998.2023.10306352
16. Devi, A.G., Thota, A., Nithya, G., Majji, S., Gopatoti, A., Dhavamani, L.: Advancement of digital image steganography using deep convolutional neural networks. In: International Interdisciplinary Humanitarian Conference for Sustainability (IIHC). Bengaluru, India, pp. 250–254 (2022). https://doi.org/10.1109/IIHC55949.2022.10060230
17. Gurumurthy, S.B., Danti, A.: Image steganography using discrete wavelet transform and convolutional NeuralNetwork. In: International Interdisciplinary Humanitarian Conference for Sustainability (IIHC). Bengaluru, India, pp. 862–868 (2022). https://doi.org/10.1109/IIHC55949.2022.10060110
18. Nadgauda, S.S., Pennamada, Y.R., Sumathi, D.: StegaNet: a deep learning model for image steganography using customized CNN and autoencoders. In: 2023 OITS International Conference on Information Technology (OCIT), Raipur, India, pp. 196–201 (2023). https://doi.org/10.1109/OCIT59427.2023.10431390
19. Yang, C., Zhang, C., Yang, X., Li, Y.: Performance study of CBAM attention mechanism in convolutional neural networks at different depths. In: 2023 IEEE 18th Conference on Industrial Electronics and Applications (ICIEA), Ningbo, China, pp. 1373–1377 (2023). https://doi.org/10.1109/ICIEA58696.2023.10241832
20. Kwak, J., Kim, K., Lee, S.-S., Jang, S.-J., Park, J.: Quantization aware training with order strategy for CNN. In: 2022 IEEE International Conference on Consumer Electronics-Asia (ICCE-Asia), Yeosu, Republic of Korea, pp. 1–3 (2022). https://doi.org/10.1109/ICCE-Asia57006.2022.9954693

Advanced Cryptography in Emerging Applications

vSPACE: Voting in a Scalable, Privacy-Aware and Confidential Election

Se Elnour, William J. Buchanan, Paul Keating, Mwrwan Abubakar, and Sirag Elnour

Abstract The *vSPACE* experimental proof-of-concept (PoC) on a TrueElect extended protocol presents a novel approach to secure, private, and scalable elections, extending the TrueElect and other related protocols with the integration of Self-Sovereign Identity (SSI) enabled Privacy-Enhancing Technologies (PETs). Zero-Trust Architecture (ZTA) principles and practices were used for a split-trust design to minimize human errors or attacks, along with using a trustworthy Infrastructure as Code (IaC) automation for staging a perfectly democratized e-governance polling; while also leveraging confidential computing, continuous authentication, multi-party computation (MPC), and Well-Architected Framework principles to address cyber-security and privacy protection challenges. Moreover, *vSPACE* integrates Distributed Ledger Technology (DLT) for immutable and certifiable audit trails, and employs Kubernetes confidential clusters as ZTA spokes for hosting decentralized applications (dapps) with a connectivity hub of an nTier-based Enterprise-Scale Landing Zone (ESLZ). The IaC model ensures rapid deployment, consistent management, and adherence to security standards, making *vSPACE* a future-proof solution for

Supported by Innovate UK.

S. Elnour (✉) · W. J. Buchanan · M. Abubakar
Blockpass ID Lab, Edinburgh Napier University, Edinburgh, UK
e-mail: vspace@c6ai.com

W. J. Buchanan
e-mail: b.buchanan@napier.ac.uk

M. Abubakar
e-mail: m.abubakar@napier.ac.uk

P. Keating
Democracy Counts, Warrington, Cheshire, UK
e-mail: paulkeating@democracycounts.co.uk

S. Elnour
TheSDEs, 8225 N FM 620, Apt 2011, Austin, TX 78726, USA
e-mail: vote@vspacewallet.com

M. G. Jaatun et al. (eds.), *Proceedings of the International Conference on Cybersecurity, Situational Awareness and Social Media*, Springer Proceedings in Complexity,
https://doi.org/10.1007/978-981-96-0401-2_4

digital voting systems. Our poster and concept note include motivations behind low-level design choices, ZTA specifications with a threat modeling assurance review, quantitative metrics on efficiency performance, and preliminary implementation IaC tooling.

Keywords Election-tech · E-governance · Privacy-enhancing technologies · Confidential computing · Cybersecurity · Continuous authentication · Artificial intelligence · Zero-trust

1 Introduction

Electronic voting is frequently employed in various decision-making elections since it is adaptable, easy to use, and inexpensive compared to traditional polls [13]. Despite this, the currently used electronic voting techniques pose the risk of excessive power and the manipulation of information, which reduces the basic levels of fairness, privacy, secrecy, anonymity, and transparency in the voting process. However, there are a variety of approaches that have been used in electronic and online voting systems. These methods use various encryption and decryption strategies to ensure that data exchanges are kept safe.

vSPACE (Voting in a Scalable, Privacy-Aware and Confidential Election) is a novel solution that extends our TrueElect (see Fig. 1), ElectAnon [10] and/or other related protocols into an extended [True***Elect***][***Anon***Creds]protocol, leveraging confidential computing, continuous authentication, and MPC; while also following the Well-Architected Framework. This paper presents an experimental PoC on SSI, using ZTA of critical IaC for cybersecurity, privacy, and ToIP protection.

The main contribution of this paper is the design and implementation of a dapps within an Election Management System (EMS) which integrates DLT for immutable and certifiable audit trails. This approach addresses the challenges of existing voting systems—offering a high level of security, privacy, and confidence for both election authorities and the electorate.

2 Related Work

The development of secure DLT-based voting systems is contingent upon meeting a comprehensive set of requirements. These criteria are essential for preserving the integrity, confidentiality, and resilience of the voting process, thereby ensuring voter confidence and the legitimacy of election outcomes.

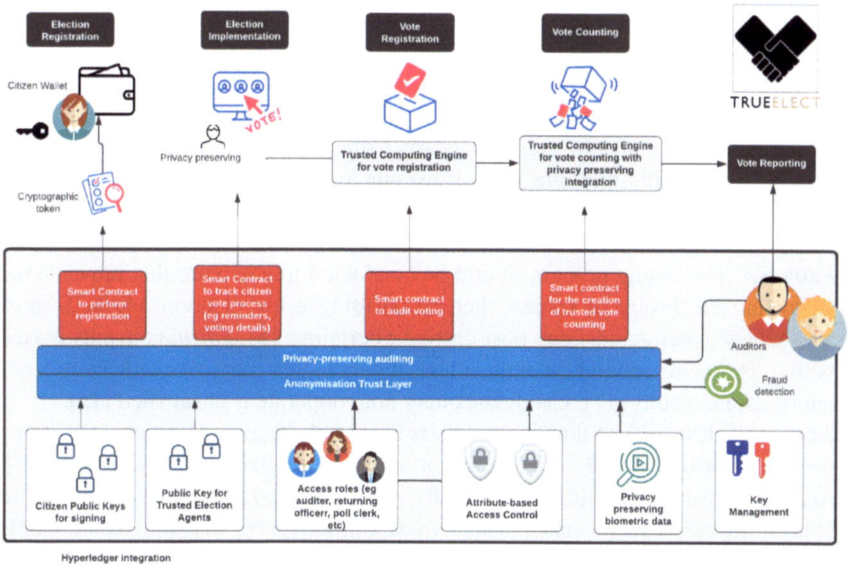

Fig. 1 TrueElect Voting Outline, credit: asecuritysite.com/blogs/graphics

Onur and Yurdakul [10] in their 2023 study, evaluated four prominent DLT-voting protocols [8, 11, 15, 17]) to understand how they meet important criteria like *Eligibility, Uniqueness, Privacy, Universal Anonymity, Fairness, Accuracy, Universal Verifiability, Individual Verifiability, Robustness, Autonomy,* and *Scalability*. These criteria are essential for secure, anonymous, fair, and accurate online voting. Our work, on the security of online elections, builds as well on earlier research by Gritzalis [5] in 2002, and Jafar et al. [6] in 2021, who also listed similar foundational requirements for trustworthy online elections using DLT.

These foundational requirements guide the design and implementation of our secure DLT-voting system. They ensure the voting process is reliable and can be trusted. The core requirements for a secure DLT-voting were highlighted as follows:

- **Eligibility**: Ensuring that only authorized individuals participate in the election is paramount. *vSPACE* must incorporate mechanisms for verifying voter eligibility without compromising their privacy [5]. Only eligible voters should vote [5]. *Eligibility* requires trusting the authority. DLT-voting is vulnerable if not properly implemented, enabling vote-buying and coercion.

- **Uniqueness**: To uphold the principle of "one person, one vote," *vSPACE* must prevent double voting by ensuring each eligible voter can cast only one ballot [5, 15]. *Uniqueness* eliminates double voting.
- **Privacy**: The confidentiality of a voter's ballot is crucial. *vSPACE* must safeguard the secrecy of the vote, preventing any party from determining a voter's choices [5]. Privacy may enable multiple votes from one voter using different keys [8, 11]. Some protocols bind keys to eligibility certificates to prevent this. No results should be available before tallying [5, 6].
- **Fairness**: The voting process should be conducted in a manner that prevents the premature disclosure of results, thereby ensuring a fair election. *vSPACE* must also secure votes against any post-casting alterations [5]. *Fairness* avoids biased voting. Protocols relying on authorities' or candidates' keys risk attacks to learn intermediate results [17]. Candidates may not cooperate if unsatisfied [15].
- **Accuracy**: It is critical that every vote is accurately recorded and counted in the final tally, with *vSPACE* being resilient against vote manipulation or fraud [5]. ***Accuracy*** discards invalid votes. Usually Zero-Knowledge Proofs (ZKPs) ensure validity. Some protocols make invalid votes impossible [8, 11]. Only valid ballots should count [1].
- **Individual Verifiability**: Electorate should have the capability to confirm that their vote has been correctly recorded and included in the final tally, enhancing their confidence in the voting process [5].
- **Universal Verifiability**: The integrity of the election process must be verifiable by any observer, ensuring that *vSPACE* provides means for third parties to independently verify the correct counting of votes [5].
- **Universal Anonymity**: In addition to privacy, *vSPACE* must ensure that a voter's identity cannot be linked to their voting behavior or other DLT activities, thus guaranteeing universal anonymity [5]; while not risking *linkage attacks* whenever making use of the electorate addresses [8, 11] or public keys [15–17].
- **Robustness**: The voting system must demonstrate resilience against a broad spectrum of potential attacks and failures, ensuring the election's smooth progression under various conditions [5].

3 Design

The [True***Elect***][***Anon***Creds]protocol is a novel approach designed to address the challenges of scalability, privacy, and confidentiality in DLT-based voting systems. Its design incorporates the principles of privacy-preserving auditing and anonymization confidence layers, as depicted in Fig. 2. It utilizes smart contracts for various stages of the election process, including voter registration, vote auditing, and the creation of trusted vote counting. The integration of a scalable DLT ensures a secure and transparent process, with public keys for citizens and trusted election agents, as well as attribute-based access control for various roles such as auditors and returning officers. Privacy-preserving biometric data and key management are also critical components of the protocol, ensuring the security and confidentiality of the voting process.

This [True***Elect***][***Anon***Creds]protocol proposal extends the capabilities of our own ***TrueElect*** older protocol, along with other complementary protocols, such as ***ElectAnon*** proposed by Onur and Yurdakul (2022); as well as the integration of the latest Hyperledger ***AnonCreds*** 1.0 alternatives for SSI to be implemented by our *vSPACE* Wallet; alongside the *vSPACE* confidential computing environment, leveraging blind signatures, MPC, and/or ZKPs to ensure voter anonymity while maintaining the integrity and verifiability of the election-tech process.

Note that the TrueElect and ElectAnon protocols were designed to facilitate secure and private voting mechanisms. Both protocols employ advanced cryptographic techniques to ensure the integrity and confidentiality of the voting process. However, they differ in their specific implementations and the flow of their respective protocols.

3.1 Overview of EMS, SSI, ZTA, IaC, and ToIP

The EMS integrated with SSI, ZTA, IaC, and ToIP represents a paradigm shift in the domain of digital elections. This integration aims to establish a certifiable, privacy-aware, and scalable framework for conducting elections, leveraging the latest advancements in cybersecurity and digital identity management.

- **Election Management System (EMS)**: The EMS is a comprehensive system designed to manage all aspects of the electoral process, from voter registration to ballot casting and result tabulation. The goal is to ensure a secure, transparent, and efficient election process that can scale to accommodate a large number of constituents while maintaining the integrity of the vote.
- **Self-Sovereign Identity (SSI)**: SSI is a user-centric approach to digital identity that gives individuals control over their personal data. In the context of EMS, SSI allows constituents to manage their digital identities securely and to interact with the election system without relinquishing personal data control to third parties.
- **Zero-Trust Architecture (ZTA)**: ZTA is a security model that operates on the principle of "never trust, always verify." It is designed to protect against internal and external threats by rigorously verifying the identity and trustworthiness of every device and user before granting access to the EMS's resources.

- **Infrastructure as Code (IaC)**: IaC is a method of managing and provisioning computing infrastructure through machine-readable definition files, rather than physical hardware configuration or interactive configuration tools. This approach enables the rapid and consistent deployment of infrastructure, which is essential for the scalability and reliability of the EMS.
- **Trust Over IP (ToIP)**: ToIP is a set of principles and standards that aim to establish a global framework for digital confidence. It combines e-governance, technology, and legal considerations to create a secure, decentralized system for digital interactions. Within the EMS, ToIP ensures that all communications and transactions are secure and that the privacy of the constituents is preserved.

3.2 Purpose and Scope of the PoC

The PoC is designed to validate the practical application and effectiveness of integrating an EMS with SSI, ZTA, IaC, and ToIP for the purpose of enhancing the security, privacy, and scalability of digital elections. The PoC aims to demonstrate the feasibility of this integrated approach and to provide a blueprint for its implementation in real-world election scenarios.

The scope of the PoC encompasses several key components:

- The creation of a digital ballot within a secure and verifiable EMS, as outlined in the Business Process Model and Notation (BPMN) Fig. 2.
- The execution of a verification algorithm that ensures the integrity of the digital ballot and the eligibility of the electorate, as depicted in the BPMN and sequence diagrams.
- The integration of a constituent digital wallet, which allows voters to securely manage their digital identities and interact with the election system using SSI principles within a ZTA framework.
- The application of IaC to facilitate the rapid and consistent deployment of the necessary infrastructure to support the EMS, while adhering to ToIP standards for secure and private digital interactions.

The PoC serves to illustrate the technical viability and the potential benefits of our adopted framework, such as improved election security, enhanced voter privacy, and the ability to scale to accommodate a large electorate. It also seeks to address the challenges associated with digital elections, including the prevention of fraud, ensuring the verifiability of votes, and maintaining the confidentiality of electorate choices.

4 Implementation

Algorithm 1 *MPC (Multi-Party Computation)* integration between business processes and distributed registry *DLT (Distributed Ledger Technology)*

Constituency Acting Returning Officers or ***POs (Presiding Officers)***:
Initiate consortium's distributed *EMS (Elections Management System)*
Finalize security tokens' *HYOK (Hold Your Own Key)* ceremony
Activate confidential *ZTA (Zero-Trust Architecture)* integrity attestation
Verify *DID (Decentralized Identity)* with *SSI (Self-Sovereign Identity)* Wallet
Enable *MFCA (Multimodal Fusion-based Continuous Authentication)*
Multi-sign constitution counterparts irrevocably as Ricardian contracts
Proceed with overseeing an anonymized *ElectionProcess* if authenticated
for each *ElectionProcess* in *EMS* **do**
Generate a smart contract for the election process
if *ElectionContractAddress* is set **then**
Generate a variable containing hard coded address
else
Generate a process contract constructor parameter for setting the election contract address
end if
end for
for each *VoterRegistration* in *EMS* **do**
Obtain voter *DID*, eligibility proof from *SSI-as-a-Service*
Generate Chaincode or Solidity code for registering voter on the *DLT*
Ensure verification of identity using AI-based biometric *MFCA*
Inject generated *DID* code to function body of registration process
end for
for each *VoteCasting* in *EMS* **do**
Obtain voter identity, ballot choice from *SSI-as-a-Service*
Generate Chaincode or Solidity code for casting vote on the *DLT*
Ensure privacy-preserving verification of voter identity and eligibility
Inject generated voting code to function body of voting process
end for
for each *VoteTallying* in *EMS* **do**
Aggregate votes using secure *MPC* and the quantum-safe *DLT*
Generate Chaincode or Solidity code for tallying votes
Ensure *ZKP (Zero-Knowledge Proof)* verification of integrity attestation
Inject generated tallying code to function body of tallying process
end for
for each *VoteCertifiability* in *EMS* **do**
Verify the certifiable audited integrity and authenticity of the votes
Use registry services *ZTA* to validate vote transactions
Update the *EMS* with the verified vote counts to be announced
Multi-sign confidential consensus by a quorum of *POs* if verified
Publish combined certifiable *ZKPs* and audit into public *DLTs*
end for

The executed verification algorithm is a pivotal component of the digital ballot creation process within the novel framework. This algorithm ensures the integrity and authenticity of the digital ballots and the eligibility of the constituents casting votes. The verification process is depicted in the sequence diagram Fig. 3 and the BPMN Fig. 2.

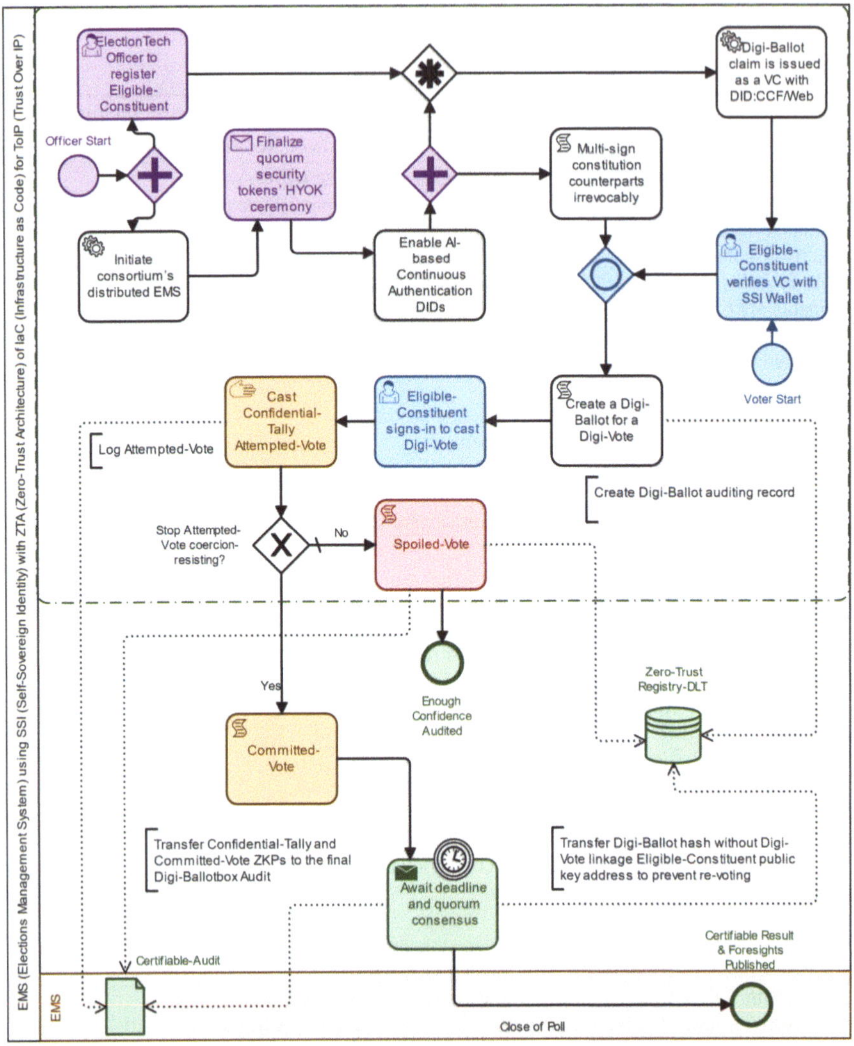

Fig. 2 Election-tech digital ballot creation BPMN

The algorithm commences with the election-tech officer's initiation of the consortium's distributed EMS. The digital ballot creation process involves several critical steps, including the finalization of a quorum security tokens' Hold Your Own Key (HYOK) ceremony and the enabling of AI-based continuous authentication Decentralized Identifiers (DIDs). Once these preliminary steps are completed, a digital ballot claim is issued as a Verifiable Credential (VC) with did:ccf/web, which is then multi-signed by the consortium's officers to establish irrevocability.

The eligible constituent, equipped with their SSI Wallet, verifies the VC. Upon successful verification, the constituent may cast a confidential-tally attempted vote. The algorithm accounts for coercion-resistance, allowing constituents to produce either a spoiled vote or a committed vote. A committed vote triggers the transfer of confidential tally and committed vote ZKPs to the final digital ballot box audit, ensuring a certifiable audit trail.

The verification algorithm concludes with the transfer of the digital ballot hash to the zero-trust registry DLT, which includes the eligible constituent public key address. The final step is the publication of certifiable results and foresights, marking the completion of a secure and verifiable voting process.

The sequence diagram Fig. 3 further elaborates on the interactions between the constituent VC-holder Backend-as-a-Service (BaaS), the SSI as a Service edge with custodianship wallet Platform-as-a-Service (PaaS), the Verification as a Service authenticator, and the zero-trust registry service with trustee DLT. These interactions are governed by our critical IaC based on the ToIP framework, ensuring a certifiable, privacy-aware, and scalable i-voting system.

4.1 Business Process Model and Notation (BPMN)

The BPMN provides a graphical representation of the election technology's digital ballot creation and the executed verification algorithm. The BPMN diagram in Fig. 2 illustrates the sequential flow and decision points that constitute the digital ballot creation process, integrating various components such as the election-tech officer, the consortium's distributed EMS, and the zero-trust registry DLT. The process begins

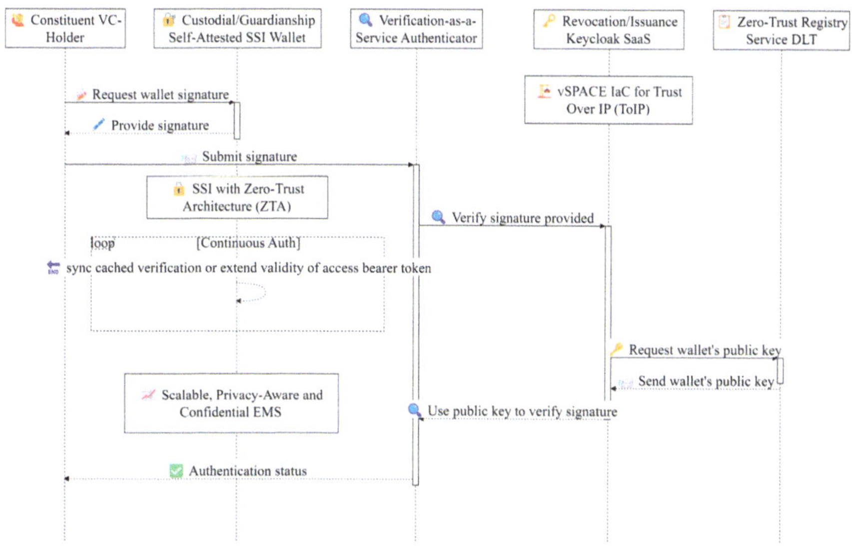

Fig. 3 Self-sovereign identity (SSI) sequence diagram

with the election-tech officer initiating the distributed EMS and proceeds through several steps, including the finalization of a quorum security tokens' HYOK ceremony, enabling AI-based continuous authentication DIDs, and the issuance of a digital ID claim as a did:ccf/web.

The BPMN Fig. 2 further describes the path taken by an eligible constituent to verify the VC with their SSI wallet and cast a confidential-tally attempted-vote. The process accounts for the possibility of vote coercion-resistance, leading to either a spoiled-vote or a committed-vote. The committed-vote then undergoes a transfer of confidential-tally and committed-vote ZKPs to the final digital ballot box audit, ensuring a certifiable audit trail. The process concludes with the transfer of the digital ballot privacy-aware hash to the zero-trust registry DLT, which includes the eligible-constituent public key address to prevent re-voting, and the publication of certifiable results and foresights.

This BPMN is critical for understanding the intricate workflow and ensuring that each step adheres to the principles of certifiability, privacy, and scalability within the context of i-voting systems. The BPMN serves as a blueprint for the implementation of the digital ballot creation process, ensuring that all actions are verifiable and auditable in a manner that upholds the integrity of the election process within our proposed framework.

4.2 Constituent Digital Wallet PoC

The constituent digital wallet PoC is a critical component of the framework, designed to demonstrate the practical implementation of a secure and privacy-preserving digital wallet system for constituents. This system is integral to the digital ballot creation process and the overall EMS, as it provides a means for constituents to interact with the election infrastructure in a manner that is both secure and verifiable.

The PoC encompasses the design and architecture of the digital wallet, which is built upon the principles of SSI and operates within a ZTA. The digital wallet serves as a secure repository for constituents' credentials and enables them to perform critical election-related actions, such as verifying their identity, receiving and casting digital ballots, and participating in the election process with a high degree of confidence in the EMS's integrity.

The digital wallet PoC demonstrates the feasibility of the constituent digital wallet system and its ability to integrate with the broader framework. It provides a tangible example of how constituents can securely manage their digital identities and participate in digital elections, thereby contributing to the advancement of certifiable, privacy-aware, and scalable i-voting systems.

4.3 DApps High-Level Design

The envisioned High-Level Design (HLD) is outlined in Fig. 4. It outlines a scalable digital wallet dapp implementation leveraging enterprise-scale IaC and best practices. By following the AWS's Well-Architected Framework and utilizing modern CI/CD and container orchestration technologies, such digital wallet dapp with an EMS aims to achieve high availability, security, and scalability for the use case of digital voting.

Our HLD choices extend the EMS by adapting coexisting DLT-based dapps and context-aware AI with anonymous Verifiable Credentials (VCs), incorporating oracle gateway with accurate bot service for FAQs and Knowledge Base (KB) inquiries. This integration is facilitated by channels APIs and logic triggers, including dynamic chaining and command language understanding (CLU) with analytics foresights via oracle connectors for the likes of weather forecasting, location mapping and travel planning, ... etc. Furthermore, the EMS utilizes anonymous verified ID role-based access control (RBAC) for a comprehensive, integrated Software as a Service (SaaS) solution. The added value proposition of our system lies in its ability to provide a confidential anonymization and privacy-preserving public engagement via its context-aware APIs gateway, which serves as a competitive edge in such critical infrastructure of polling election-tech.

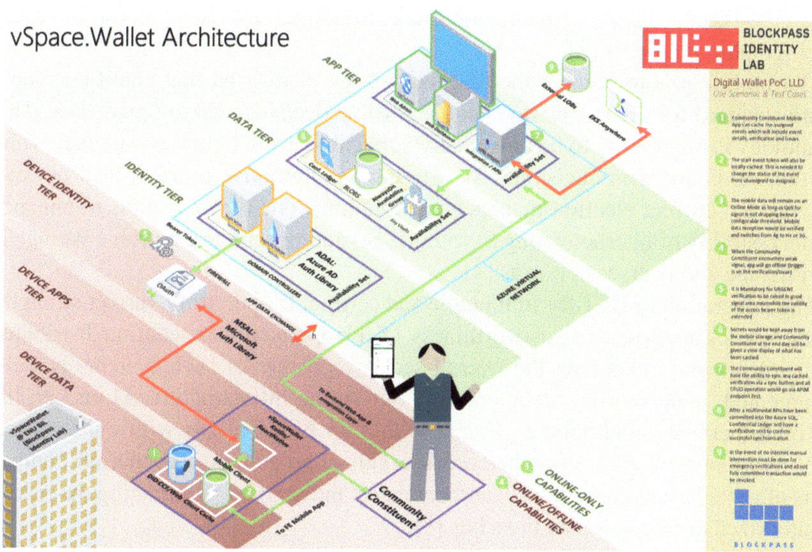

Fig. 4 Constituent digital wallet DApp IaC

5 Evaluation Analysis

ElectAnon is implemented with smart contracts and a ZKP gadget, and it has been observed to run safely for 1,000,000 electorate. While the [True***Elect***][***Anon***Creds]protocol's current evaluation and security analysis are further provided in this section, advanced analysis and more detailed results are pending final insights on our envisioned *vSPACE* R&D pilot [Certifiable Consensus using Combined Confidential Computing and Continuous Authentication with Integrity], which aims to achieve certifiable consensus through the integration of confidential computing and continuous authentication with integrity attestation. This system is designed as a SERVE (Secure Electronic Registration and Voting Election-tech) inspired IaC. This approach is critical for ensuring the security, privacy, and integrity of the voting process, addressing the challenges of insider threats, network vulnerabilities, and auditing difficulties that have been identified in previous studies [4].

In addition to these foundational elements, our research is deeply invested in exploring advanced cryptographic techniques, notably ZKPs. ZKPs enable a party to prove the truth of a statement without revealing any information beyond the validity of the statement itself. This cryptographic method is essential for a voting infrastructure that prioritizes privacy, verifiability, and transparency while eliminating associated risks. The seminal works of Fiat and Shamir (1986), followed by Goldreich and Oren (1994), laid the groundwork for ZKPs, highlighting their potential in secure cryptographic protocols [3].

Our ambition is to pioneer the first technology of its kind that could become a future standard for secure, private, and accessible decision-making processes. This technology is envisioned to be universally applicable, from public engagements and elections to boardroom decisions. By offering an alternative to traditional by-mail systems and current election management solutions, our *vSPACE* aims to enhance the accessibility of internet voting (i-voting) [14] and/or remote electronic voting [7] for absentee and overseas citizens [4].

The integration of DLT for immutable and certifiable audit trails further strengthens the EMS's integrity and trustworthiness. DLT's application in securing distributed deployments and enhancing the overall security posture of such systems is well-documented, underscoring its potential in our voting infrastructure. Moreover, the use of Kubernetes confidential clusters within an ESLZ aligns with our commitment to adopting a ZTA for cybersecurity. This approach ensures that all components of *vSPACE* are continuously authenticated and verified, minimizing the risk of unauthorized access and data breaches [4].

5.1 Security Analysis

The security analysis of the PoC on the [True***Elect***][***Anon***Creds]protocol is critical in ensuring the integrity, confidentiality, and scalability of digital elections. This analysis leverages the integration of SSI within a ZTA, employing confidential computing, continuous authentication, and Well-Architected Framework principles.

The *vSPACE* framework employs a Kubernetes confidential cluster within an ESLZ, integrating DLT for immutable and certifiable audit trails. The IaC model ensures rapid deployment, consistent management, and adherence to security standards, making *vSPACE* a future-proof solution for digital voting systems.

The security assurance focuses on several key areas:

- **Certifiable Confidential Computing**: Ensuring that data is encrypted at rest, in transit, and in use. The use of confidential computing technologies, such as CNCF-certified Constellation Kubernetes with aTLS (attested TLS) protocol, provides end-to-end certifiable confidentiality.
- **Privacy-aware Continuous Authentication with SSI**: Leveraging continuous authentication mechanisms, to ensure that only authorized individuals can access the voting system. This is further enhanced by the integration of Entra Verified ID or Keycloak for SSI OpenID Connect-based authentication.
- **Secure Multi-Party Computation (MPC)**: Utilizing MPC to enable secure computation over encrypted data, ensuring the privacy of electorate choices while allowing for the accurate tallying of votes.
- **Distributed Ledger Technology (DLT) Auditability**: Integrating DLT for creating immutable and certifiable audit trails, enhancing the transparency and verifiability of the election process.
- **Well-Architected Framework principles**: Adhering to the Well-Architected Framework principles for operational excellence, security, reliability, performance efficiency, cost optimization, and sustainability. This includes defending *vSPACE* against various cyberattacks.
- **ZTA Infrastructure as Code (IaC) compliance**: Employing IaC for rapid deployment and consistent management of the voting infrastructure, ensuring compliance with security standards.

5.2 Protocol Evaluation Comparative Analysis

Table 1 compares [True***Elect***][***Anon***Creds]in contrast to the related protocols of McCorry et al. (2017), Chaintegrity (2019), Yang et al. (2020), Panja et al. (2020), Priscore, and ElectAnon; as evaluated by Onur and Yurdakul [10]:

- **Scalability**: *vSPACE* must efficiently manage a large number of electorate without compromising performance or security, enabling scalability according to election needs [5].

Table 1 Evaluation summary of related DLT-voting studies

DLT-voting work	Eligibility requirement	Uniqueness requirement	Privacy requirement	Universal anonymity Req.	Fairness requirement Req.	Accuracy requirement	Universal-verifiability Req.	Individual-verifiability Req.	Robustness requirement	Self-tallying	Scalability	Certifiable confidentiality	DIDs/SSI continuous Auth.	Consensus AI hybridization	Integrity audit autonomy
McCorry et al.[8]	✓	x	✓	x	✓	✓	✓	✓	o	✓	x	x	x	x	x
Chaintegrity [17]	✓	✓	✓	x	x	✓	✓	✓	x	x	x	x	x	x	x
Yang et al. [15]	✓	✓	✓	x	x	✓	o	✓	x	o	x	x	x	x	x
Panja et al.[11]	✓	x	✓	x	✓	✓	✓	✓	o	✓	x	x	x	x	x
Priscore [16]	✓	✓	o	x	✓	✓	✓	✓	o	o	x	x	x	x	x
ElectAnon [10]	✓	✓	✓	✓	✓	✓	✓	✓	✓	✓	✓	x	x	x	x
[True***Elect***][***Anon***Creds](this work)	✓	✓	✓	✓	✓	✓	✓	✓	✓	✓	✓	✓	✓	✓	✓

✓: implemented, x: not implemented, o: partially implemented

- **Certifiable Confidentiality**: Providing immutable and certifiable audit trails through the integration of off-chain confidential computing and/or for the confidential cluster of an on-chain DLT implementation.
- **DIDs/SSI Continuous Authentication**: Utilizing SSI for Multimodal Fusion-based Continuous Authentication (MFCA).
- **Consensus AI Hybridization**: Leveraging AI techniques (such as: feature extraction and anomaly detection) for the likes of privacy-preserving biometric authentication, enhanced consensus mechanisms and/or 51% attack prevention.
- **Integrity Audit Autonomy**: Once initiated, *vSPACE* should function independently without the need for manual intervention, ensuring an uninterrupted election process; while also ensuring the integrity of the voting process through comprehensive auditability.
- **Coercion Resistance and Receipt Freeness**: Last, but not least, *vSPACE* complies with the requirement for a voting process to be free from undue influence and manipulation [2]. *Coercion Resistance* ensures that electorate cannot be compelled to vote in a particular way. *Receipt Freeness* ensures that electorate cannot obtain any proof that could be used to sell their vote.

6 Conclusions

This research white paper has outlined the *vSPACE* experimental PoC on the [True***Elect***][***Anon***Creds]protocol, adopting a modular approach to extend TrueElect and related protocols by integrating SSI-empowered Privacy-Enhancing Technologies (PETs). The added value included scalability, integrated privacy, and confidentiality; not to mention the pending research directions of innovating the credentialing cybersecurity through quantum-robust and AI-based privacy-preserving biometric-enabled verified ID with Multimodal Fusion-based Continuous Authentication (MFCA), to be strengthened with swappable libraries, such as QR-code integration [9] and an efficient Blockene-like [12] auditability; while allowing a further robustification of *vSPACE* by adapting a collusion-proof, constrained, cellular-data-friendly, citizen-powered consensus with coexisting AI and DLT-based dapps in such zero-trust environment, to minimize the potential threat of any split-trust assumption on local consortium candidates or voting authority sub-committees.

References

1. Anane, R., Freeland, R., Theodoropoulos, G.: e-Voting requirements and implementation. In: Proceedings of the 9th IEEE International Conference on E-Commerce Technology (CEC 2007), pp. 382–392. IEEE Computer Society (2007). https://doi.org/10.1109/CEC-EEE.2007.42
2. Benaloh, J., Tuinstra, D.: Receipt-free secret-ballot elections. In: Proceedings of the Twenty-Sixth Annual ACM Symposium on Theory of Computing (STOC '94), pp. 544–553. ACM (1994). https://doi.org/10.1145/195058.195407
3. Feige, U., Fiat, A., Shamir, A.: Zero-knowledge proofs of identity. J. Cryptol. **1**, 77–94 (1988). https://doi.org/10.1007/BF02351717
4. Gibson, J.P., Krimmer, R., Teague, V., Pomares, J.: A review of E-voting: the past, present and future. Ann. Telecommun. **71**(7), 279–286 (2016). https://doi.org/10.1007/s12243-016-0525-8
5. Gritzalis, D.A.: Principles and requirements for a secure e-voting system. Comput. Secur. **21**(6), 539–556 (2002). https://doi.org/10.1016/S0167-4048(02)01014-3
6. Jafar, U., Aziz, M.J.A., Shukur, Z.: Blockchain for electronic voting system-review and open research challenges. Jafar2021 **21**(17), 5874 (2021). https://doi.org/10.3390/s21175874
7. Kohno, T., Stubblefield, A., Rubin, A.D., Wallach, D.S.: Analysis of an electronic voting system. Proc. IEEE Symp. Secur. Priv. **2004**, 27–40 (2004). https://doi.org/10.1109/SECPRI.2004.1301313
8. McCorry, P., Shahandashti, S.F., Hao, F.: A smart contract for boardroom voting with maximum voter privacy. In: Proceedings of the International Conference on Financial Cryptography and Data Security, pp. 357–375. Springer (2017)
9. Olaniyi, O.M., Folorunso, T.A., Ahmed, A., Joseph, O.: Design of secure electronic voting system using fingerprint biometrics and crypto-watermarking approach. Int. J. Inform. Eng. Electron. Business (IJIEEB) **8**(5), 9–17 (2016). https://doi.org/10.5815/ijieeb.2016.05.02
10. Onur, C., Yurdakul, A.: ElectAnon: a blockchain-based, anonymous, robust, and scalable ranked-choice voting protocol. Distrib. Ledger Technol. **2**(3), 19 (2023). https://doi.org/10.1145/3598302
11. Panja, S., Bag, S., Hao, F., Roy, B.: A smart contract system for decentralized Borda count voting. IEEE Trans Eng. Manage. **67**(4), 1323–1339 (2020). https://doi.org/10.1109/TEM.2020.2986371
12. Satija, S., Mehra, A., Singanamalla, S., Grover, K., Sivathanu, M., Chandran, N., Gupta, D., Lokam, S.: Blockene: a high-throughput blockchain over mobile devices. In: Proceedings of the 14th USENIX Symposium on Operating Systems Design and Implementation (OSDI 20), pp. 567–582. USENIX Association (2020). https://doi.org/10.48550/arXiv.2010.07277
13. Singh, A., Yadav, A., Kumar, A., Singh, K.: A review of smart electronic voting machine. In: Poonia, R.C., Singh, V., Singh Jat, D., Diván, M.J., Khan, M.S. (eds.) Proceedings of Third International Conference on Sustainable Computing, pp. 85–95. Springer Nature Singapore, Singapore (2022). ISBN: 978-981-16-4538-9

14. Vassil, K., Solvak, M., Vinkel, P., Trechsel, A.H., Alvarez, R.M.: The diffusion of internet voting. Usage patterns of internet voting in Estonia between 2005 and 2015. Gov. Inform. Quart. **33**(3), 453–459 (2016). https://doi.org/10.1016/j.giq.2016.06.007
15. Yang, X., Yi, X., Nepal, S., Kelarev, A., Han, F.: Blockchain voting: publicly verifiable online voting protocol without trusted tallying authorities. Future Gener. Comput. Syst. **112**, 859–874 (2020). https://doi.org/10.1016/j.future.2020.06.051
16. Yang, Y., Guan, Z., Wan, Z., Weng, J., Pang, H.H., Deng, R.H.: PriScore: blockchain-based self-tallying election system supporting score voting. IEEE Trans. Inform. Forensics Secur. **16**, 4705–4720 (2021). Available online: www.Semanticscholar.org/CorpusID:238416976
17. Zhang, S., Wang, L., Xiong, H.: Chaintegrity: blockchain-enabled large-scale e-voting system with robustness and universal verifiability. Int. J. Inform. Secur. **19**(3), 323–341 (2020). https://doi.org/10.1007/s10207-019-00465-8

Design of Pairing Free Attribute-Based Encryption for Smart Grid Applications

Anjali Kumari, Sreevallabh Karanam, Swejan Annabathini, and Kamalakanta Sethi

Abstract Smart grids are integrated systems utilizing lightweight and intelligent device to monitor, manage, and optimize the production, distribution, and consumption of electricity more effectively. Smart grids employ smart meters which continuously measure the incoming and outgoing flow of electricity and shared those power consumption data to other connected devices. Here, confidential data is shared with many stakeholders like the users (the actual user of electricity), phasor units *etc.* Therefore, the data security here plays a crucial role. Initially, Pairing based Ciphertext-Policy Attribute-Based Encryption (PB CP-ABE) was used in order to provide fine grain access control, data integrity and confidentiality. But the computationally complexity of this cryptographic technique added over-head to the system reducing its performance. To improve the performance matrix of smart grids, we propose the Pairing Free CP-ABE (PF CP-ABE) which replaces the complex bilinear pairing with simpler scalar multiplication. Finally, we have analyzed the security, and collusion resistance properties of the proposed cryptosystem along with it's performance analysis.

A. Kumari · S. Karanam · S. Annabathini · K. Sethi (✉)
Indian Institute of Information Technology, Sricity, India
e-mail: kamalakanta.s@iiits.in

A. Kumari
e-mail: anjali.k21@iiits.in

S. Karanam
e-mail: sreevallabh.k21@iiits.in

S. Annabathini
e-mail: swejan.a21@iiits.in

M. G. Jaatun et al. (eds.), *Proceedings of the International Conference on Cybersecurity, Situational Awareness and Social Media*, Springer Proceedings in Complexity,
https://doi.org/10.1007/978-981-96-0401-2_5

1 Introduction

Smart grids, are the next generation of electricity networks [1] that uses complex and integrated systems to deliver electricity from it's generation point to different stakeholders. It is a Vast network of interconnected components that work together to keep your lights on and your devices powered. Smart grids leverage sensors and communication across the entire network, enabling real-time data exchange for optimized electricity delivery. Smart grids connect generators, grid operators, consumers, and market players. This collaboration optimizes the entire electricity system for minimal cost, reduced environmental impact, and maximized reliability, resilience, flexibility, and stability.

Smart grids [1] collect a wealth of data from smart meters and other sources. This data includes details about the consumers private information along with its energy usage patterns. Security breaches can expose this data to unauthorized access, leading to privacy violations and misuse. Also, hackers could exploit vulnerabilities to inject false data, disrupting power delivery or even causing outages. This makes secure sharing of data across the smart grid is crucial. HoweVer, concerns arise about who has access to this data and how it's used. Breaches in access control systems could grant unauthorized entities access to sensitive data.

Traditionally, smart meters send data to utility companies using a system that relies on certificates to verify the receiver [2]. However, this method has vulnerabilities. Hackers could steal or forge certificates, compromising data and causing financial losses for the grid. To address these concerns, the data is stored in secure, third-party repositories with stringent access controls. However, a major challenge is ensuring the security of these repositories, as a breach could expose critical data. In order to address this problem, a solution called Attribute-Based Encryption (ABE) [3] was introduced to improve security. In Ciphertext-Attribute-Based Encryption (CP-ABE) [4], only authorized users with specific attributes (like location or permission level) can access encrypted data. This prevents unauthorized access and protects sensitive information.

While ABE offers a promising solution for securing data in smart grids, the traditional CP-ABE [4] approach has a significant drawback in this context: its reliance on bilinear pairings. The bilinear pairings are complex mathematical calculations that act as a secure lock on encrypted data. While they ensure only authorized users with the appropriate "key" (attributes) can access the information, these calculations can be computationally expensive. This becomes a problem for smart meters in smart grid networks. These devices are often designed to be low-power and have limited processing capabilities. The complex calculations required by CP-ABE pairings can be a burden for these resource-constrained devices, potentially slowing down communication and hindering overall efficiency.

This paper proposes an innovative solution of Pairing-free CP-ABE (PF CP-ABE) [2, 5] to this bottleneck. As the name suggests, it eliminates the need for complex bilinear pairings operations entirely. This significantly reduces the computational workload required to encrypt and decrypt data. Therefore, it is a novel and more efficient cryptographic solution that achieves the same level of security without requiring the same level of processing power. This makes pairing-free CP-ABE a more suitable choice for resource-constrained devices like smart meters in a smart grid environment.

The rest of the paper is structured as follows. Section 2 offers a comprehensive review of relevant attribute-based encryption schemes used for data security and smart grids. Section 3 establishes the foundational concepts, including access structures, linear secret sharing schemes, and elliptic curve cryptography. Section 4 dives into the details of our proposed pairing-free cryptosystem utilizing elliptic curve cryptography. Section 5 rigorously analyzes the security of the scheme, while Section 6 evaluates its efficiency. Sect. 6 presents experimental validation of the performance analysis. Finally, Sect. 7 concludes the paper by summarizing our contributions and outlining potential future research directions.

2 Related Work

Attribute-Based Encryption (ABE), introduced by Sahai and Waters in 2006 [4], offers a powerful mechanism for fine-grained access control. It operates under a one-to-many encryption paradigm, where decryption is successful only if a user's attributes satisfy a predefined access policy. ABE schemes come in two primary flavors: Key-Policy ABE (KP-ABE) and Ciphertext-Policy ABE (CP-ABE) [6]. In KP-ABE, ciphertexts are associated with attributes, while decryption keys are issued based on user attributes. Conversely, CP-ABE grants users attribute-based keys and associates access policies with the ciphertexts. Following the introduction of CP-ABE, numerous schemes have been proposed that rely on a central authority for attribute management and key distribution [7–10].

In the field of smart grid security, Hur et al. [2, 11, 12] introduced a CP-ABE scheme designed to ensure policy privacy by obscuring the access structure within the ciphertext. However, this approach suffers from computational inefficiencies and limited expressiveness. To achieve constant-size ciphertexts, Odelu et al. [5, 9] proposed a CP-ABE scheme offering a compact ciphertext. Nevertheless, these advancements come at the cost of restricted expressiveness, as their access structures are limited to AND gates with wildcards.

In the field of ABE using Elliptic Curve Cryptography (ECC), recent works are focused on achieving lightweight solutions for resource-constrained IoT (Internet of Things) devices. Building on this, Odelu et al. [5, 9] and Alsheri et al. [13] proposed CP-ABE schemes with constant-size secret keys and ciphertexts, specifically designed to address the limitations of mobile devices in IoT environments. Merlino et al. [2] proposed detection of DDoS attacks in Smart Grids. Their work exemplifies the ongoing effort to make ABE more efficient and applicable to practical resource-constrained scenarios.

In the field of lightweight CP-ABE, the focus is on improving efficiency and security for resource-constrained devices. One approach, Cooperative CP-ABE, distributes encryption tasks among devices to ease the load. Additionally, using hardware like FPGAs with lightweight algorithms like Blowfish shows promise in reducing processing time. However, challenges remain in achieving strong security against modern attacks while optimizing resource usage. Further research is needed to overcome these limitations and create more efficient and secure lightweight CP-ABE solutions [13].

3 Background

This section lays the groundwork for our proposed cryptosystem by explaining the mathematical concepts and security assumptions behind it.

3.1 Access Structure

Definition 1 *(Access Structure)*:

Let $\mathcal{A} = \{A_1, A_2, \ldots, A_n\}$ be a finite set of attributes. A collection $\mathbb{S} \subseteq 2^{\mathcal{A}}$ is a monotone access structure if it satisfies the following properties:

Monotonicity: $\forall B, C \in \mathbb{S}$: $B \subseteq C \rightarrow C \in \mathbb{S}$

An access structure $\mathbb{S}$ is a collection of non-empty subsets of $\mathcal{A}$, *i.e.*, $\mathbb{S} \subseteq 2^{\mathcal{A}} \setminus \{\emptyset\}$. The sets in $\mathbb{S}$ are called the "authorized sets", representing combinations of attributes that allow decryption. The remaining sets in $2^{\mathcal{A}} \setminus \mathbb{S}$ are "unauthorized sets".

Within ABE systems, access control is governed by a pre-defined policy. This policy, represented as an access structure, dictates the specific attributes, a user must possess to decrypt a ciphertext. For instance, an access structure expressed as the boolean formula $A \wedge B \wedge (C \vee D)$ signifies that successful decryption requires the user to hold attributes A, B, and either C or D. In simpler terms as shown in Fig. 1, the user must satisfy all of the conditions (A and B) and at least one of the sub-conditions (C or D) within the policy to decrypt the information.

3.2 Linear Secret Sharing (LSS) Schemes

LSSS matrices offer a mechanism to realize intricate monotone access policies within CP-ABE schemes.

Definition 2 *(LSSS)*: A secret sharing scheme Π on an attribute universe U is called linear over $\mathbb{Z}_p$ if the following conditions are satisfied:

Fig. 1 Access tree

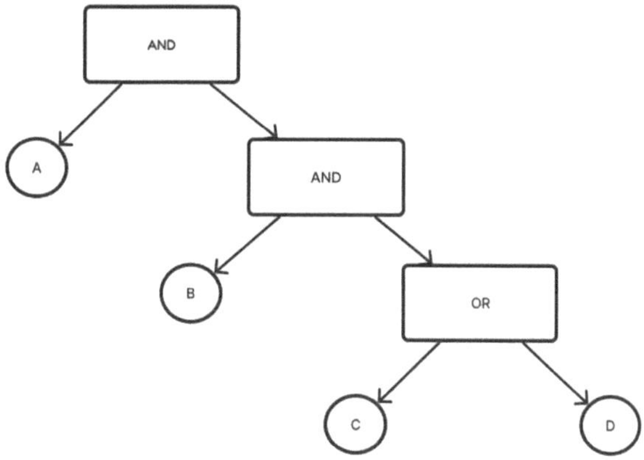

1. A vector over $\mathbb{Z}_p$ is generated by the shares for each party $s \in \mathbb{Z}_p$.
2. For each access structure $\mathbb{A}$ on U, there exists a matrix A of order $l \times n$ called the "share-generating matrix" and a function ρ such that $\rho : \{1, ..., l\} \rightarrow U$.
3. A column vector $v = (s, r_2, ..., r_l)$ chooses s as its first element and randomly chooses the rest $r_2, ..., r_l \in \mathbb{Z}_p$. Then, according to the LSS Scheme, $\boldsymbol{\lambda} = A\mathbf{v}$ is the vector of l shares of the secret s. It is to state that the share λ_x belongs to $\rho(x)$.

If a linear secret sharing scheme is defined as above, it also satisfies the linear reconstruction property. To be specific, let $\Pi \in \mathbb{A}$ be any authorized set, where $\mathbb{A}$ is the access structure, and let I be the corresponding set of row number $\{x : \rho(x) \in \Pi\}$. Then there must exist constants $\{c_x \in \mathbb{Z}_p\}_{x \in I}$, such that if $\{\lambda_x\}$ are the shares of secret s, the secret can be recovered by computing $\sum_{x \in I} c_x \lambda_x = s$. The pair (A, ρ) represents the access policy in our mathematical construction.

The LSSS matrix generation process originates from algorithm [7]. This algorithm takes as input an access tree that represents a monotone Boolean formula. The access tree itself consists of internal nodes representing logical *AND* or *OR* gates, and leaf nodes representing individual attributes. The algorithm operates by traversing the access tree in a top-down manner, as follows:

- For *OR* gate:
  - Both the child nodes assigned with the same vector v,
  - The forechosen constant c, for the share now acts as an counter and is kept unchanged;
- For *AND* gate:
  - Appends the vector v with 0 to change its length to c,
  - To denote concatenation, the *right child node* is labeled with the vector $v \| 1, \|$,
  - The *left child node* is labeled with $v_\theta = (0, ..., 0) \| - 1$, where $|v_\theta| = c$,
  - Increase the value of c by 1.

The algorithm assigns labels to the tree nodes as depicted in Fig. 2. These labels are then converted into a matrix, denoted as LSSS, where each row corresponds

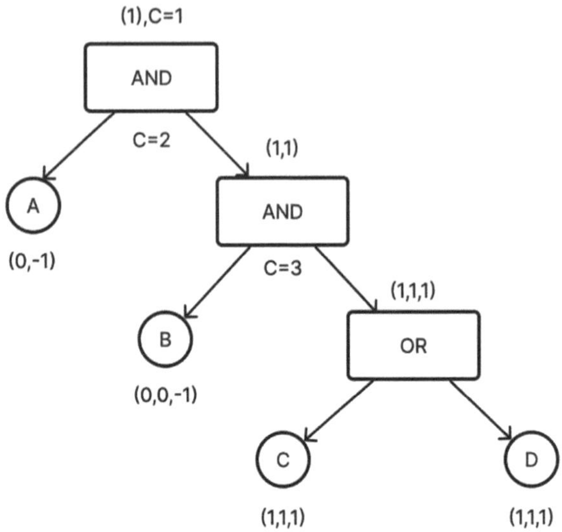

Fig. 2 Labeled access tree

to a leaf node. If the label vectors associated with the leaves have varying lengths, the algorithm employs zero-padding. This process ensures that all vectors within the LSSS matrix share a uniform length, corresponding to the length of the longest vector.

The LSSS matrix of the access tree in Fig. 1, after applying the algorithm:

$$A = \begin{bmatrix} 0 & -1 & 0 \\ 0 & 0 & -1 \\ 1 & 1 & 1 \\ 1 & 1 & 1 \end{bmatrix} \begin{matrix} \rho(1) = A \\ \rho(2) = B \\ \rho(3) = C \\ \rho(4) = D \end{matrix}$$

The function denoted by ρ establishes a one-to-one correspondence between each row of the LSSS matrix and a specific attribute (A, B, C, or D) as illustrated in the above formula. This criterion requires that the linear span of the matrix rows corresponding to attributes within S must encompass the vector $(1, 0, ..., 0)$. In simpler terms, a combination of these attribute-labeled rows must be able to reconstruct the vector $(1, 0, ..., 0)$.

3.3 *Elliptic Curve Cryptography*

Elliptic curves are described by the set of solutions to certain equations in two variables. Elliptic curves defined modulo a prime p are of central importance in public-key cryptography.

Definition 3 *(Elliptic Curve)*:

Let $a, b \in \mathbb{R}$ be constants such that $4a^3 + 27b^2 \neq 0$. A ***non-singular elliptic curve*** is the set $\mathcal{E}$ of solutions $(x, y) \in \mathbb{R} \times \mathbb{R}$ to the equation.

$$y^2 = x^3 + ax + b$$

together with a special point $\mathcal{O}$ called the ***point at infinity***. The following is the definition of *Elliptic Curve Discrete Logarithm Problem (ECDLP)* which secures ECC:

Let $\mathcal{E}$ be an elliptic curve defined over a finite field $\mathbb{F}_q$ and let P be a point of order n such that $P \in \mathcal{E}(\mathbb{F}_q)$. Given $Q \in \langle P \rangle$, the *ECDLP* is to find the integer l in range $l \in [0, n-1]$, such that $Q = lP$.

Elliptic Curve Cryptography ECC offers a compelling alternative to RSA. It achieves a comparable level of security with significantly smaller key sizes. This advantage stems from the inherent difficulty of solving the Elliptic Curve Discrete Logarithm Problem ECDLP compared to factoring large integers in RSA. An ECC encryption protocol typically follows a three-step process. First, the plaintext message undergoes a conversion into a point Q on a specific elliptic curve. Subsequently, the actual encryption protocol involving two parties, conventionally denoted as Alice and Bob, takes place.

1. *Key Generation*:
   - Alice and Bob use DLP and agree on a same elliptic curve $y^2 = x^3 + ax + b(mod\ p)$, with a generator point G.
   - Alice selects an integer $X_A \in \mathbb{A}_p$ as private and computes it's public key $P_A = X_A G$.
   - Bob selects an integer $X_B \in \mathbb{B}_p$ as private and computes it's public key $P_B = X_B G$.
2. *Encryption*: To encrypt Q, Alice randomly selects an integer $k \in \mathbb{Z}_p$ and computes the two parts of the ciphertext, $C_1 = kG$ and $C_2 = Q + kP_b$. Alice sends (C_1, C_2) to Bob.
3. *Decryption*: After receiving the ciphertext, Bob may multiply C_1 by his private key X_B and subtracts it from C_2. That is:

$$\begin{aligned} C_2 - X_B C_1 &= (Q + kP_b) - X_B(kG) \\ &= (Q + kX_B G) - X_B kG \\ &= Q \end{aligned}$$

In this way, the message Q is transferred from Alice to Bob.

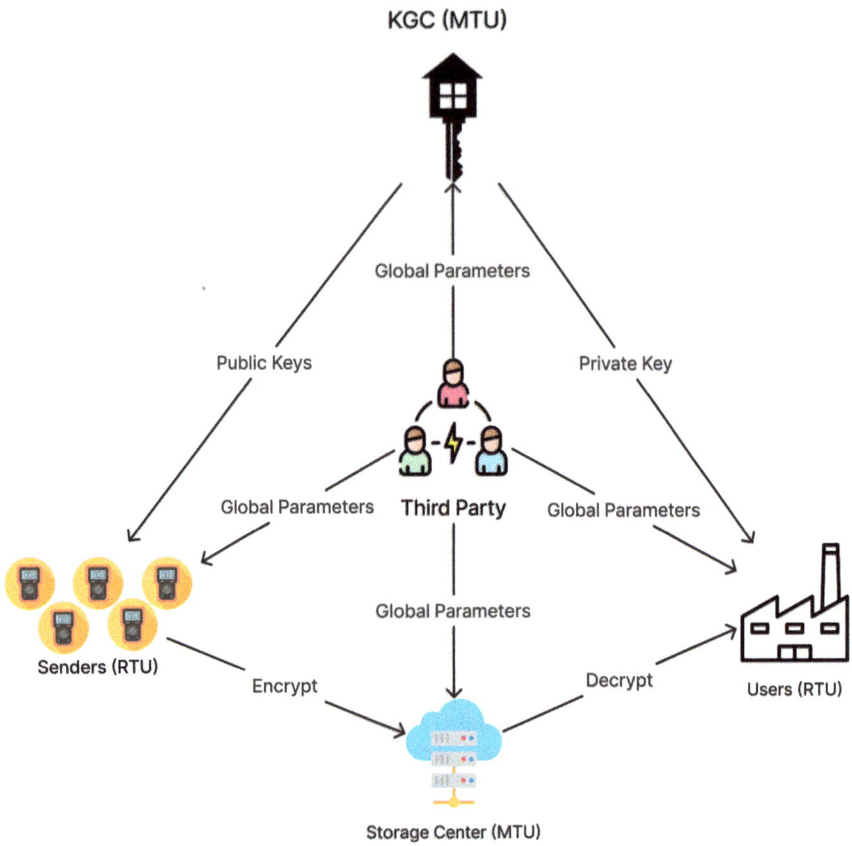

Fig. 3 SCADA architecture

4 The Proposed Pairing Free Cryptosystem

In this section, we will discuss our proposed pairing free cryptosystem in details for smart grid applications. Here, we present the system model followed by it's mathematical construction.

4.1 System Model

This system model adopts a data-centric perspective for the smart grid architecture. By doing so, the cryptographic system can be specifically designed to address the unique data security challenges within this framework. Our proposed cryptosystem is shown in Fig. 3 and the workflow is presented as follows.

Our initial focus is on Supervisory Control and Data Acquisition (SCADA) systems as a foundation for secure and reliable smart grid management and maintenance. Within the SCADA architecture, the master terminal unit (MTU) serves as the central hub, collecting and disseminating data to all subordinate control entities within the utility.

Individual substations utilize Remote Terminal Units (RTUs) to facilitate data collection and reporting from field sensors and phasor measurement units (PMUs) to the MTU. This data is then accessible to human operators via a human-machine interface (HMI). Notably, a decentralized grid infrastructure enables independent management of distinct grid entities by various operators within their respective trust domains. We can categorize the various entities in smart grid from a security perspective in the following manner:

1. The **Third Party** generates the global parameters and distributes the global parameters to all other entities. It does not play role after generating global parameters.
2. The **key generation center (KGC)** serves as the central authority responsible for generating cryptographic keys and parameters for all participants within the smart grid. This secure key management system leverages an entity-based access control approach, where the KGC assigns cryptographic credentials based on the specific attributes of each entity. Importantly, the KGC itself maintains no ability to decrypt the protected information within the smart grid, ensuring data confidentiality.
3. The **storage center (SC)** acts as a secure data repository within the smart grid, managing user access control. In the context of the SCADA model, the MTU assumes the role of the storage center, housing data collected from RTUs and regulating access for authorized users.
4. The **sender** represents an entity responsible for generating and transmitting data to the storage center. Within our model, the sender defines an access policy based on attributes, encrypts the data using this policy and sends the encrypted data to the Storage Center. For example, an RTU (remote terminal unit) or a power metering device can function as a sender. In this scenario, the sender would define an access policy for its metering data, restricting access only to authorized entities. This policy might be structured as: ((“Department: Billing” OR “Department: Anomaly Detection”) AND (“City: Cambridge” OR “City: Princeton”)).
5. **Users** are the people working in the power supply companies and produce important information like the billing charges and energy feedbacks. They communicate these information with customers via smart meters. They interact with the HMI to retrieve information from the storage center. A user can decrypt only those ciphertexts stored in the storage center where their attributes satisfy the pre-defined access policy associated with the ciphertext.

4.2 Mathematical Construction

In this sub-section, we present the detailed mathematical construction of our proposed cryptosystem for smart grid applications. Our cryptographic scheme consists the set of algorithms, which is divided into four process groups: **Setup**, **Key Generation**, **Encryption** and **Decryption**.

4.2.1 Setup

This process involves establishing essential parameters and generating public-private key pairs for various entities. First, a trusted third party executes a *GlobalSetup* algorithm that generates all the cryptographic parameters required for the system to function. The third party's role concludes after the creation of initial parameters, and they do not participate in the further activites. Following this, two internal entities, the Key Generation Center (KGC) and the Storage Key Center (SC), run their dedicated setup algorithms (*KGCSetup* and *SCSetup*) utilizing the established global parameters. The *KGCSetup* algorithm generates the KGC's own public-private key pair and facilitates the generation of public-private key pairs for end users. The *SCSetup* algorithm, on the other hand, generates a dedicated public-private key pair specifically for the storage center (SC).

GlobalSetup $(q) \rightarrow (GP)$: Let $\mathbb{F}_q$ be a finite field of prime order q and $\mathcal{E}$ be an elliptic cure defined over the finite field $\mathbb{F}_q$. We fix $G \in \mathcal{E}$ with prime order r ($rG = \mathcal{O}$; $\mathcal{O}$ is the point of infinity). The Elliptic Curve Discrete Logarithm Problem (ECDLP) is assumed intractable. Given a point $Q \in \mathcal{E}$ known to be a scalar multiple of G (*i.e.*, $Q = kG$ for some integer $k \in \mathbb{Z}_r^*$), finding k is computationally infeasible. This hardness underpins the system's security. Additionally, a cryptographic hash function $H : \{0, 1\}^* \rightarrow \mathbb{Z}_r$ is chosen. This function maps arbitrary bit strings to elements in the same modular group as G, enabling further cryptographic operations while ensuring fixed-size outputs regardless of the input length. Finally we all variables are included to produce a Global parameter(GP) as:

$$GP = \{\mathcal{E}, G, H\}$$

KGCSetup $(GP) \rightarrow (PK_K, SK_K)$: The Key Generation Center (KGC) executes this algorithm, utilizing global parameters (GP) as input. Following this execution, the KGC generates a public key, denoted as PK_K, and a corresponding secret key, denoted as SK_K. In this step, the KGC chooses a random number $n \in \mathbb{Z}_r$ as the master key and publishes nG as its public key.

$$SK_K = n, PK_K = nG$$

SCSetup $(GP) \rightarrow (PK_S, SK_S)$: The Storage Center (SC) executes this algorithim, utilizing global parameters (GP) as input. Following this execution, the SC generates a public key, denoted as PK_S, and a corresponding secret key, denoted as SK_S. In this step, the SC chooses a random number $m \in \mathbb{Z}_r$ as the secret key and publishes mG as its public key.

$$SK_S = m, PK_S = mG$$

4.2.2 Key Generation

The *KeyGen* algorithm is invoked by the Key Generation Center (KGC) to provide the Public key of each attribute and Secret Key components corresponding to the attributes possessed by the user. Let A be the set consisting of n attributes, then for each attribute $i \in A$, we choose $k_i \in \mathbb{Z}_r$ randomly and derive it's public key $PK_i = k_i G$.

KeyGen $(GP, GID, i, SK_K) \rightarrow (SK_{i,GID})$: The key generation center computes the secret key SK_i for an attribute attribute i of a user, having the global identifier as GID as follows:

$$SK_{i,GID} = k_i + H(GID)n$$

Subsequently, KGC records this attribute i on its corresponding attribute list. Hence the private key of user with GID based on the attributes will be:

$$SK_{GID} = \{SK_{i,GID}\}_{i \in \{1,2,3,\ldots,n\}}$$

4.2.3 Encryption

The encryption process takes as input a message (J), an access control structure defined by an access matrix (A) and a threshold value (ρ), and the public keys corresponding to all attributes referenced within the access policy. The algorithm then generates a ciphertext (CT) that embodies the specified access control requirements.

Encrypt $(GP, M, (A, \rho), PK_i) \rightarrow (CT)$: Plaintext M is mapped to a point J on curve $\mathcal{E}$. We choose random number $s \in \mathbb{Z}_r$ in order to calculate the first component of the ciphertext $C_0 = J + sG$. The access policy chosen by the data owner is used to generate LSSS access structure which is fed into the encryption algorithm. The output after encryption is an $n \times l$ access matrix A with ρ mapping its rows to attributes.

A random vector $v \in \mathbb{Z}_r^l$ is chosen, with s as its first entry. Let us denote $\lambda_x = v \cdot A_x$, where A_x is the xth row of A. Another vector $u \in \mathbb{Z}_r^l$ is chosen such that 0 is its first entry and ω_x denoted as $u \cdot A_x$. The sender computes ciphertext CT as follows and sends (CT, GID) to the storage center.

$$\left.\begin{aligned} C_{1,x} &= \lambda_x G + \omega_x PK_{\rho(x)} \\ C_{2,x} &= \omega_x G \end{aligned}\right\} \forall x \in A$$

$$CT = \{C_0, \{C_{1,x}, C_{2,x}\}_{x \in \{1,2,,3\ldots,n\}}\}$$

4.2.4 Decryption

Successful decryption of ciphertext CT is contingent upon the following two factors:

1. The user must possess a valid set of attribute secret keys,
2. These keys must align with the access control matrix governing the ciphertext.

The receiver receives the encrypted ciphertext from the storage center (SC) and decrypts using the following procedure.

Decrypt $(GP, CT, SK_{i,GID}) \rightarrow (J)$: In order to decrypt the ciphertext CT, the user finds out the set of rows I_x from A such that each row x is in the form of $(1, 0, \cdots, 0)$. Then the user then submits its *GID* along with $(C_{2,x}, \rho(x))$ for each $x \in I_x$ to the storage center. Subsequently, the storage center verifies the attributes of the user against the attribute list. If the request is successfully verified, then the storage center computes the mathematical expression given below and sends the results to the user for further computation.

$$\begin{aligned}\sum_x C_{2,x} \cdot SK_{\rho(x),GID} &= \sum_x (\omega_x G(k_{\rho(x)} + H(GID)n)) \\ &= \sum_x (\omega_x k_{\rho(x)} G + \omega_x H(GID)nG)\end{aligned}$$

For each row x in I_x, the user computes the following expression:

$$\begin{aligned}&\sum_x C_{1,x} - \sum_x (C_{2,x} \cdot SK_{\rho(x),GID}) \\ &= \sum_x (\lambda_x G + \omega_x PK_{\rho(x)}) - \sum_x (\omega_x k_{\rho(x)} G + \omega_x H(GID)nG) \\ &= \sum_x (\lambda_x G + \omega_x H(GID)nG)\end{aligned}$$

A constant $y_x \in \mathbb{Z}_r$ is chosen by the user such that $\sum_x y_x x = (1, 0, ..., 0)$. Then, the user computes $\sum_x y_x(\lambda_x G - \omega_x H(GID)nG)$. Since $v \cdot (1, 0, ..., 0) = s$ and $u \cdot (1, 0, ..., 0) = 0$, therefore

$$\sum_x y_x(\lambda_x G - \omega_x H(GID)nG) = sG$$

Now, the user computes $C_0 - sG = J$. Finally, J is mapped back to the plaintext M.

5 Security Analysis

In this section we prove that our Pairing Free CP-ABE scheme for smart grid systems is secure. The notations used for security and performance analyses is shown in Table 1.

5.1 *Data Security*

Our proposed PF-CP-ABE scheme leverages elliptic curve cryptography to ensure the confidentiality of secret keys associated with user attributes. The system strictly controls access to these keys. Only authorized users possessing a specific attribute are eligible to receive the corresponding secret key k_i from the designated attribute authority. The underlying mathematical hardness of the ECDLP guarantees the security of this process. In simpler terms, even if an unauthorized user doesn't have the required attribute, they cannot computationally derive any information about the secret key k_i solely by observing the corresponding public key $K_i G$ within a reasonable time frame.

Any ciphertext C_0 has only one implicit message. Assuming message M such that it can be mapped to mG, where $m \in \mathbb{Z}_r$, since s is always randomly chosen by the data owner, $C_0 = (m + s)G$ is just a random point in the perspective of an attacker. The ECDLP ensures that an attacker cannot gain any meaningful information about the secret message M without possessing the corresponding secret key s. This secret key is further protected using a secret sharing scheme *i.e.*, LSSS. Here, the secret key is further divided into λ_x shares. Recovering the complete secret key requires a qualified subset of these shares, so only a data user with the appropriate attributes can decrypt ciphertext. In essence, both ECDLP and the secret sharing scheme con-

Table 1 Notations

Abbreviation		Description
N_a	$\rightarrow$	Number of attributes in the cryptosystem
n_r	$\rightarrow$	Number of rows in the LSSS matrix A
C_G	$\rightarrow$	Multiplication in a group
C_e	$\rightarrow$	Bilinear pairing
C_{Z_N}	$\rightarrow$	Field exponentiation Z_N where $\|N\| = 1024$
D_a	$\rightarrow$	Minimum number of attributes that satisfy the access policy
U_a	$\rightarrow$	Set of user attributes

tribute to a robust security architecture by isolating sensitive data and access control. This ensures data confidentiality and it is achieved by ensuring that unauthorized users, lacking the attributes specified in the access policy, cannot derive the necessary attributes *wrt,* any row x ($x \in A$). Consequently, for such users, the equation $\sum_x c_x \lambda_x = (1, 0, ..., 0)$. This inherently prevents the calculation of the secret key s along with the initial element of vector v and this effectively prevents the unauthorized access to the encrypted data.

5.2 Collusion Resistance

Our proposed PF CP-ABE Schema prevents collusion attacks by guaranteeing data integrity. Here, collusion resistance signifies that even if multiple users collaborate, they should be unable to decrypt the ciphertexts unless at least one of them possesses the individual decryption capability. To achieve this, the scheme employs Group Identifier (GID) which is linked to a user's attributes. This binding prevents unauthorized combinations of attributes from different users during the decryption process, upholding the system's security.

For example, assuming the access policy $\mathbb{S} \subseteq 2^{\mathcal{A}} \setminus \{\emptyset\}$ and Alice wants to collude with Bob to decrypt a ciphertext. Independent access to the encrypted data is not feasible for either party. While Alice possesses attributes α and β, and Bob holds attribute γ, the decryption process requires possession of a specific combination of these attributes which are not currently held by either individual. Assuming that the collusion, is possible then Alice's and Bob's keys will look like the expressions given below for some x:

$$\lambda_x G - H(GID_{\text{Alice}})\omega_x nG \tag{1}$$

$$\lambda_x G - H(GID_{Bob})\omega_x nG \tag{2}$$

For any authorized user, he/she can choose $c_x \in \mathbb{Z}_r$ in such a way that $\sum_x c_x x = (1, 0, ..., 0)$ and also compute sG. Since Alice and Bob have different *GIDs* their hash outputs will not be the same *i.e.,* $H(GID_{\text{Alice}}) \neq H(GID_{Bob})$ and $\sum_x H(GID)\omega_x G \neq 0$. From the above equations (Eqs. 1 and 2), we can say that they cannot recover sG and this concludes that the cryptosystem is collusion resistant.

6 Performance Analysis

This section commences by analyzing the storage requirements and communication costs associated with our proposed scheme. Subsequently, an experimental evaluation is conducted to compare the computational efficiency of our scheme against existing alternatives. To ensure clarity during the comparison, a table (Table 1) has been included to define any new notations introduced for this purpose.

6.1 Storage Overhead

Smart grids rely on lightweight IoT devices with limited resources. Therefore, storage efficiency becomes a critical factor when analyzing performance. This analysis examines storage overhead across three key entities: the Key Generation Center (KGC), the Storage Center (SC), and the users.

6.1.1 Key Generation Center

The KGC issues and revokes attribute keys for users. It also stores a list of attributes associated with each user. Consequently, the KGC's storage requirements grow linearly with the total number of attributes and users in the system.

6.1.2 Storage Center

Similar to traditional servers, the SC bears the majority of storage overhead. It stores user IDs along with encrypted data (ciphertexts). Each ciphertext comprises two parts $(C_0, C_{1,x}, C_{2,x})$, and the overall storage requirement is directly proportional to the number of attributes used for encryption.

6.1.3 User

Unlike traditional cryptosystems, users in this system are relieved of the burden of storing and managing keys locally. The KGC handles this task. Users only need to retain their own private keys for decryption.

6.2 Communication Overhead

Our system introduces increased communication overhead compared to traditional approaches. This is because users rely on the Key Generation Center (KGC) to decrypt data, as the KGC holds the secret keys for user attributes. This necessitates the user to communicate with the KGC during decryption. However, a key advantage of our system lies in attribute revocation. Unlike traditional systems where revocation might require updating multiple user keys, our system allows the KGC to simply modify the attribute for a single user. This significantly reduces the overall communication overhead associated with revocation. Hence, despite the rise in communication cost for decryption, the simplified revocation process leads to an overall gain in communication efficiency.

6.3 Computation Overhead

In this section, we compare the performance of the proposed PF-CP-ABE with the state-of-the-art works in terms of time complexity. Table 2 shows how our PF-CP-ABE is computationally efficient with existing schemes [5, 7, 9, 10]. Table 3 gives you a rough idea of how much time each kind of cryptographic operation typically takes. Although CP-ABE schemes presented in [9, 10] are computationally efficient because they use very small secret keys, this efficiency only works for complex access policies. On the other hand, Scheme [5] offers efficient encryption and decryption, its restricted access structure (limited to "AND" gates only) and reliance on long RSA keys (min. size 2048 bits) make it unsuitable for resource-constrained smart grid environments. While the scheme [7] uses LSSS, it's encryption time scales linearly with the number of attributes.

6.4 Experimental Analysis

We implemented both our scheme and schemes [5, 7] on a system running Ubuntu Linux 16.04LTS. The system has an Intel I7 CPU at 2.60 GHz and 2GB of RAM.

Table 2 Time complexity comparison of 'our scheme' with existing schemes

Scheme	Encryption	Decryption
[7]	$C_e + (7N_r + 2)C_g$	$2D_a(C_e + 2C_g)$
[9]	$(N_a - \|A\| + 2)C_g$	$(N_a - \|A\| + 3)C_g$
[5]	$3C_{Z_N}$	$3C_{Z_N}$
[10]	$(2(N_a - \|A\|) + 2)C_g$	$(2(N_a - \|A\|) + 2)C_g + 3C_e$
Our scheme	$(4N_r + 1)C_g$	$(D_a + 1)C_g$

Table 3 Execution time of various operations used in the experiment

C_e	C_G	C_{Z_N}
4.12 ms	2.19 ms	1.67 ms

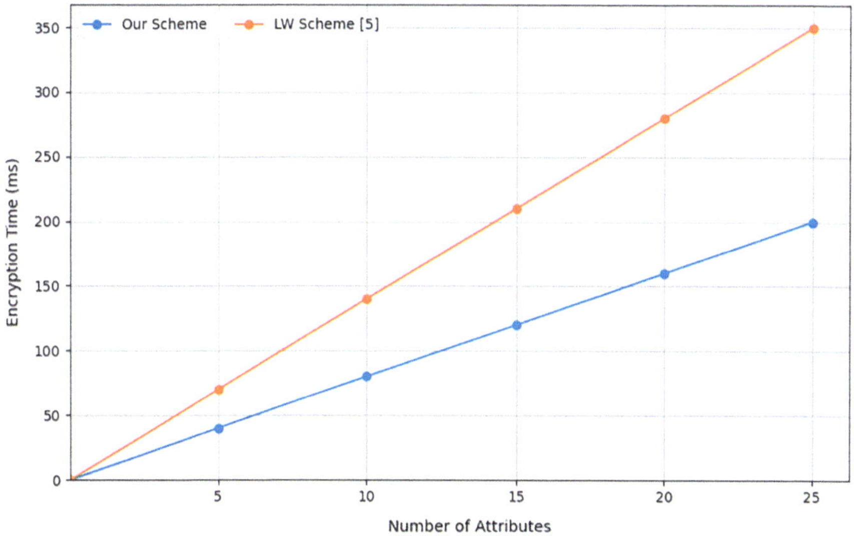

Fig. 4 Comparisons of encryption-time of 'our scheme' versus LW scheme [7]

Our implementation relies on a well-established cryptography library (charm, version 0.50) that uses a special mathematical structure called a 160-bit elliptic curve group. This curve helps ensure the security of the system. For efficiency, the curve is built on a foundation of calculations involving very large numbers (512-bit finite field). This setup aims to achieve a security level equivalent to protecting information with an 80-bit key. To ensure reliable results, all our experiments were repeated 30 times, and the reported values represent the average outcome.

Our proposed pairing-free CP-ABE scheme exhibits consistently faster encryption times compared to LW encryption, as shown in Fig. 4. This efficiency improvement implies that our scheme offers a simpler cryptographic solution, potentially leveraging optimized algorithms and reduced computational overhead. Furthermore, the consistent performance difference between the two approaches highlights the scalability of our technique. This demonstrates, even if the message size increases, the encryption time maintains its consistency.

The results depicted in Fig. 5 affirm the dominance of our pairing-free CP-ABE encryption scheme over LW Scheme [7] in terms of decryption time across diverse attribute scenarios. Notably, Our Scheme consistently outperforms Scheme [7] as the number of attributes increases, indicating the better efficiency of our decryption processes. The sustained performance gap between our scheme and LW encryption across varying the number of attributes, highlights the scalability of our CP-ABE scheme. This ensures that it maintains efficiency advantages as decryption tasks become more complex.

In Fig. 6 we analyze the key generation times of Our Scheme with schmes [5, 7] when the number of attributes increases. It can be observed from the figure that

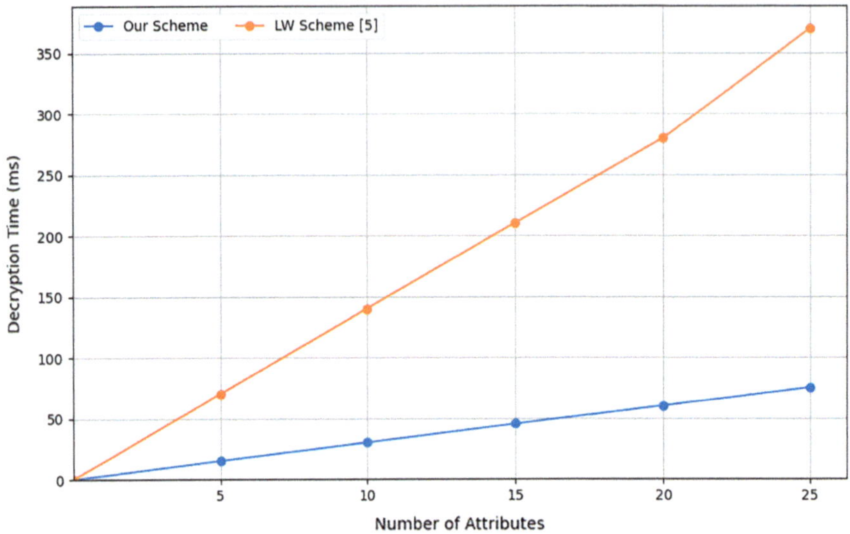

Fig. 5 Comparisons of decryption-time of 'our scheme' versus LW scheme [7]

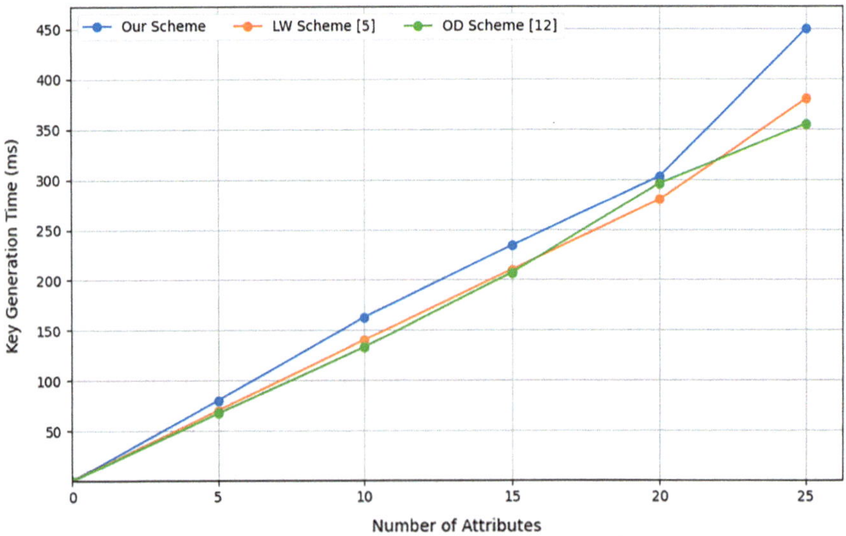

Fig. 6 Camparison of *KeyGen* times for 'our scheme' with LW scheme [7] and OD Scheme [5]

when the number of attributes increases, key generation time generally increases for all the schemes indicating the escalating computational complexity associated with cryptographic operations. We can conclude from the figure that the better the performance of the machine (higher number of threads, larger RAM), the lower time it takes for key generation.

One major drawback of scheme [7] is that users need to combine their secret keys with encrypted data (ciphertexts) during decryption, which can be computationally expensive. Our approach offers a significant advantage by leveraging the Elliptic Curve Cryptography (ECC) algorithm. We replace complex bilinear pairings with simpler scalar multiplications on elliptic curves. This not only reduces calculation times but also saves overall processing overhead. Additionally, our scheme cleverly offloads some of the decryption workload to the attribute authority. This minimizes the computational burden on individual users. In essence, our PF-CP-ABE scheme delivers superior efficiency, making it a much better fit for practical applications in smart grid systems.

7 Conclusion

In this paper, a novel data security system leveraging Pairing-Free Ciphertext-Policy Attribute-Based Encryption (PF-CP-ABE) is proposed to significantly improve data security for Smart Grid Applications. This system replaces the complex CP-ABE scheme used for smart grid data security with a more streamlined and efficient Paring Free CP-ABE approach. This transition not only improves efficiency, performance, and fault tolerance in securing data within Smart Grid architectures but also ensures robust data security and collusion resistance. The system's foundation lies in elliptic curve cryptography, which safeguards the secret keys, restricting access according to the user attributes, and preventing unauthorized decryption of ciphertexts. Furthermore, the scheme's inherent resilience against collusion attacks reinforces data integrity within Smart Grids. Overall, the proposed system offers a promising solution to address the evolving security challenges in Smart Grid applications, providing a comprehensive and effective approach to data protection in the context of Smart Grid architecture.

References

1. Abrahamsen, F.E., Yun, A., Cheffena, M.: Communication technologies for smart grid: a comprehensive survey. Sensors **21**(23), 8087 (2021)
2. Merlino, J.C., Asiri, M., Saxena, N.: Ddos cyber-incident detection in smart grids. Sustainability **14**(5), 2730 (2022)
3. Horváth, M.: Attribute-based encryption optimized for cloud computing. Infocommun. J. **7**(2), 1–9 (2015)

4. Bethencourt, J., Sahai, A., Waters, B.: Ciphertext-policy attribute-based encryption. In: Proceedings of IEEE Symposium on Security and Privacy, pp. 321–334 (2007)
5. Odelu, V., Das, A.K., Khan, M.K., Choo, K.-K.R., Jo, M.: Expressive CP-ABE scheme for mobile devices in IoT satisfying constant-size keys and ciphertexts. IEEE Access **5**, 3273–3283 (2017)
6. Goyal, V., Pandey, O., Sahai, A., Waters, B.: Attibute-based encryption for fine-grained access control of encrypted data. In: Proceedings of ACM Conferences on Computer and Communications Security (ACM CCS), pp. 89–98, Virginia, USA
7. Lewko, A., Waters, B.: Decentralizing attribute-based encryption. In: Annual International Conference on the Theory and Applications of Cryptographic Techniques, pp. 568–588 (2011)
8. Liu, Y., Ning, P., Reiter, M.: False data injection attacks against state estimation in electric power grids. ACM Trans. Inf. Syst. Secur. (TISSEC) **14**(1), 13 (2011)
9. Odelu, V., Das, A.K.: Design of a New CP-ABE with Constant-Size Secret Keys for Lightweight Devices Using Elliptic Curve Cryptography. Wiley, Hoboken, NJ, USA (2016)
10. Guo, F., Mu, Y., Susilo, W., Wong, D.S., Varadharajan, V.: CP-ABE with constant-size keys for lightweight devices. IEEE Trans. Inf. Forensics Secur. **9**(5), 763–771 (2014)
11. Hur, J.: Attribute-based secure data sharing with hidden policies in smart grid. IEEE Trans. Parallel Distrib. Syst. **24**(11), 2171–2180 (2013)
12. Hu, C., Yu, J., Cheng, X., Tian, Z., Akkaya., K., Sun, L.: CP-ABSC: an attribute-based signcryption scheme to secure multicast communications in smart grids. Math. Found. Comput. (2018)
13. Alsheri, A., Khan, I.: Types of lightweight cryptographies in current developments for resource constrained machine type communication devices: challenges and opportunities. IEEE Trans. Cryptogr. (2022)

Critical Infrastructures in the Cloud

Martin Gilje Jaatun and Geir Kjetil Hanssen

Abstract Cloud computing is increasingly being used not only to support critical infrastructure applications, but actually forms a vital part of them. This paper discusses challenges faced by custodians of critical infrastructures when moving to the cloud, and outlines some security requirements that are relevant to apply to critical infrastructure cloud applications.

1 Introduction

Depending on who you ask, Cloud Computing was either something that emerged as a paradigm shift about a decade ago, or alternatively it is just the last in a long line of incremental developments in how we do computing since the 1950-ies. Be that as it may, it is nonetheless clear that the cloud is not in your basement; if you are using cloud computing, your data is being processed and stored on somebody else's computer—and possibly in another legislative domain.

There has been a fair bit of concern regarding end-user privacy and holding cloud providers to account for how they manage personal data in the cloud [1], but for critical infrastructure it's less about privacy and more about societal impact. However, some critical infrastructure services are also tightly interwoven with the lives of citizens, and so the privacy aspect can often not be ignored.

Traditional critical infrastructures include electric power distribution, telecommunications, and water distribution networks [2]. Another example is the exploration and production of oil- and gas resources in challenging environments, like at offshore installations [3]. With the possible exception of telecommunications, these have been considered rather conservative and staid institutions, more characterized by "business as usual" than "agility".

M. G. Jaatun (✉) · G. K. Hanssen
SINTEF Digital, Trondheim, Norway
e-mail: martin.g.jaatun@sintef.no

G. K. Hanssen
e-mail: geir.k.hanssen@sintef.no

M. G. Jaatun et al. (eds.), *Proceedings of the International Conference on Cybersecurity, Situational Awareness and Social Media*, Springer Proceedings in Complexity,
https://doi.org/10.1007/978-981-96-0401-2_6

Frequently, driven by a growing need for more computational services, critical infrastructure custodians find themselves "inadvertently" procuring cloud services through a third party that is not nominally a cloud service provider, as illustrated in Fig. 1. The third party will typically provide a Software-as-a-Service (SaaS) application that in turn uses other cloud processing or storage services in a provider chain as indicated in the figure. This is sometimes referred to as "mashups" or "federated services", where service elements are dynamically composed to form a complex internet service [4]. This has previously been identified as an accountability challenge [1], but is just as relevant when critical societal functions may be impacted. The solutions most commonly seen typically involve export of time series data from, e.g., SCADA systems for off-site "number crunching", and it is easy to assume that as long as the export is "secure" (i.e., that there is no way for an attacker to abuse the communication path out of the critical infrastructure in order to gain access from the outside), the security risk is negligible. However, the point of doing the analysis is invariably to support decisions regarding operation of the critical infrastructure, and thus tampering with the data once exported may cause real damage further down the line.

This potential weakness in the flow of data *out* of a the organization, to an external cloud-based service provider, and back *into the organization*, was highlighted within a study initiated by the Norwegian Petroleum Authority in 2020 [3]. The hypothetical (but realistic) case was that data is gathered from, e.g., sensors or edge-devices at the so-called 'sharp end' within the operational organization, e.g. at the drilling deck on an oil rig. Then, data are transferred outside the organization to a cloud-based service

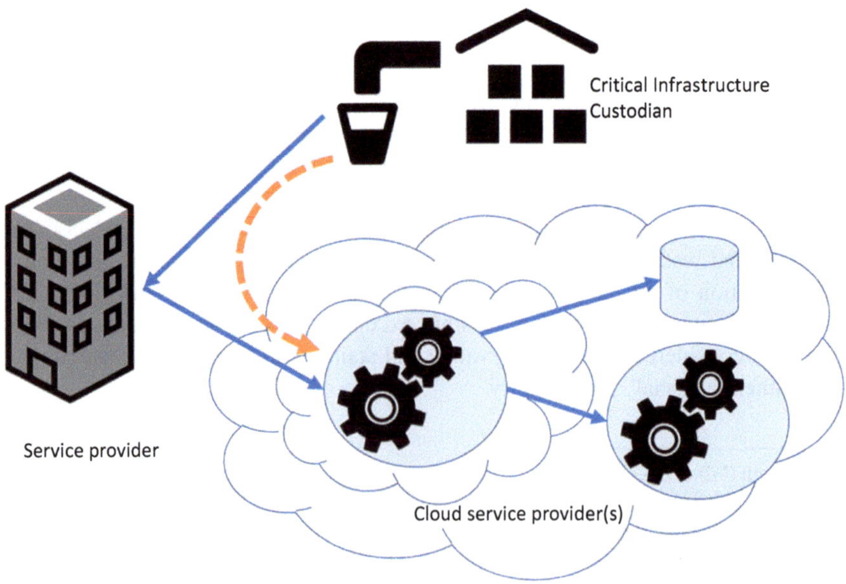

Fig. 1 Conceptual model of offering cloud services to critical infrastructure [5]

provider, e.g. providing operator-support services, that in return, is fed back into the organization and for example, made accessible to the operator via a tablet. Then, the operator uses this information to make decisions about the activities at the drilling deck. Such a scenario may seem secure and safe as the data may be well protected within the organization; oil and gas installations meet very high cybersecurity standards and have numerous security measures, e.g. separation between information technology—IT, and operational technology—OT. Furthermore, the scenario may seem technically sound and safe since the services that are provided (by the external cloud-based service provider) have no technical interface with any of the internal systems. However, because the data to and from the cloud-based provider travel via the Internet, there is a theoretical probability that someone may tamper with the data, either out- or in-going. And reaching all the way to the sharp- and critical end of the operational organization, wrong data (or operator support) may lead to hazardous events, potentially life threatening.

The sceptical reader may argue that this constructed scenario is exaggerated and not realistic. However, we believe it illustrates the general trend within critical domains and infrastructures, where the access to rich data is growing rapidly. This is followed by new technology and computational capacity to exploit data, typical via cloud-based resources and AI solutions. In short, data can be used to digitally transform and enhance businesses—in search for more efficient and profitable operations. An example taken from the oil- and gas domain, is that "intelligent oil fields" can increase productivity by 2–8%, with 2–6% better extraction [6].

With this backdrop, we look into the case of security in critical infrastructures that interface with the cloud. Our aim is to highlight the need for well-considered cybersecurity measures and strategies, and to invite the research community in developing an understanding of this emerging challenge, and the countermeasures that should compensate for the risks.

2 Background

There are a number of standards and good practice documents that provide requirements and guidance on general cloud security. ISO/IEC 27017 [7] has been co-developed with the International Telecommunication Union (ITU), and provides additional controls for applying ISO/IEC 27002 [8] to cloud services. The Cloud Security Alliance (CSA) has created the Cloud Controls Matrix (CCM) [9] which covers both of the former, and more.

In Europe, the General Data Protection Regulation [10] received far more attention than the EU NIS Directive [11] (also known as NIS1). NIS1 only mentions "cloud" in a couple of places. The NIS Directive indicates that public administrations should be able to ask cloud service providers "additional security measures beyond what [they] would normally offer", but does not go into details what these measures might be. The directive also states that member states can enforce their own national security requirements on cloud services.

The second version of the NIS-directive (NIS2) [12] mentions cloud specifically in recital 33, 34, 35, 113, 114, 116, and 117. These mentions group cloud providers with any other service providers, and specify that for a cloud provider, the competent national authority will be the EU country where the cloud provider has their main establishment.

The current situation is that the majority of the large cloud providers are not based in Europe, and this makes it increasingly likely that unless great care is taken, a large part of the cloud provider chain illustrated in Fig. 1 will physically reside outside Europe. This represents a jurisdictional challenge, and recent events [13] have sown doubts about to what extent US cloud providers can handle personal data of European citizens. Critical infrastructure providers thus need to think carefully about the requirements that they will pose to cloud service providers.

Most critical infrastructures have primarily been focused on Operational Technology (OT) rather than Information Technology (IT), and the premier OT security standards can be found in the IEC 62443 standards series [14]. IEC 62443 is still under development and does not actually discuss cloud solutions, but IEC 62443-3-2 [15] mentions that "the [System Under Consideration] may include [...] emerging technologies such as [...] cloud-based solutions".

3 Cloud Security Requirements

Based on the previous section, it is clear that it is expected that cloud providers offering services to a critical infrastructure will be presented with additional security requirements, but it is not immediately obvious what these requirements might be. Røstum and Jaatun [5] provide a selection of requirements (based on a by now partly outdated report by Bernsmed et al. [16]) that could be relevant for a water network operator. In the following (see Sect. 3.1, 3.2, 3.3 and 3.4) we will present a selection of these requirements, focusing on those that have relevance to a broader set of European critical infrastructures.

Note that the requirements are not specific on, e.g., what encryption algorithms or key lengths are mandated. Each custodian of a critical infrastructure needs to establish what represents current good practice—currently, 128-bit AES remains a reasonable choice for symmetric encryption [17], but this is a moving target, and everything may change quickly once quantum computing[1] becomes generally available [18].

This also holds true for requirements specifying frequency of, e.g., backups; the actual frequency is highly dependent on the type of critical infrastructure and type of cloud service.

[1] Generally, the public-key algorithms based on the hard problems of integer factorization and discrete log (RSA, Diffie-Hellman) will be trivially broken, and most symmetric algorithms (such as AES) will need to double their key size.

3.1 Requirements Related to Data Processing

- Isolation—*Ensure that all data is isolated from other customers' data*
  - All in-memory data shall be segregated from data belonging to other customers
  - The cloud provider must implement mechanisms that ensure that different virtual machines do not influence each other
  - Data sent to the cloud service related to a specific request are not visible to other users of the service
- Monitoring—*Ensuring that breaches of permissible use agreements are detected*
  - Behaviour of running virtual machines (VMs) shall be monitored continuously
- Physical location—*Ensure data is processed in a specific geographic location*
  - As a rule, all data should be processed in data centres based in Europe[2]
- Migration—*Ensure that migration between different physical servers is performed securely*
  - All VMs must be encrypted during migration

3.2 Requirements Related to Data Transfer

- Encryption—*Ensure that data is not transferred in clear text*
  - Up- and downloading of data to/from the cloud service is encrypted
  - All communication stages should be encrypted
  - End-to-end encryption shall be used whenever possible
- Integrity—*Ensure correctness and consistency in customer data*
  - Up- and downloading of data to/from the cloud service is integrity protected *Integrity protection can be performed using different means, such as digital signatures, message authentication codes, and other combination of encryption mechanisms and hash functions.*
- Isolation—*Ensure that all data is isolated from other customers' data*
  - The cloud service provider offers network isolation between customers, ensuring that no data traffic to/from one customer can be eavesdropped on by another

[2] This assumes that the critical infrastructure in question is based in Europe—in the general case, it will be an advantage that the cloud provider is located in the same jurisdiction as the critical infrastructure.

3.3 Requirements Related to Access Control

- Access control for administration—*Ensure secure access to the cloud administrative interface (dashboard)*
  - The cloud provider shall enforce a good practice password policy, focused on length and complexity of passwords
  - The cloud provider shall support multi-factor authentication
  - The cloud provider shall support third-party authentication solutions for simple login (SAML [19]/OpenID [20])
- Access control for users—*Ensure secure access for cloud users*
  - The cloud provider shall provide a system for creating, updating, suspending and deleting user accounts, to remove access of employees when they leave the organization
  - All cloud users should have unique user accounts; no joint accounts are to be used
  - Access to cloud services should be role-based
  - The principle of least privilege should be applied when assigning privileges to roles

3.4 Requirements Related to Data Storage

- Encryption—*Ensure that data is not stored in clear text when not in use*
  - All data is encrypted when stored. Disk encryption is sufficient (including virtual disks)
  - Data from each infrastructure custodian must be encrypted with separate encryption keys
- Physical location—*Ensure data is stored in a specific geographic location*
  - As a rule, cloud providers based in Europe should be preferred (see also requirements to data storage)
- Isolation—*Ensure that all data is isolated from other customers' data*
  - Information must be segmented in such a manner that all data from a given critical infrastructure provider is segregated from data belonging to other customers
- Ownership—*Ensure that the customer retains ownership of own data*
  - All data stored in the cloud solution remains the property of the critical infrastructure provider

  - A data processing agreement shall be entered into with the supplier. This can be with a third party that develops services using a cloud service provider, or directly with the cloud service provider itself
  - The cloud service provider may not use data from the critical infrastructure provider for the former's own purposes

- Portability—*Ensuring portability of customer data*

  - Data must not be locked in on the cloud provider's platform, but must be exportable to a pre-agreed (preferably open) format on demand

- Integrity—*Ensure correctness and consistency in customer data*

  - Integrity is maintained for all data stored in the cloud solution (see data transfer requirements above)

- Deletion—*Ensure proper deletion of all data upon customer request*

  - All replicated data shall be deleted within a specified deadline when requested by customer

- Backup—*Ensure backup is performed and maintained in a proper manner*

  - The cloud solution shall create backups at least [daily][3]
  - A local (off-cloud) backup of the cloud data shall be performed at least [weekly][3] – this should be usable also when the cloud service is not available.
  - A detailed scheme for how long backups should be retained must be devised. E.g., daily backup: 21 days; weekly backup: 12 weeks; monthly backup: 6 months; quarterly backup: two years
  - Backups stored in the cloud must be checked by restoring to shadow system at least monthly
  - Local (off-cloud) backup must be checked [weekly] (when it is created)
  - The cloud provider shall commit to a guaranteed maximum time for restoring of backup copy
  - Backup copies must be stored geo-redundant with respect to where they are normally stored

4 Discussion

A cloud service provider will implement several layers of security that will protect against different types of attack, as illustrated in Fig. 2. Normally, any cloud provider that adheres to good practice according to, e.g., CSA [9] or ISO [7] will have these covered, but critical infrastructure providers that make unverified assumptions with respect to the compliance of their cloud service provider do so at their peril. In many critical infrastructures, the concept of safety barriers has a long history, but with

[3] The actual frequency must be decided by the critical infrastructure custodian.

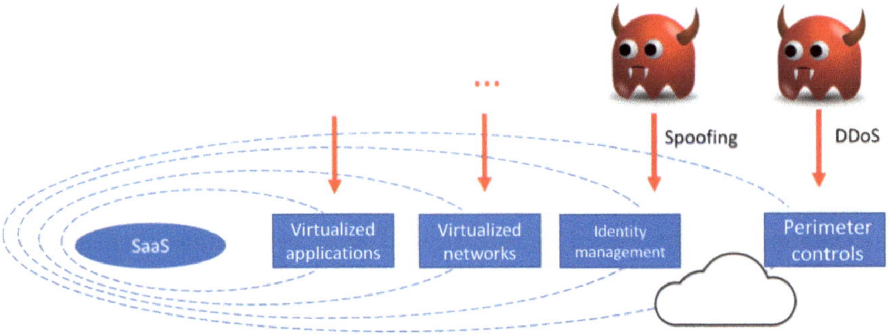

Fig. 2 Layers of security controls in the cloud [5]

increased connectivity a need for specific cybersecurity barriers [21] has emerged. How such cybersecurity barriers can interact with cloud security controls remain a topic for future study.

Depending on the cloud service model that is applied, there will be different degrees of shared responsibility for cloud security between the cloud customer (i.e., the critical infrastructure custodian) and the cloud provider [22]. For the most part, we have assumed a Software-as-a-Service (SaaS) service model, in which case the responsibility rests solely on the cloud provider, but if a Platform-as-a-Service (PaaS) or Infrastructure-as-a-Service (IaaS) service model has been chosen, the cloud customer has to take increasing responsibility, as illustrated in Fig. 3.

The NIS1 Directive [11] indicated that national security requirements may be imposed on cloud service providers, but the jurisdictional challenges represented by overseas providers are likely to make this interesting, to say the least. Geopolitical events and shifting regimes have highlighted the challenges that are inherent in putting all your cloud eggs in a single foreign basket—after Russia's full-on invasion

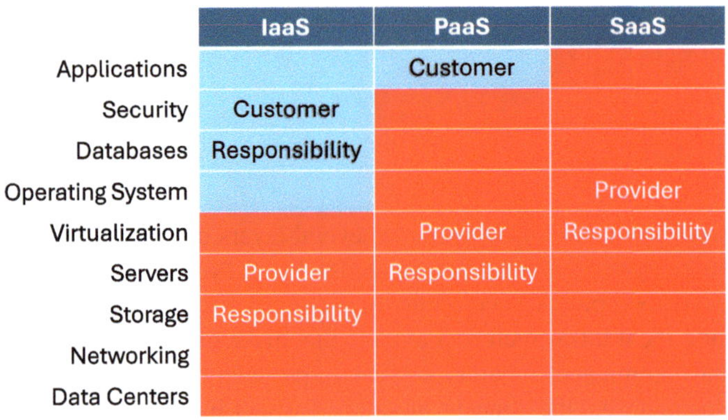

Fig. 3 Shared responsibility between cloud provider and cloud customer

of Ukraine in February 2022, many organizations rushed to shift their operations from data centres in Ukraine to Western Europe. Many European players are also wary about recent developments in US politics, with fears that the next president will take the country in a non-democratic direction, with ensuing disregard for international law and existing bilateral agreements. This is effectively eroding the traditional trust that Europeans have had in US-based cloud services, and has fuelled the search for European[4] alternatives.

Considering that the cloud-specific guidance in official EU documents [11, 12] is so vague, there might be a market for more concrete guidelines tailored to various critical infrastructures. The suggestions provided here barely scratch the surface, but they could provide a starting point in conjunction with the good practice publications mentioned above.

5 Conclusion

Cloud services are global in nature, and a cloud provider chain that starts in one country is likely to cross several borders, both inside Europe and beyond. Critical infrastructures are already using the cloud for a myriad of tasks, and this will only increase in the future. The relevant players need to embrace this fact, and work actively toward standards and guidelines that allow them to use the cloud in an acceptably secure manner.

Acknowledgements This paper is based on research funded by the Research Council of Norway through the Cybersecurity Barrier Management project, grant no. 326717; the Norwegian Petroleum Safety Authority; Norwegian Water; and the European Commission through the STOP-IT project, grant no. 740610.

References

1. Jaatun, M.G., Pearson, S., Gittler, F., Leenes, R., Niezen, M.: Enhancing accountability in the cloud. Int. J. Inf. Manag. (2016). https://doi.org/10.1016/j.ijinfomgt.2016.03.004. http://www.sciencedirect.com/science/article/pii/S0268401216301475
2. George, R.: Critical infrastructure protection. Int. J. Crit. Infrastruct. Protect. **1**, 4–5 (2008). https://doi.org/10.1016/j.ijcip.2008.08.010. https://www.sciencedirect.com/science/article/pii/S1874548208000061
3. Hanssen, G.K., Onshus, T., Jaatun, M.G., Myklebust, T., Ottermo, M.V., Lundteigen, M.A.: Principles of digitalisation and IT-OT integration. Technical Report SINTEF Digital (2023). https://ntnuopen.ntnu.no/ntnu-xmlui/handle/11250/3127881
4. Meland, P.H.: Composite services with dynamic behaviour. In: Secure and Trustworthy Service Composition: The Aniketos Approach, pp. 1–9. Springer (2014)

[4] For all practical purposes, this means the European Union and the European Economic Area (EU/EEA).

5. Røstum, J., Jaatun, M.G.: Informasjonssikkerhet og skybaserte tjenester for vannbransjen (in Norwegian). Technical Report, Norsk Vann (2018). https://norskvann.no/index.php/kompetanse/va-bokhandelen/produkt/681-a238-informasjonssikkerhet-og-skybaserte-tjenester-for-vannbransjen. Rapport 238/2018
6. Lu, H., Guo, L., Azimi, M., Huang, K.: Oil and gas 4.0 era: a systematic review and outlook. Comput. Ind. **111**, 68–90 (2019)
7. ISO: Information technology—security techniques—code of practice for information security controls based on ISO/IEC 27002 for cloud services. ISO/IEC Standard 27017:2015 (2018). https://www.iso.org/standard/43757.html
8. ISO: Information technology—security techniques—code of practice for information security controls. ISO/IEC Standard 27002:2013 (2013). https://www.iso.org/standard/54533.html
9. Cloud Security Alliance: Cloud Controls Matrix. https://cloudsecurityalliance.org/artifacts/cloud-controls-matrix-v3-0-1/
10. European Commission: Regulation (EU) 2016/679 of the European Parliament and of the Council of 27 April 2016 on the protection of natural persons with regard to the processing of personal data and on the free movement of such data, and repealing Directive 95/46/EC (General Data Protection Regulation) (2016). https://eur-lex.europa.eu/eli/reg/2016/679/oj
11. European Commission: Directive (EU) 2016/1148 of the European Parliament and of the Council of 6 July 2016 concerning measures for a high common level of security of network and information systems across the Union (2016). https://eur-lex.europa.eu/eli/dir/2016/1148/oj
12. European Commission: DIRECTIVE (EU) 2022/2555 OF THE EUROPEAN PARLIAMENT AND OF THE COUNCIL of 14 December 2022 on measures for a high common level of cybersecurity across the union, amending regulation (EU) no 910/2014 and directive (EU) 2018/1972, and repealing directive (EU) 2016/1148 (NIS 2 Directive) (2022). https://eur-lex.europa.eu/legal-content/EN/TXT/?uri=CELEX:32022L2555
13. Gilbert, F.: What Schrems 2 Means for your Privacy Shield Program. CSA Blog. https://cloudsecurityalliance.org/blog/2020/08/10/what-schrems-2-means-for-your-privacy-shield-program/
14. IEC: IEC/TS 62443-1-1:2009 Security for industrial automation and control systems—Part 1-1: Terminology, concepts and models (2009)
15. IEC: IEC 62443-3-2:2020 Security for industrial automation and control systems—Part 3-2: Security Risk assessment for system design (2020). https://webstore.iec.ch/publication/30727
16. Bernsmed, K., Meland, P.H., Jaatun, M.G.: Cloud security requirements. Technical Report, SINTEF ICT (2015). https://infosec.sintef.no/wp-content/uploads/2015/08/Cloud-Security-Requirements-v2.0.pdf. Report number A27131
17. Barker, E., Roginsky, A.: Transitioning the use of cryptographic algorithms and key lengths. Technical Report National Institute of Standards and Technology (2018). https://nvlpubs.nist.gov/nistpubs/SpecialPublications/NIST.SP.800-131Ar2.pdf
18. Mavroeidis, V., Vishi, K., Zych, M.D., Jøsang, A.: The impact of quantum computing on present cryptography. Int. J. Adv. Comput. Sci. Appl. **9**(3) (2018). https://doi.org/10.14569/IJACSA.2018.090354. http://dx.doi.org/10.14569/IJACSA.2018.090354
19. OASIS: Security Assertion Markup Language (SAML) v2.0 (2005). https://www.oasis-open.org/standard/saml/
20. OpenID: OpenID Foundation (2023). https://openid.net/
21. Øien, K., Hauge, S., Jaatun, M.G., Flå, L., Bodsberg, L.: A survey on cybersecurity barrier management in process control environments. In: 2022 IEEE International Conference on Cloud Computing Technology and Science (CloudCom), pp. 113–120 (2022). https://doi.org/10.1109/CloudCom55334.2022.00026
22. GSA: Cloud basics | cloud information centre. https://cic.gsa.gov/basics/cloud-basics

Security Operations Centre (SOC) and Advanced Malware Detection Techniques

An Exemplar Incident Response Plan for Security Operations Centre Analysts

Enoch Agyepong and Cyril Onwubiko

Abstract The significance of a cyber security incident response plan cannot be overemphasised, especially at a time when cyberattacks are becoming increasingly prevalent and sophisticated. Organisations need to have a focused and coordinated approach when responding to cyber incidents. Security Operations Centres (SOCs) need a well-defined incident response plan to facilitate detection and response activities in the event of a cyberattack. While there are several well-known incident response plans, these plans typically have a broad scope and are designed to assist teams and organisations to develop their own incident handling processes. Previous studies propose a tailored incident response plan for the Computer Security Incident Response Team (CSIRT). This paper presents an example incident response plan for SOC analysts. The proposed approach leverages the United States National Institute of Standards and Technology (NIST) incident response framework. The proposed approach will be beneficial to analysts with varying levels of experience/ expertise, especially junior and early-career analysts, since research indicates that these categories of analysts are burdened by the complexity of security incident analysis.

1 Introduction

A cybersecurity incident response plan (IR plan), also known as an incident handling process, can be defined as a set of processes and artefacts designed to detect, respond, and mitigate the impact of cybersecurity incidents to an organisation [1]. The goal is to reduce the negative impact resulting from incidents, and to facilitate reinstation to normal business operations as quickly as possible when an incident occurs, and to

E. Agyepong (✉)
Airbus UK, Quadrant House, Celtic Springs, Coedkernew, Newport NP10 8FZ, UK
e-mail: enoch.agyepong@airbus.com

C. Onwubiko
Cyber Security Intelligence, E-Security Group, Research Series, London, UK
e-mail: cyril@research-series.com

M. G. Jaatun et al. (eds.), *Proceedings of the International Conference on Cybersecurity, Situational Awareness and Social Media*, Springer Proceedings in Complexity,
https://doi.org/10.1007/978-981-96-0401-2_7

learn from the incident to improve future response efforts. Incident response is a core function of security operations centres (SOCs) [2]. A SOC is a centralised unit within an organisation that is responsible for monitoring and managing security-related events, alerts and incidents. The SOC function can also be provided by managed SOC providers, in the case of outsourced SOCs, and in some cases, organisations may adopt a hybrid approach, where some responsibilities of the SOC are performed in-house (insource), while the other responsibilities are provided by an external managed SOC provider (outsource). Analysts in SOCs are responsible for detecting, analysing, responding, and mitigating security incidents [3].

An incident is an event (or alert) that is not part of normal operations that could lead to loss or harm [4]. It is an occurrence that actually or potentially jeopardises the confidentiality, integrity, or availability of an information systems or the information the system processes, stores, or transmits or that constitutes a violation or imminent threat of violation of security policies, procedures, or acceptable use policies. An incident can occur in the following scenarios: when a user receives a malicious file; when a denial-of-service attack is launched on an organisation's web servers; when an employee violates a security policy; when attackers steal sensitive information; and when organised criminals launch ransomware attacks on an organisation. Incidents occur as a result of deliberate or accidental malicious events. An event is any observable occurrence that originates from a computer system. Systems produce events that might not always be malicious or harmful to a computer system. An event could be a user connecting to a file; a system crash; an unauthorised use of system privileges; or an unauthorised access to sensitive information [5].

It is important to note that incidents can still occur even when extensive security measures are in place to safeguard an organisation from both internal/inside and external/outside threats. Hence, an effective incident response plan ensures that when cyber incidents happen, an organisation can respond in a manner that will reduce or mitigate the damage [6]. The European Union Agency for Cybersecurity, ENISA [7], has proposed an incident handling guide for Computer Security Incident Response Teams (CSIRTs). It is pertinent to note that a SOC is not a CSIRT, although their functions are closely related [8, 9]. The activities of a CSIRT are a subset of the functions of a SOC [10]. SOC analysts would benefit from an incident response plan based on best practices derived from reputable organisations such as SANS and NIST to guide their operations. Additionally, a tailored incident response plan for SOC analysts will help junior and/or inexperienced analysts who suffer from the complexities of incident response [11]. Moreover, studies show that SOCs tend to suffer from tacit knowledge, where the knowledge, skills, and abilities an individual gains through experience are often difficult to put into words or otherwise communicate across the team [2, 12]. Having cues to support analysts during incident handling will greatly benefit both entry-level and junior SOC analysts.

This paper makes two key contributions:

1. *We propose an incident response (IR) plan for the SOC analysts using the NIST framework.*
2. *We provide an example incident handling checklist for SOC analysts.*

The remainder of this paper is organised as follows: Sect. 2 presents the common incident response plans. Sect. 3 presents the proposed incident handling process and the checklist to support SOC analysts. Discussions are presented in Sect. 4, and conclusions and future work in Sect. 5.

2 Incidents Response Plans

Two of the most popular incident response plans, one developed by NIST and the other by SANS are compared in Fig. 1.

NIST [13] suggests four phases for incident handling: (1) preparation; (2) detection and analysis; (3) containment, eradication, and recovery; and (4) post-incident activities. Similarly, SANS [5] suggests six steps for incident handling (see Fig. 1): (1) preparations; (2) identification; (3) containment; (4) eradication; (5) recovery; and (6) lessons learned. Both the NIST and SANS approaches serve as guides to assist organisations in establishing cyber incident response capabilities [14].

A noteworthy difference between the two frameworks is that while NIST combines containment, eradication, and recovery, SANS separates them as distinct phases of the IR process. Post-incident activity and lessons learned are comparable in both

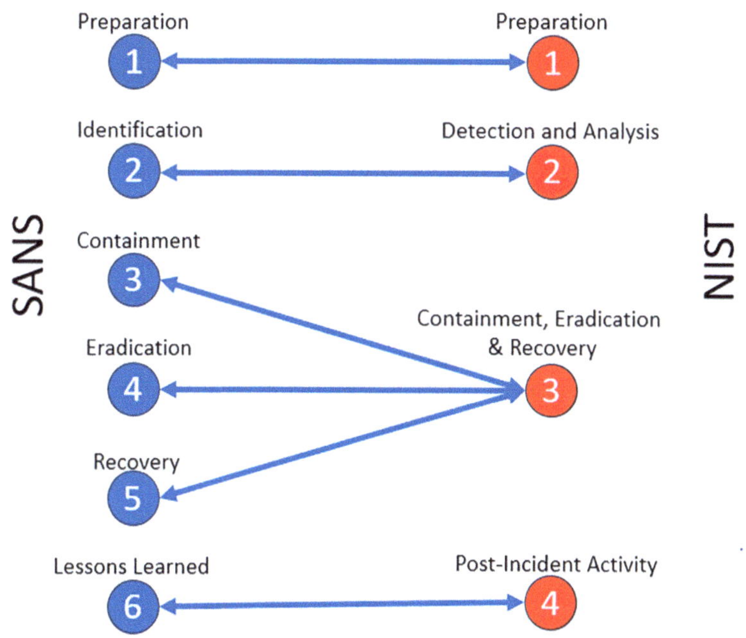

Fig. 1 Common security incident response plans

frameworks. The preparation, identification, detection, and analysis stages have the same implications (see Fig. 1).

The different phases are discussed:

Preparation: Preparation is the proactive planning and coordination of internal and external resources to control, contain, and mitigate incidents [4]. Preparation starts with a security policy that has the support of management and includes having the required people, processes, and technology to prevent and respond to incidents [15]. The preparation phase identifies key individuals who will be involved in the incident response such as incident responders, senior responsible officers (SROs), legal, and tech SMEs. This phase also encompasses IR playbook creation, readiness, tabletop exercises (TTX), and external call-out contract with an IR retainer and digital forensics company, and in addition to insurance underwriters. The preparation must also define the monitoring procedure, along with the notification and escalation processes.

Identification: This phase entails spotting abnormalities, signals, telemetries and threats from events that could endanger the organisation. It involves monitoring the organisation's enterprise to detect symptomatic events before a security breach occurs. To ascertain if an event qualifies as an incident, this stage necessitates gathering and correlating events from a variety of sources, including log files, error messages, and events/alerts from other controls like intrusion detection systems and firewalls that can yield high-fidelity proof [15]. Once certain events are determined to indicate an incident through a process called triaging, it is crucial to report it promptly to give the IR team sufficient time to collect evidence and prepare for the preceding steps. At this stage of an incident, the IR team should coordinate communication with system administrators and management teams about the impacted services based on the 'need to know' principle. Communication and coordination between SOC (and management) are critical, even more so if the scope of the incident is perceived to have a significant impact to the organisation.

Containment: The purpose of this phase is to limit further damage to the organisation and equally important, to reduce and control the attacker's ability to expand or extend the current incident [15]. Containment can be as simple as isolating an infected network segment, shutting down hacked production servers, and redirecting traffic to failover servers. Containment can also include unplugging the affected system's network cable or turning off switches and/or routers in broad sections of the network to isolate compromised systems from those that have not been hacked. As a result, the issue would be separated from the rest of the production network, minimising the blast radius.

Eradication: This stage entails erasing any artefacts left by the attacker, such as backdoors, compromise paths etc. The information gathered during the identification and containment phases will be used to determine the causes of the occurrence. SANS [5] asserts that this is the most difficult part of incident handling activity because no one can be certain if they have completely and safely removed malicious code left by attackers or closed all backdoors or compromise paths introduced as a result of the incident. Reformatting and re-installing operating systems are both useful techniques that can be used to eradicate malicious artifacts left behind by an attacker. When seeking to eliminate harmful code, the course of action must include investigating what triggered the issue, identifying the vector of infection, and taking steps to prevent this from happening again [5].

Recovery: The goal of this phase is to reintroduce affected systems that have now been cleansed into the production environment so that another incident does not occur. Testing, monitoring, and validating the systems being reintroduced into production is critical to prevent malware reinfection or any other compromise. The objective should be to safely restore any impacted systems and data to the organisation's enterprise. The decision to reintroduce systems, infrastructure/data into the enterprise must be decided by the business or system owner [5] and, in conjunction, by the organisation's Chief Information Security Officer (CISO) or whomever is accountable for the cybersecurity of the organisation. Analysts can give advice and make recommendations, but the ultimate responsibility to restore a system rest with the CISO or system owner. Recovery requires restoring all operations and services and ensuring that operations are performing as expected. Restoring all services in some cases may not be feasible; in cases where data may have been lost, compromised, or corrupted, this becomes impossible. SANS [5] recommends a retest of restored systems to ascertain they are back to their original state post-incident.

Lessons learned/Post-Incident Activity: The final step in the incident response process is to document what happened and improve processes to prevent it from happening again. Learning from the past and documenting the lessons gained are essential components of incident management and all effective engineering processes. It may help to avoid similar events in the future [16]. Analysts and engineers could use lesson-learned documents to improve their work methods.

2.1 The ENISA Incident Handling Process

ENISA [7] outlines four components of incident handling designed for CSIRT (see Fig. 2).

- Detection of the incident.
- Triage, which involves the assessment of the incident.
- Analysis, which gathers more information about the incident.
- Incident response, which is taking the necessary actions to remediate the incident.

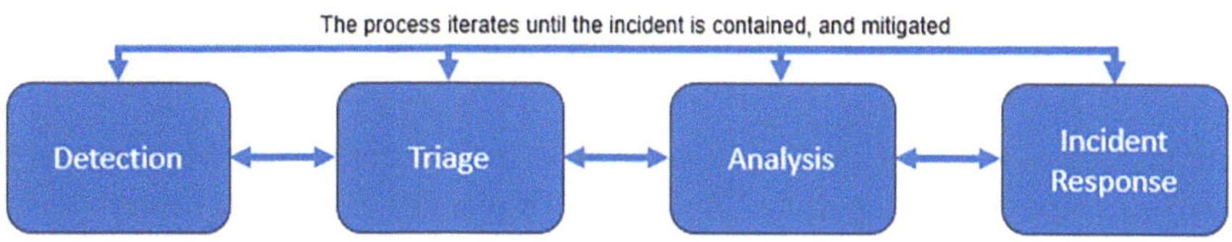

Fig. 2 Incident handling for CSIRT [7]

The ENISA approach differs from the NIST and SANS frameworks in that it does not place a strong emphasis on planning and lessons learned as part of the incident handling process. A possible reason for this could be that it is assumed that an incident management plan will capture the preparation and make provisions for capturing lessons learned.

The ISO (International Organisation for Standardisation) also has a similar standard to ENISA known as the ISO/IEC 27,035 Information security incident management that offers guidelines for managing information security incidents effectively [17]. It encompasses the entire incident management process, from detection and reporting to response and recovery. A major difference between the ENISA approach and ISO/IEC 27,035 is that ISO/IEC 27,035 provides a broader framework for managing information security incidents, covering various aspects such as detection, reporting, response, and recovery. In contrast, the ENISA guidelines, as the name suggests, are developed by the European Union Agency for Cybersecurity and are specifically tailored for CSIRTs and for the needs and requirements of European organisations and institutions [18].

Given the close relationship between CSIRT and SOC, it is rational to argue that the incident handling process proposed for CSIRT can also be used for SOC; however, it is vital to note that the two are not the same. SOCs are responsible for a larger range of issues than CSIRTs. Jacobs et al. [19] highlight the differences between CSIRT and SOC.

Notable difference between CSIRT and SOC are shown in Fig. 3. Among them are real-time monitoring, reporting of incidents, the development of signatures and use cases, as well as sensor tuning, which do not fall under the ambit of CSIRT. A SOC that does not have a process in place to improve their use cases or signatures faces the challenges of a huge number of false positives and, even worse, false negatives. While a false positive represents a false alert, a false negative denotes an instance in which a security tool intended to detect a particular threat fails to do so. This could be a result of a lack of use cases or failing to use the sensors.

Our work is motivated by a desire to enhance incident response activities conducted by SOCs, as existing frameworks are broad. We opine that a tailored incident response plan for analysts will provide more relevant guidance and actionable strategies tailored to the unique nature of SOCs. Additionally, it is our contention that a novel incident response plan that is more targeted and streamlined would enable SOCs to deal with cyber incidents with greater efficiency and effectiveness.

3 Proposed Incident Response Plan for SOC Analysts

The primary functions of a SOC as reported in the literature are protective monitoring, analysis, response and reporting [20–24]. An IR plan for SOC could be derived by mapping the core functions of SOC to the NIST incident response framework (see Table 1).

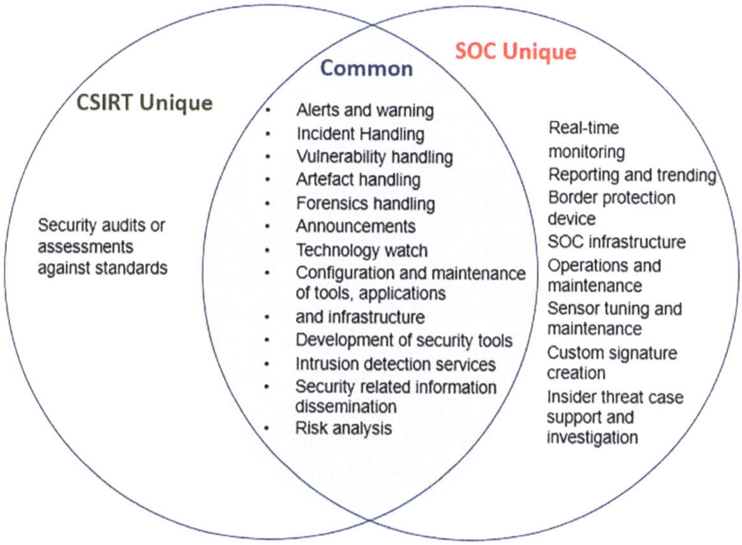

Fig. 3 The difference between CSIRT and SOC [19]

Table 1 An incident response plan for SOC analysts

SOC analysts' function as reported in the literature	Incident response phases for analysts	NIST framework
Preparatory Activities *Use case development and writing of signatures for intrusion detection systems* [25] *Training and Tuning sensors* [10, 19] *Creation of detection rules, dashboards and IR playbooks* ***Protective Monitoring Function*** *Continuous monitoring and detection*[20–24, 26]	*Preparation*	*Preparation*
Analysis Function *Event correlation, triage and analysing*[20–24, 26]	*Incident Discovery, Analysis and Confirmation*	*Detection & Analysis*
Response and Reporting Function [20–24, 26]	*Containment, Eradication & Recovery*	*Containment, Eradication & Recovery*
–	*Post-Incident Activity*	*Post-Incident Activity*

Using the incident response phases for analysts presented in Table 1, we design Fig. 4 to show the activities of an analyst under each of the phases.

3.1 Preparation

Analysts tasked with protectively monitoring an organisation's enterprise must be provided with the relevant role-specific training needed to react effectively and swiftly when there is an attack. In addition, analysts must have access to relevant SOC documentation such as IR playbooks and plan that will inform their operations [27], contact details of relevant stakeholders they need to contact to escalate and report an incident.

Other preparatory activities include development of protective monitoring artefacts such as use case development, security monitoring dashboard, log parsers and integration, detection rules, and IR playbook creation etc.

Protective Monitoring (a.k.a. Continuous Monitoring)

We opine that continuous monitoring and detection of security events is a vital step in the incident handling process for analysts. This is because without an effective continuous monitoring and detection then the incident response process will always lag in time. Effective detection means the mean time to detect (MTTD) is reduced, which invariably reduces the mean time to respond (MTTR).

Effective continuous monitoring is underpinned on having efficient continuous monitoring tooling in place, such as well configured and baselined endpoint detection and response (EDR) and security incident and event monitoring (SIEM), and automation using security orchestration, automation and response (SOAR) to swiftly detect threats and incidents. Since events are only symptomatic of an incident, and multiple streams of events from several hundreds of endpoints in the enterprise can be overwhelming to process manually, it is important that these several thousands of event stream are automatically analysed by correlating them using a SIEM. Furthermore, using a SIEM, and provided events from all endpoints in the enterprise are forwarded in real-time to the SIEM, will allow the SOC to triage all security events and hence determine whether they are indicative of a security incident. Since attack indicators are rarely isolated events on a particular system component or system, continuous correlation of streams of events from multiple endpoints is vital in detecting security incidents [25].

Analysts should be well-trained to monitor IT systems for deviations from regular operations and determine whether they constitute true security incidents. Analysts must have access to log data from numerous devices, such as firewalls, antivirus software, and intrusion detection systems, to enable them to check for signs of attacks across an organisation's enterprise. When an incident is detected, analysts must gather all relevant information as detailed under Sect. 3.2, to determine the nature and severity of the occurrence. Incidents can be reported by users; hence incidents

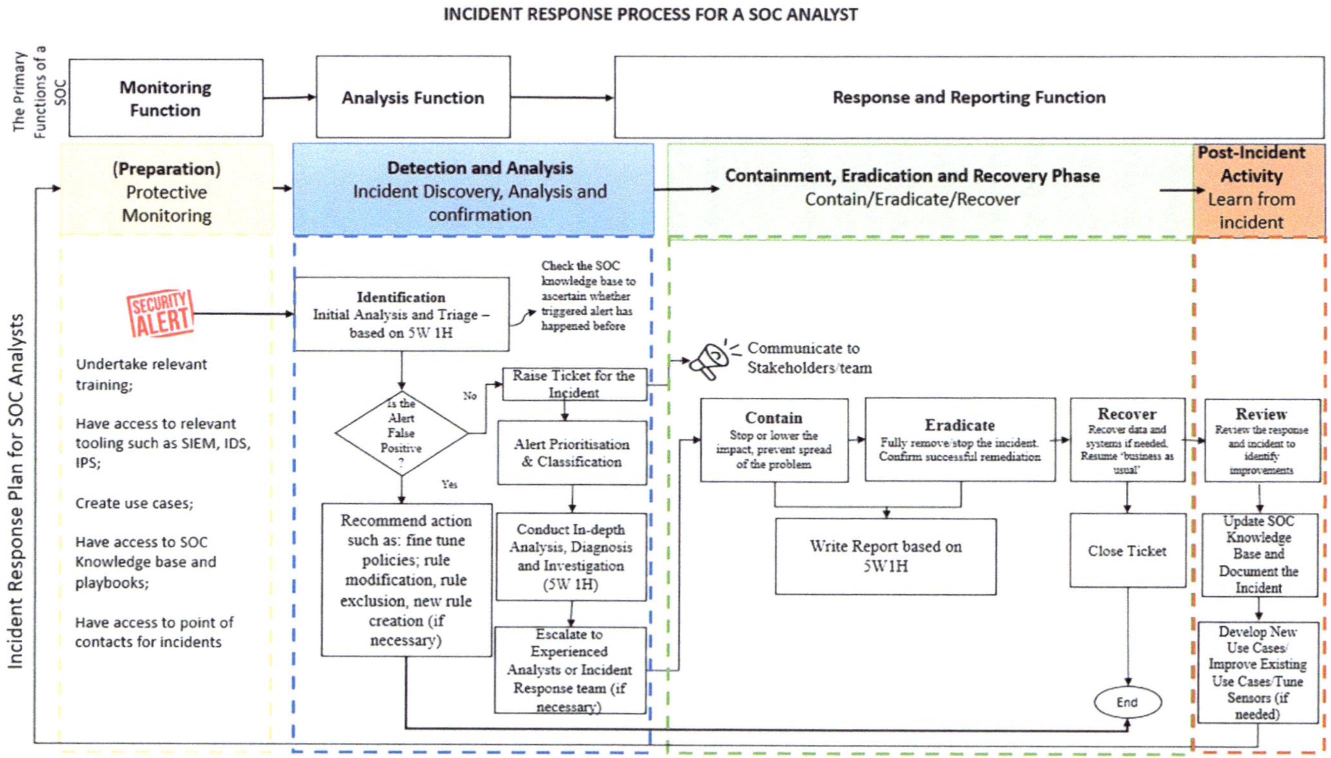

Fig. 4 Proposed incident response plan for SOC analysts

can be reported via a variety of communication channels, including email, phone call, walk-in report, web forms and API.

3.2 *Incident Discovery, Analysis and Confirmation*

Analysis

As shown in Fig. 4 once an alert is triggered, it should be analysed and confirmed to determine whether it poses a threat. The analysis function starts with assessment to ascertain whether triggered event is a real event or a false positive—a process that can be described as the triage of an alert, as described below. The 5W1H (*who, where, when, what, why and how*) serves as a guide during the analysis (see Table 2).

Triaging an Alert—Initial Analysis

When doing an analysis, an analyst's main goal is to determine whether it is a false positive or a true positive as shown in Fig. 4. Triage is typically the initial stage in determining the impact of an incident and its outcomes. Analysts would typically use an in-house knowledge database to validate and determine whether an alert is a "real incident" or a false positive. Based on this, an analyst will decide how to respond to the alert. If it is a false positive, an analyst would normally recommend actions such as tuning out or altering the signature to improve detection [11]. The analysis of a security event should result in it being qualified as a true incident or a false positive [25].

Confirmed False Positives

If the alert is deemed a false positive, an analyst must determine whether there is a need to fine-tune the sensor to improve future detection. They could also decide whether existing signatures, rules, or use cases should be modified to improve future detection.

True Positive -Real Incident

If an alert is deemed to be a malicious event (i.e., an incident), analysts should notify and communicate this to other team members for support (if required). A confirmed incident must be reported to the SOC manager, who will decide on whether other key stakeholders should also be notified. Analysts then move on to the next phase of their analysis, which involves the prioritisation and classification of the alert.

Classification and Prioritisation

If an incident is observed, the analyst will start with the prioritisation of the incident. Alerts and incidents processed by analysts are typically characterised as high incidents, medium incidents, and low incidents [4, 25]. After triage and verification, an analyst needs to classify an incident based on its severity. It is classified according to a classification schema, which is defined by the SOC. Classifying the incident

Table 2 The 5W1H [25]

Cues for analysing an incident and writing technical incident report					
Who	Where	When	What	Why	How
Who are the potential attackers/ adversaries?	Where is the attacker or adversary targeting or likely to target and exploit?	When was the attack or incident first noticed?	What does the attacker already know? What capability do the attacker have?	Why is this incident of interest?	How was the potential attack discovered?
– Attack Path (External threat or Insiders?) – Source IP Ad- dress/ Attacker IP – Address Source Port/ Service – Source MAC Address – Attacker Username (if internal) – Attacker Host Name – Attacker User Agent (if applicable) – Is the source IP informa- tion known? – Is the destination IP infor- mation known? – Is the username of the attacker known?	– Impacted Host/ Application – Destination IP Address/ Attacker IP Address – Destination Port/Servic – Destination MAC Address – Location of Detection – Has network or client machine in- volved been identified? – Have the log sources been identified? – Has the source of the event been identified?	– Date and Time of Detection including time zone – Reporting Device Detection time – Manager Receipt Time – Is the event still continuing?	– Name of Alert/ Trigger File/Email/URL – Domain Name Asset Name – User Account IPS Signature/ Use Case Event ID/Type/OS Breach Type – Incident Severity/ Classification – Is offence about reconnaissance? – Is offence about attack delivery? – Is offence about exploitation? – Is the offence about system compromise?	– Risk Context and geographical – Information and threat description – System criticality – Potential Impact – Have the rules associated with the incident been identified? – Has the first rule that triggered the incident been identified? – Was the rule updated recently? – Were there any rules testing in place?	– Method of Detection – Mitigation Factors – Playbook used – Did the attacker download any executable? – Is it a known attacker? – Did the attacker access malicious website? – Does the computer have any vulnerability?

will also allow for incidents to be prioritised. A ticket is raised for the incident. A ticketing system would be used to capture all the details of an incident, documenting the history of the incident from registration to closure.

3.3 Containment, Eradication and Recovery

Incident containment, eradication, and recovery are critical steps in the incident response process for SOCs. This phase is aimed at limiting the damage caused by the incident, eliminating the cause of the incident, and restoring affected systems and services to their normal state [15].

Containment

The response will be determined by a pre-planned strategy for containing the incident and ensuring that the organisation can continue with its usual business operations. An analyst's goal would be to stop or reduce the impact of the problem and prevent it from spreading. The SOC could draw on a playbook [27, 28]. A SOC playbook is a step-by-step guide designed to help security analysts navigate and manage security incidents effectively. SOC analysts should aim to work with other members of the SOC team or relevant parties to perform short-term containment, for example, by isolating the network segment that is under attack. Then focus on long-term containment, which involves temporary fixes to allow systems to be used in production while rebuilding clean systems. Analysts may draw on other teams to support containment and remediation action. Analysts could draw on the network operations team, for example, to help reconfigure a firewall. They may also draw on a forensic or malware team for malware analysis.

Eradication

Analysts should also collaborate with the relevant teams and experts, such as malware or forensic specialists, to remove malware from all affected systems, establishing the root cause of the attack (although this is usually a very difficult task, and often, exact root cause is indeterministic), and take steps to prevent similar attacks in the future. The goal here is to completely remove or halt the occurrence. Analysts should also confirm successful remediation.

Recovery

The fundamental purpose of the incident handling process is to resolve and recover from the incident as quickly and efficiently as possible. Analysts should aim to bring affected production systems back online carefully to prevent further attacks. The introduction of the system should be done carefully to avoid reinfection by malware where data is being restored. The process should be to test, verify, and monitor affected systems to ensure they are back to normal activity. The objective of this is to recover data and systems (where possible).

3.4 *Post-Incident Activity*

Post-incident activities provide the opportunity to review the IR process and plan to identify improvements. The lessons learned process takes some time due to their very nature, involving all the relevant stakeholders, retrospective, root cause analysis (RCA), and finalisation of the report. Analysts may have to close the ticket associated with an incident once it has been remedied and resolved due to service level agreements (SLAs), even before the lessons learned are concluded. Notwithstanding, lessons learned sessions play a pivotal role in the continuous improvement of incident handling processes. Lessons learned could also help organisations enhance their overall cybersecurity posture. As part of the lessons learned, analysts may develop new use cases and amend existing ones to improve future detection, for instance, if the SIEM or EDR missed detecting the events that led to the incident, or if the events/incident were wrongly classified, such as true negative, which were not detected or false negative, which were not alerted even though the system should have, then the SIEM and EDR should be re-tuned and re-baselined. Furthermore, analysts may work with the SOC engineering team to tune sensors or detection systems to improve future detection.

3.5 *Incident Response Checklist for SOC Analysts*

Table 3 presents a checklist that can be used by SOC analysts during incident response.

4 Discussion

Cyber incidents, such as denial-of-service attacks, ransomware or malware, can disrupt normal business operations and lead to substantial financial and administrative burden to the organisation if not resolved swiftly. Analysts need well-defined incident handling processes to support their operations, restore services and maintain business continuity. Having an effective incident handling process would allow analysts to respond rapidly to cyber incidents. A prompt response can help contain the incident and prevent further damage. Efficient incident handling can help minimise the impact of a cyber incident, reducing the potential damage to data, systems, and overall business operations.

A SOC without a clear incident response plan is exposing itself to various risks and negative consequences. Indeed, without a formal incident response plan, SOCs may struggle to effectively manage and mitigate the impact of security incidents. This can result in longer response times, increased damage to systems and data, and significant financial losses. Response time is a key performance indicator for an

Table 3 Incident response checklist for SOC analysts

Incident response phase	Actions for analysts
Preparation	• Ensure that analysts have access to playbooks. A SOC playbook is a detailed guideline designed to assist security analysts in properly navigating and dealing with security concerns. Analysts, for example, could have access to playbooks for dealing with phishing or ransomware attacks • Ensure that analysts have access to contact information for all key stakeholders • Ensure that analysts have access to protective monitoring tooling • Ensure that analysts have access to policy for outside third-party notifications **——— —Protective Monitoring —- —** • Analysts must have the tools to be able to monitor an organisation's enterprise, looking for signs of malicious or unusual activity • Configure the monitoring solutions, such as intrusion detection systems, firewalls, and Security Information and Event Management (SIEM), to trigger real-time alerts of suspicious behavior • Configure sensors and SIEM to detect unusual privilege account activity in comparison to a baseline and trigger an investigation
Incident discovery, analysis and confirmation	• Use the 5W1H to guide the detection of incidents, analysis, and confirmation Table 2: The 5W1H [25] • What is the scope of the incident and how will it affect business? • Have the source(s) of the incident been located? If so, where, when, and what are they? • Describe how the SOC first learned of the attack • Analyse audit logs and security applications to identify unusual or suspicious account behavior or activities that indicate a likely attack and confirm an attack has occurred • Analysts must have awareness of the sensitivity around data type, particularly personally identifiable information. This is important because there are specific reporting requirements and timeframes for data breaches concerning personal data • Describe the possible attacker's capabilities, activities, and motivations, as well as any known or expected capabilities • Prepare an incident timeline to keep an ongoing record of when the attack occurred and subsequent milestones in analysis and response • Check applications for signatures, IP address ranges, file hashes, processes, executable names, URLs, and domain names of known malicious websites • Collect meeting notes in a central repository to use in preparing communications with stakeholders • Inform other team members about the discovery • Analyse incident indicators of compromise (IOCs) with threat intelligence tools • Determine the scope of an incident by evaluating which privileged accounts were recently used, whether any passwords were changed, and what programs were per- formed. A snapshot of the audit logs can also be taken for comparison • Check applications for signatures, IP address ranges, file hashes, processes, executable names, URLs, and domain names of known malicious websites using real- time threat intelligence • Analyse audit logs and security applications to identify unusual or suspicious account behavior or activities that indicate a likely attack and confirm an attack has occurred

(continued)

Table 3 (continued)

Incident response phase	Actions for analysts
Containment	• Determine whether the incident can be isolated. If this is the case, isolate the impacted systems. If not, collaborate with system owners and/or managers to decide the next steps to take to contain the situation • The technical and security teams should be able to swiftly access and monitor systems by using temporary privileged accounts • Keep the evidence safe. Before performing any steps that might compromise data integrity on the original media, back up any affected systems as soon as possible • Passwords should be changed for all users, services, apps, and network accounts • To prevent the attacker from spreading malicious mal- ware, increase the sensitivity of application security controls (allowing, denying, and restricting) • Remove systems from the production network or take systems offline if necessary • Analyse, record, and confirm any instances of potential data exfiltration occurrences across the network • Enable additional, temporary permissions if needed for the technical and security teams to quickly access and monitor systems • Change passwords for all users, service, application, and network accounts immediately and automatically—without preventing them from accessing the systems they use to complete their jobs • Increase privileged account and system monitoring by restricting access to sensitive systems, requiring additional approval processes for privileged access, or mandating multi-factor authentication for privileged accounts
Eradication	• Re-image the system if possible, and then harden it with patches and/or other countermeasures to minimise or lessen the chance of an attack? If not, please explain why • Check that the attackers' malware and other artifacts have been removed and that the compromised systems have been hardened against attacks in the future. If not, please explain why not • Using readily customisable reports, compare data before and after the occurrence to swiftly detect which privileged accounts may be malicious and audit their life cycle • Monitor and audit any privileged account activity to ensure that they are returning to normal expected usage • Close all firewall and network connections • Test devices and programs to ensure that any malicious code has been removed • To ensure that systems are effectively reset, compare data before and after the occurrence • Potentially share information on eradication tactics with others

(continued)

Table 3 (continued)

Incident response phase	Actions for analysts
Recovery	• Check to ensure that the impacted system(s) have been patched and hardened against both the recent and potential future attacks • Check to see what day and time it would be possible to get the impacted systems back online • Make certain that the tools required to monitor and verify that the systems being returned to production are not hacked by the same means that caused the original incident are available • Download and install security patches • Close network access and change passwords • Conduct a vulnerability analysis • Return to production any systems that were taken down • Examine the forensic evidence that has been gathered
Post-incident activity	• If necessary, create new use cases and signatures to improve future detection • To improve detection, consider tuning sensors such as intrusion detection systems and intrusion prevention systems • If the nature of the incident requires it, provide further training to SOC team members • Determine what worked and what needs to be improved • Consider whether existing playbooks or processes need to be updated and make any necessary changes to reflect new policies and needs • Share information externally if possible, so that others in the organisation's intelligence-sharing group can benefit • Report the final outcome to all stakeholders • Check to ensure that all the necessary documentation from the incident has been written • If the incident response report has been completed, ensure that it answers the following questions: (Who? What? Where? Why? When? and How?)

effective SOC [20, 29] Furthermore, in the absence of a predefined incident response plan, response efforts by any SOC will lack coordination and structure. This can lead to confusion among analysts regarding their roles and responsibilities during an incident, potentially causing delays in response activities and allowing the incident to escalate further.

Additionally, the business receiving the SOC services may suffer reputational damage if they are unable to response to cyber-attacks in an effective and efficient manner. Moreover, in the event of a security incident or data breach, organisations without a robust incident response plan will struggle to effectively communicate with stakeholders, including customers, partners, and the public. An incident response plan provides a framework for documenting and analysing security incidents, allowing organisations to identify root causes, lessons learned, and areas for improvement. A SOC analyst incident response plan would also ensure that key knowledge and best practices for dealing with issues are not forgotten over time and that lessons learned are steadily incorporated. This will be useful in addressing the problem with tacit knowledge [2]. The process will also offer a repeatable process in the way that the SOC conducts incident response.

Without a structured approach to incident response, SOCs may miss vital opportunities to learn from past incidents and enhance their security posture over time. It is imperative to rehearse incident handling processes using realistic tabletop exercise scenarios.

5 Conclusions, Limitations and Future Work

The importance of an incident response plan cannot be overstated in today's rapidly evolving threat landscape. First, security incidents are completely unavoidable even against well-protected enterprises, this is because incidents happen for several reasons—accidentally or deliberately. On one hand, the bar to launch cyberattacks is ever so low with organised criminals now operating ransomware as a business,[1] botnets are rented as a service for as little as $100 per use per day.[2] On the other hand, regulatory directives are becoming more stringent, mandating speedy breach disclosure, and where penalties for data breaches are becoming punitive. As organisations face an ever-increasing number of security incidents and data breaches, having a continuous and comprehensive incident response plan is essential for mitigating risks, minimising the impact of incidents, and maintaining business continuity.

While there are many recognised cyber security incident handling strategies, these existing plans are often overly general, serving primarily as frameworks for teams or organisations to develop their tailored incident plans. In contrast, our suggested incident response plan, as detailed in Fig. 4, is tailored specifically for SOCs. In addition, we have provided an example incident handling checklist for SOC analysts with cues that can be used by analysts. The paper also guides practitioners how to customise the existing framework to their specific situation.

By implementing an incident response plan, SOCs streamline their response efforts, and facilitate timely and effective coordination among the analysts operating in the SOC. Furthermore, a structured incident response plan enables SOCs to comply with regulatory requirements, protect sensitive data, and safeguard the reputation of the organisation receiving the SOC services in the face of cyber-attacks. Beyond immediate incident containment and remediation, an incident response plan provides a framework for continuous improvement and learning. By documenting and analysing security incidents, SOCs can identify vulnerabilities, assess their security posture, and implement proactive measures to enhance resilience and readiness for future threats.

[1] Ransomware as a business service is a business model between ransomware operators and affiliates in which affiliates pay to launch ransomware attacks developed by operators - https://www.crowdstrike.com/cybersecurity-101/ransomware/ransomware-as-a-service-raas/.

[2] Botnet booter rentals for as little as $100 per day - https://www.darkreading.com/cloud-security/cybercrime-a-black-market-price-list-from-the-dark-web.

The proposed SOC IR plan draws on insight from existing work and experiential knowledge of the authors to recommend the cues on the incident handling checklist, but a drawback is that the proposed cues have not yet been tested in an organisational SOC. Hence, a potential future work would see the proposed cues used in an organisational SOC and using this process to drive further improvements of the model.

References

1. Education Service District (EDS).: 6550-P13 Procedure—Information Security Incident Management, Policy (2023)
2. Reeves, A., Ashenden, D.: Understanding decision making in security operations centres: building the case for cyber deception technology. Front. Psychol. **14**, (2023)
3. Kearney, P., Abdelsamea, M., Schmoor, X., Shah, F., Vickers, I.: Combating alert fatigue in the security operations centre
4. Onwubiko, C., Ouazzane, K.: SOTER: A playbook for cyber security incident management. IEEE Trans. Eng. Manag. 1–22 2019
5. Institute, S.A.N.S.: SEC504: Hacker Techniques, Exploits, and Incident Handling. The SANS Institute, Boston (2024)
6. Grigaliūnas, Š., Schmidt, M., Brūzgienė, R., Smyrli, P., Bidikov, V.: Leveraging taxonomical engineering for security baseline compliance in international regulatory frameworks. Futur. Internet **15**(10), 2023. https://doi.org/10.3390/fi15100330
7. European Network and Information Security Agency (ENISA).: Good practice guide for incident management (2010)
8. Aijaz, L., Aslam, B., Umar, K.: Security operations center—a need for an academic environment. In World Symposium on Computer Networks and Information Security (WSCNIS), pp. 1–7 (2015)
9. Paz, S.: Cybersecurity standards and frameworks. IEEE Technol. Eng. Manag. Soc. Body Knowl. 397–416 (2023)
10. Jacobs, P., Arnab, A., Irwin, B.: Classification of security operation centers. In: 2013 Information Security for South Africa, pp. 1–7. IEEE, https://doi.org/10.1109/ISSA.2013.6641054 (2013)
11. Zhong, C., Yen, J., Liu, P., Erbacher, RF.: Learning from experts expereince: toward automated cyber security data triage. IEEE Syst. J. 1, 1–12 (2018). https://doi.org/10.1109/JSYST.2018.2828832
12. Vermeer, M., Kadenko, N., van Eeten, M., Gañán, C., Parkin, S.: Alert alchemy: SOC workflows and decisions in the management of NIDS rules. In: Proceedings of the 2023 ACM SIGSAC Conference on Computer and Communications Security, ACM, pp. 2770–278 (2023)
13. Cichonski, P., Millar, T., Grance, T., Scarfone, K.: Computer security incident handling guide recommendations of the national institute of standards and technology (2012). https://doi.org/10.6028/NIST.SP.800-61r2
14. Cynet.: NIST incident response. Cynet
15. Huda M.: Intrusion detection & incident response for information security: technologies and challenges. In: 2022 2nd International Conference on Advance Computing and Innovative Technologies in Engineering, ICACITE 2022, Institute of Electrical and Electronics Engineers Inc, pp. 1472–1476. (2022). https://doi.org/10.1109/ICACITE53722.2022.9823454.
16. Patterson, C.M., Nurse, J.R.C., Franqueira, V.N.L.: I don't think we're there yet': The practices and challenges of organisational learning from cyber security incidents. Comput. Secur. 139. https://doi.org/10.1016/j.cose.2023.103699

17. ISO (International Organization for Standardization), ISO/IEC 27035–1:2023 Information technology Information security incident management (2023)
18. ENISA, European Union Agency for Cybersecurity (ENISA), https://european-union.europa.eu/institutions-law-budget/institutions-and-bodies/search-all-eu-institutions-and-bodies/european-union-agency-cybersecurity-enisa_en
19. Jacobs, P., Grobler, M., Von Solms, S.: E-CMIRC: Towards a model for the integration of services between SOCs and CSIRTs (2016). https://www.researchgate.net/publication/306506311
20. Kokulu, F.B., Bao, T., Doupe, A., Shoshitaishvili, Y., Ahn, G.J., Zhao, Z.: Matched and mismatched SOCs : a qualitative study on security operations center issues. Association of Computing Machinery (ACM) (2019)
21. Lif, P., Sommestad, T.: Human factors related to the performance of intrusion detection operators. In HAISA, Lesvos, Greece, pp. 265–275 (2015). Accessed 29 Nov 2018. https://pdfs.semanticscholar.org/9ec5/14d9705aa4e85356fb5385f04932e68cdb6c.pdf
22. Majid, M.A., Ariffi, K.A.Z.: Success factors for cyber security operation center (SOC) establishment. In: International Conference on Informatics, Engineering, Science and Technology, Bandung: European Alliance for Innovation (EAI), 2019. https://doi.org/10.4108/eai.18-7-2019.2287841
23. Onwubiko, C.: Cyber security operations centre: security monitoring for protecting business and supporting cyber defense strategy. In: 2015 International Conference on Cyber Situational Awareness, Data Analytics and Assessment (CyberSA), London, UK, pp. 1–10. IEEE, https://doi.org/10.1109/CyberSA.2015.7166125(2015)
24. Schinagl, S., Schoon, K., Paans, R.: A framework for designing a security operations centre (SOC). In: 2015 48th Hawaii International Conference on System Sciences, pp. 2253–2262. IEEE, https://doi.org/10.1109/HICSS.2015.270 (2015)
25. Agyepong, E., Cherdantseva, Y., Reinecke, P., Burnap, P.: A systematic method for measuring the performance of a cyber security operations centre analyst. Comput. Secur. **124**, 102959 (2023). https://doi.org/10.1016/J.COSE.2022.102959
26. Agyepong, E., Cherdantseva, Y., Reinecke, P., Burnap, P.: Towards a framework for measuring the performance of a security operations center analyst. In: 2020 International Conference on Cyber Security and Protection of Digital Services (Cyber Security), Dublin, pp. 1–8. IEEE, (2020)
27. Shaked, A., Cherdantseva, Y., Burnap, P., Maynard, P.: Operations-informed incident response playbooks. Comput Secur 134 (2023)
28. Shaked, A., Cherdantseva, Y., Burnap, P.: Model-Based incident response playbooks. In: Proceedings of the 17th International Conference on Availability, Reliability and Security, pp. 1–7 (2022)
29. Onwubiko, C., Onwubiko, A.: Cyber KPI for return on security investment. In: 2019 International Conference on Cyber Situational Awareness, Data Analytics and Assessment, Cyber SA, pp. 1–8 (2019). https://doi.org/10.1109/CyberSA.2019.8899375

Malware Detection Using Machine Learning Algorithms

Nureni Ayofe Azeez, Ogechukwu Juliet Nzeribe, Charles Van der Vyver, and Ademola Philip Abidoye

Abstract Recent years have seen a surge in malware attacks, which have caused considerable financial and operational damage to individuals and businesses alike. As malware techniques continue to evolve, it is essential to be proactive and adaptive in order to detect and reduce the risk of such attacks. To address this issue, this research is focused on creating a sophisticated malware detection system that utilizes machine learning algorithms to detect Malware attacks. With this technique, a comparative assessment of the algorithms used was carried out. The models were trained using four datasets. The machine learning algorithms used are Decision Trees, Xgboost, Logistic Regression, Naïve Bayes and Random Forest, are tested for their ability to detect Malware behaviors. A comparison of the results was carried using four evaluation metrics namely: Accuracy, Precision, Recall and F1-score. At the end of the experiments - For Dataset 1, XGBoost had the highest accuracy of 0.982456. For Dataset 2, Random Forest had the highest accuracy of 0.990432. For Dataset 3, XGBoost, Random Forest and Decision Tree had the highest accuracy of 1.0 meaning they achieved a perfect accuracy of 100%. For Dataset 4, both Random Forest and XGBoost had the highest accuracy with the values respectively 0.628474 and 0.628135. A bar chart was used to graphically represent our results for each evaluation metrics. Boxplot used to represent results showing the minimum, maximum, and quartile ranges.

N. A. Azeez (✉) · O. J. Nzeribe
Department of Computer Sciences, Faculty of Science, University of Lagos, Lagos, Nigeria
e-mail: nazeez@unilag.edu.ng

O. J. Nzeribe
e-mail: ogechrisn@gmail.com

C. Van der Vyver
School of Computer Science and Information Systems, Faculty of Natural and Agricultural Sciences, Vaal Triangle Campus, North-West University, Vanderbijlpark, South Africa
e-mail: Charles.VanDerVyver@nwu.ac.za

A. P. Abidoye
Kennesaw State University, Atlanta, USA
e-mail: ademola.abidoye@gmail.com

M. G. Jaatun et al. (eds.), *Proceedings of the International Conference on Cybersecurity, Situational Awareness and Social Media*, Springer Proceedings in Complexity,
https://doi.org/10.1007/978-981-96-0401-2_8

1 Introduction

Malware is a malicious software that causes files on your computer or network to be locked up until you pay a ransom. It's become a major cyber threat in recent years, affecting people, companies, and even major infrastructure. When it comes to Malware detection, you need to understand how Malware attacks are becoming more common, what the attackers are using, and how to detect and stop them. As the world moves more and more into digital space, data storage has become the norm for businesses of all sizes. In the information age, businesses want to be able to quickly access their data to communicate with customers, collaborate with partners, and do everyday tasks. When a Malware attack happens, it can have a huge impact on a company. It can mean losing access to their digital assets forever, having to completely shut down their business, losing millions in revenue, and damaging their reputation. What is Malware? Malware, or malicious software, is any program or file that is intentionally harmful to a computer, network or server. [1].

Malware can infect networks and devices and is designed to harm those devices, networks and/or their users in some way. Depending on the type of malware and its goal, this harm may present itself differently to the user or endpoint. In some cases, the effect malware has is relatively mild and benign, and in others, it can be disastrous. No matter the method, all types of malwares are designed to exploit devices at the expense of the user and to the benefit of the hacker – the person who has designed and/or deployed it. Malware can cause a lot of damage, like financial losses, data loss, and even damage to your reputation. Malware is becoming more and more sophisticated and targeted. Cybercriminals are using encryption algorithms and other tricks to hide their identity and make it harder for people to find them [7, 10].

Malware is a form of malicious software that has become increasingly prevalent in recent years, affecting individuals, businesses and organizations alike. It has caused an estimated $1 trillion in economic losses globally. It can be spread through a variety of vectors, such as email phishing schemes, malicious web pages, drive by downloads, exploitation of software or operating system vulnerabilities, and social engineering techniques. Among the most prominent Malware attacks are WannaCry (Malware), Petya (Malware) and Ryuk (Malware). These attacks have affected a variety of organizations, including healthcare providers, financial institutions, government agencies, and critical infrastructure.

As Malware attacks become increasingly sophisticated, it is increasingly difficult to detect them and mitigate them effectively. Malware detection methods and technologies are not up to the task of quickly spotting and stopping Malware attacks, which can cause huge financial losses, data leaks, and disruption to operations. Keep an eye on new and changing Malware types and reduce their impact to protect the security of data and systems.

2 A Summary of Related Works

A survey on malware was conducted by [1], which focused on the threats, attacks, and preventive measures related to Malware. The survey analyzed various aspects of Malware, including its history, types, propagation techniques, and attack vectors. The authors also discussed the impact of Malware on individuals, businesses, and organizations, as well as the current state of Malware prevention and mitigation techniques. The survey highlighted the increasing prevalence and sophistication of Malware attacks, as well as the growing threat to critical infrastructure and national security.

[1] The authors identified various preventive measures, including user education, backup and recovery solutions, and security software, as well as the importance of collaboration between industry, government, and law enforcement in combating Malware. Overall, the survey provided a comprehensive overview of malware and its impact on society, as well as the current state of prevention and mitigation techniques. It identified the need for continued research and development in the area of malware prevention and detection, as well as the importance of user education and awareness in protecting against these types of attacks.

[2] provided a comprehensive overview of Malware attacks, their evolution, and current trends in Malware. The survey covers various aspects of Malware, including its history, types of Malware, propagation methods, encryption techniques, and ransom payment methods. The survey highlights the increasing sophistication of Malware attacks, with attackers using advanced encryption techniques and social engineering tactics to trick users into downloading and installing malware. The survey also notes the emergence of new types of malware, including hybrid malware that combines features of both malware and other types of malware. In terms of prevention and mitigation strategies, the survey highlights the importance of regular backups, user awareness training, and antivirus software. The survey also notes that there is a need for more research into effective prevention and mitigation strategies, as well as improved collaboration between security researchers, law enforcement agencies, and other stakeholders. Overall, the survey provides valuable insights into the evolving threat landscape of malware and highlights the need for continued research and development of effective prevention and mitigation strategies.

[3] provided an overview of the history, types, and characteristics of Malware. It also discussed the various attack vectors used by malware, including email, malvertising, and exploit kits. The survey also examined existing malware detection and prevention techniques, such as signature-based detection, behavior-based detection, and data backup. It highlighted the limitations of these techniques and the need for more advanced detection methods. Furthermore, the survey discussed the economic impact of malware on organizations and individuals. It highlighted the increasing frequency and sophistication of malware attacks and the need for a proactive approach to malware prevention and mitigation. Overall, the survey provided a comprehensive overview of malware and the various methods for detecting and preventing it.

Brik et al. [4] was centered on the evolution of malware and the challenges associated with detecting and mitigating malware attacks [4] also examined the economic impact of malware attacks, including the costs associated with data loss and system downtime. The study emphasized the need for proactive measures to prevent malware attacks, including regular backups, security awareness training, and vulnerability management. The authors also discussed the importance of incident response planning and the need for effective detection and mitigation strategies.

The survey by [4] provides a comprehensive overview of the Malware landscape and highlights the economic impact and challenges associated with Malware attacks. The study emphasizes the importance of preventive measures and incident response planning in mitigating the effects of Malware attacks. The authors also discuss the limitations of current detection and mitigation strategies and the need for continued research and development in this area.

Meidan et al. [5] conducted the survey focused on the technical aspects of Malware, including the different families of Malware, their encryption techniques, and the methods used for ransom payment and recovery. The study also discussed the evolution of Malware, including the shift from simple screen-locking Malware to more sophisticated file-encrypting Malware [5] also highlighted the challenges of detecting Malware, particularly the difficulty of distinguishing Malware from legitimate encryption processes. The authors discussed various detection methods, including signature-based detection, behavior-based detection, and machine learning-based detection.

The study emphasized the need for a multi-layered approach to Malware detection, combining different detection techniques to increase detection accuracy. The survey by [5] provided a technical perspective on Malware, highlighting the encryption techniques and evolution of Malware families. The study also provided insights into the challenges of Malware detection and the importance of a multi-layered approach to detection. The authors emphasized the need for continued research and development in the field of Malware detection and prevention.

Alharbi et al. [6] examined the state-of-the-art techniques for Malware detection, prevention, and recovery. The study emphasized the need for a multi-layered approach to Malware defense, including network and endpoint security, backup and recovery systems, and user awareness and education [6] discussed various methods for Malware detection, including signature-based detection, behavior-based detection, and machine learning-based detection. The study highlighted the limitations of signature-based detection, which can be ineffective against new and unknown Malware variants. The authors also discussed the challenges of behavior-based detection, which can generate false positives and may not be effective against advanced Malware attacks.

The study by [6] emphasized the potential of machine learning-based detection for Malware, particularly deep learning and neural network approaches. The authors highlighted the need for large, diverse datasets for training machine learning models and the importance of feature selection and data pre-processing in improving detection accuracy. Overall, the survey by [6] provides valuable insights into the state-of-the-art techniques for Malware detection and defense. The study highlights the need for a comprehensive approach to Malware defense and the potential of machine learning-based detection for improving detection accuracy and efficiency.

Singh and Jain [7] focused on the different types of Malware, their propagation techniques, and the impact of Malware attacks on organizations. The study identified several types of Malware, including crypto Malware, screen locker Malware, and mobile Malware. The authors also discussed the various propagation techniques used by Malware attackers, such as phishing emails, malicious attachments, and exploit kits [7] also highlighted the impact of Malware attacks on organizations, including financial losses, data breaches, and reputational damage.

The study emphasized the importance of Malware prevention and detection measures, including regular backups, patching systems and software, and implementing endpoint security solutions. The survey by [7] provides valuable insights into the different types of Malware and their propagation techniques, as well as the impact of Malware attacks on organizations. The study emphasizes the importance of Malware prevention and detection measures and provides useful recommendations for organizations to protect themselves against Malware attacks.

Zohaib et al. [8] was fascinated about the impact of Malware on businesses and examined the various types of Malware, including file-encrypting, screen-locking, and mobile Malware. The study highlighted the significant financial and reputational damage caused by Malware attacks, as well as the potential legal and regulatory implications for affected organizations [8] also discussed the methods used by attackers to distribute Malware, including spam emails, malvertising, and exploit kits. The study emphasized the importance of preventive measures, including regular backups, security awareness training, and effective security controls.

The authors also highlighted the need for incident response plans to ensure effective response and recovery from Malware attacks. The survey by [8] provides valuable insights into the impact of Malware attacks on businesses and the importance of preventive measures and incident response plans. While the study did not specifically focus on Malware detection, it highlighted the need for organizations to take a comprehensive approach to malware prevention and mitigation.

Zawoad et al. [11] was interested in the detection and mitigation of Malware attacks. The survey provided an overview of the different types of Malware attacks and their characteristics, as well as the methods used for Malware detection and mitigation. The study identified several detection methods, including signature-based detection, behavior-based detection, and hybrid detection. The authors discussed the advantages and disadvantages of each method and highlighted the importance of using multiple detection methods to improve detection accuracy [11] also discussed the challenges of Malware mitigation, including the difficulty of recovering encrypted data and the risk of paying the ransom.

Singh et al. [10] conducted the survey focused on the impact of Malware attacks on organizations and individuals and the various prevention and mitigation strategies that can be employed. The study identified several types of Malware attacks, including locker Malware, encrypting Malware, and master boot record (MBR) Malware. The authors discussed the methods used by Malware attackers, including phishing emails, social engineering, and drive-by downloads. They also highlighted the importance of timely detection and response to malware attacks to minimize their impact [10] discussed various preventive measures against Malware attacks, including regular backups, user education and awareness, endpoint protection, and network segmentation. The study also highlighted the challenges of detecting and mitigating malware attacks, including the difficulty of identifying new and evolving variants of malware and the increasing sophistication of attack methods.

Overall, the survey by [10] provides a comprehensive overview of the Malware landscape and the various prevention and mitigation strategies that can be employed. While the study did not specifically focus on Malware detection, it provided valuable insights into the impact of malware attacks and the importance of preventive measures in protecting against these attacks.

Another previous survey on malware was conducted by [12], which focused on the technical aspects of malware and the challenges of detecting and mitigating these attacks. The study provided an overview of malware techniques, including encryption-based malware, screen-locking malware, and data exfiltration-based Malware. The survey also discussed the methods used by attackers to distribute Malware, such as spam emails, exploit kits, and malicious websites [12] also discussed the challenges of detecting and mitigating Malware attacks, including the difficulty of identifying Malware in encrypted traffic, the need for timely detection, and the importance of effective response and recovery strategies.

The authors highlighted the role of machine learning and artificial intelligence in malware detection and prevention, and the need for collaboration between security researchers, industry, and law enforcement. The study by [12] provided valuable insights into the technical aspects of malware and the challenges of detecting and mitigating these attacks. The study emphasized the importance of proactive measures, such as backup and recovery systems, and the need for continuous monitoring and analysis of network traffic to detect potential malware attacks.

Connolly L et al. (2020) discovered that within the observed sample, organization size, by itself, did not affect the severity of attacks. Prior findings and opinions on the relationship between organization size and the incidence of Malware attacks are rather inconsistent, with some saying that Malware is mainly a problem for large enterprises and others saying that small and medium enterprises (SMEs) make up the bulk of the victims. Of the organizations observed, SMEs and large organizations were similarly impacted by Malware attacks and in most cases the impact felt was of high severity. Malware does not target organizations of a particular size. All organizations, small, medium and large, are equally affected. It was observed several large organizations that experienced severe consequences of crypto-Malware attacks as well as SMEs. Therefore, regardless of how large or small an organization is, there is no room for complacency. Cyberattacks on profit-driven organizations normally lead to substantial financial losses, reputational damage and loss of customers.

Moore C. (2021) discussed that due to evolution of more sophisticated Malwares, packet inspection method is not sufficient to rely on. To detect the Malware, a network monitoring system should be deployed. A network detection system monitors the network to detect the malicious activity. A network-based Malware detection system was proposed. Honeypots were created on each workstation to observe the system behavior. The workstations were observed to find the malicious behavior. Once a malicious behavior was observed the network administrator disabled the user account and shut down the system to isolate it with the system. This detection system did not prevent the launch of the attack and lead to lose of data if an attack is launched.

3 Methodology

Developing a methodology for malware detection involves a combination of techniques and strategies to identify, prevent, and mitigate malware attacks. The steps include:

Developing a methodology for malware detection involves a combination of techniques and strategies to identify, prevent, and mitigate malware attacks. The steps include:

3.1 Data Collection

The summary of the datasets used is presented in Table 1.

3.1.1 Dataset 1

The first dataset was gotten from Github website (GitHub 2023). ClaMP (Classification of Malware with PE headers) - A Malware classifier dataset built with header fields' values of Portable Executable files. The total samples are 5184 - malware (2683) and benign (2501). The dataset contain the information publicly available as https://raw.githubusercontent.com/urwithajit9/ClaMP/master/dataset/ClaMP_Raw-5184.csv.

Table 1 Dataset source

Dataset	URL	Source
Dataset 1	https://raw.githubusercontent.com/urwithajit9/ClaMP/master/dataset/ClaMP_Raw-5184.csv	Github
Dataset 2	https://www.kaggle.com/competitions/malware-detection/data	Kaggle
Dataset 3	https://www.kaggle.com/datasets/nsaravana/malware-detection	Kaggle
Dataset 4	https://figshare.com/articles/dataset/Windows_Malware_Detection_Dataset/21608262	Figshare

3.1.2 Dataset 2

The second dataset was gotten from Kaggle. The dataset has 216,352 entries of malware (140,849) and benign (75,503). The raw data here was obtained from the malware security partner of Meraz' 18 - Annual Techno Cultural festival of IIT Bhilai, the said raw data constituted malware and legitimate files. The dataset contained five files and Kaggle-data.csv was used. The dataset is publicly available as https://www.kaggle.com/competitions/malware-detection/data.

3.1.3 Dataset 3

The third dataset was gotten from Kaggle website. The dataset has malware and benign of both 50,000 each. It is publicly available as https://www.kaggle.com/datasets/nsaravana/malware-detection.

3.1.4 Dataset 4

The fourth dataset was gotten from Figshare. A dataset for Windows Portable Executable Samples with four feature sets. It contains four CSV files, one CSV file per feature set. The first feature set (DLLs_Imported.csv file) contains the DLLs imported by each malware family. The dataset is a multiclass classification meaning it has more than two class label. Malware types here includes Benign (0), Redline Stealer (1), Downloader (2), RAT (3), Banking Trojan (4), SnakeKeyLogger (5) and Spyware (6). The dataset is publicly available as https://figshare.com/articles/dataset/Windows_Malware_Detection_Dataset/21608262.

3.2 *Machine Learning Algorithms*

In this research work, four (4) different Algorithms (classifiers) were used in this research while training the data set in the course of the research. The algorithms include.

3.2.1 Logistic Regression

Logistic regression, also known as supervised machine learning, is a type of statistical algorithm that analyzes the relationship between a collection of independent variables and their dependent binary variables. It models the relationship between features and the likelihood of a positive class using sigmoid functions. The linear combination of the input features is mapped to a probability value from 0 to 1. The goal of the algorithm is to find optimal coefficients that maximize the likelihood of observed data for the model.

We know the equation of the straight line can be written as:

$$y = b_0 + b_1x_1 + b_2x_2 + b_3x_3 + \ldots + b_nx_n \tag{1}$$

In Logistic Regression y can be between 0 and 1 only, so for this let's divide the above equation by $(1 - y)$:

$$\frac{y}{(1-y)}; 0 for y = 0 and infinity for y = 1 \tag{2}$$

But we need a range between - [infinity] to + [infinity], then take the logarithm of the equation it will become:

$$loglog\left[\frac{y}{1-y}\right] = b_0 + b_1x_1 + b_2x_2 + b_3x_3 + \ldots + b_nx_n \tag{3}$$

The above equation is the final equation for Logistic Regression.

The mathematical formula involves using the sigmoid to model the relationship between the input features and the probability of the binary outcome. The sigmoid function (σ) is used to map the linear combination (z) to a probability value (p) between 0 and 1:

$$p = \sigma(z) = \frac{1}{\left(1 + e^{(-z)}\right)} \tag{4}$$

where e is equal to the base of the natural logarithm (Euler's number).

3.2.2 Decision Trees

Decision trees are used for both classifying and regressing tasks. They are supervised learning methods that make decisions by dividing data into sub-data sets according to the values of the input features. The sub-data sets represent nodes in the tree and the leaves correspond to the predicted output. The algorithm selects the most suitable feature and value for each sub-data set based on a partitioning criterion, which is often used for classifying tasks. The process of recreating sub-data sets continues until a given depth is reached. For classification tasks, the majority of classes in a leaf node are assigned as the predictions. To avoid overfitting, a maximum depth has been set for decision trees.

$$E(S) = \sum_{i=1}^{c} -p_1p_1 \tag{5}$$

$$G(S, C) = E(S) - \sum_{w \in values(C)} \frac{S_w}{S} E(S_w) \tag{6}$$

where:

E – Entropy is a logarithmic algorithm for exploration analysis.

G – Gini for false reduction classification.

3.2.3 Random Forest

Random Forest involves the combination of multiple decision trees to enhance the accuracy and robustness of predictions. The "forest" is the set of decision trees. The "random" part comes from introducing randomness into the training. The process involves the training of a set of decision trees, with the final prediction being determined by the aggregation of the results of the individual trees. Their ability to minimize overfitting, work with big data, and make precise predictions has made them a go-to choice for data scientists and professionals. Renowned for its high performance and scalability making it a valuable particularly when used in combination with decision trees. After training, Predictions for an unseen sample x^i can be made by averaging the predictions from all the individual classification trees on x^i

$$f^i = \frac{1}{B}\left(\sum_{b=1}^{B} f_b\left(fx^i\right)\right) \tag{7}$$

$$RFfi_i = \frac{\sum_j normfi_{ij}}{\sum_{j \in allfeatures, k \in alltrees} normfi_{jk}} \tag{8}$$

3.2.4 XGBoost (Extreme Gradient Boosting)

A machine learning algorithm that utilizes a gradient boosting framework to enhance its performance and scalability. It is an ensemble learning method that combines the predictions of multiple weak models to produce a stronger prediction. Renowned for its superior predictive capabilities and scalability, as well as its capacity to process large datasets. Tree pruning is a new technique that XGBoost uses to reduce the size of decision trees in training. This helps avoid overfitting and makes it easier to use the system's resources. It also helps XGBoost handle any missing values in the data and figure out the best way to use them during training.

The objective function (loss function and regularization) at iteration t

$$L^{(t)} = \sum_{i=1}^{n} l\left(y_i, \hat{y}_i^{(t-1)} + f_t(x_i)\right) + \Omega(f_t) \quad (9)$$

Can be seen as

$$f(X + \Delta X) where X = \hat{y}_i^{(t-1)} \quad (10)$$

It is easy to see that the XGBoost objective is a function of functions (i.e. l is a function of CART learners, a sum of the current and previous additive trees).

XGBoost objective using second-order Taylor approximation.

F(x) is the loss function l, while a is the previous step (t-1) predicted value and Δx is the new learner we need to add in step t.

$$f(x) \approx f(a) + f'(a)(x-a) + \frac{1}{2}f''(a)(x-a)^2 \quad (11)$$

$$L^{(t)} \approx \Sigma_{i=1}^{n}\left[l\left(y_i, \hat{y}^{(t-1)}\right) + g_i f_t(x_i)\right] + \Omega(f_t) \quad (12)$$

where:

$$g_i = \partial_{\hat{y}^{(t-1)}} l\left(y_i, \hat{y}^{(t-1)}\right) \text{ and } h_i = \partial^2_{\hat{y}^{(t-1)}} l\left(y_i, \hat{y}^{(t-1)}\right) \quad (13)$$

The above is first and second order gradient statistics of the loss function.

XGBoost simplified objective

$$\tilde{l}^{(t)} = \Sigma_{i=1}^{n}\left[g_i f_t(x_i) + \frac{1}{2} h_i f_t^2(x_i)\right] + \Omega(f_t) \quad (14)$$

3.2.5 Naïve Bayes

One of the most popular and straightforward probabilistic classification algorithms based on Bayes' theorem with the assumption of feature independence. Naive Bayes algorithm is called "naive" because it assumes that a feature's presence in a class doesn't depend on the presence of the other features in the class. Basically, it believes that all features are equal and contribute independently to a class's probability. The basic concept involves predicting the probability of a class for a set of features using Bayes' conditional probability. The formula for Bayes' theorem is as follows:

$$p\left(\frac{y}{x}\right) = \frac{\left(p\left(\frac{x}{y}\right) * p(y)\right)}{p(x)} \quad (15)$$

where:

$p(\frac{y}{x})$ is the probability of class y given the features x \left(the posterior probability\right).
$p\left(\frac{x}{y}\right)$ is the probability of features x given class y (likelihood).
$p(y)$is the probability of class y in the dataset \left(prior probability\right).
$p(x)$is the probability of features x in the dataset.

3.3 Evaluation Metrics

They are quantitative measures that measure a model or system's performance. These metrics are used to measure a model's performance, compare models, and make decisions about improvements. Four metrics were used in this project to measure the performance of these classifiers namely: Accuracy, Precision, Recall and F1-score.

4 Discussion

4.1 Results and Analysis

4.1.1 Data Analysis for Dataset 1

Table 2 shows the result of the machine learning algorithm for Dataset 1. XGBoost and Random Forest both had the highest accuracy of 0.982456 and 0.980263 respectively followed by Decision Tree with the accuracy of 0.970760 then Logistics Regression with accuracy of 0.899122. The lowest accuracy is Naïve Bayes with 0.633771.

Figure 1 gives a clear graphical representation of the evaluation metrics and the algorithms using Dataset 1. The bar chart represents each of the machine learning algorithms and the results they produced across each of the evaluation metrics used (Table 3).

Table 2 Results obtained for dataset 1

Classifier	Accuracy	Precision	Recall	F1 Score
XGBoost	0.982456	0.985893	0.976708	0.981279
Logistic regression	0.899122	0.901587	0.881987	0.891679
Decision tree	0.970760	0.977848	0.959627	0.968652
Random forest	0.980263	0.982785	0.975155	0.978955
Naïve bayes	0.633771	0.938650	0.237577	0.379182

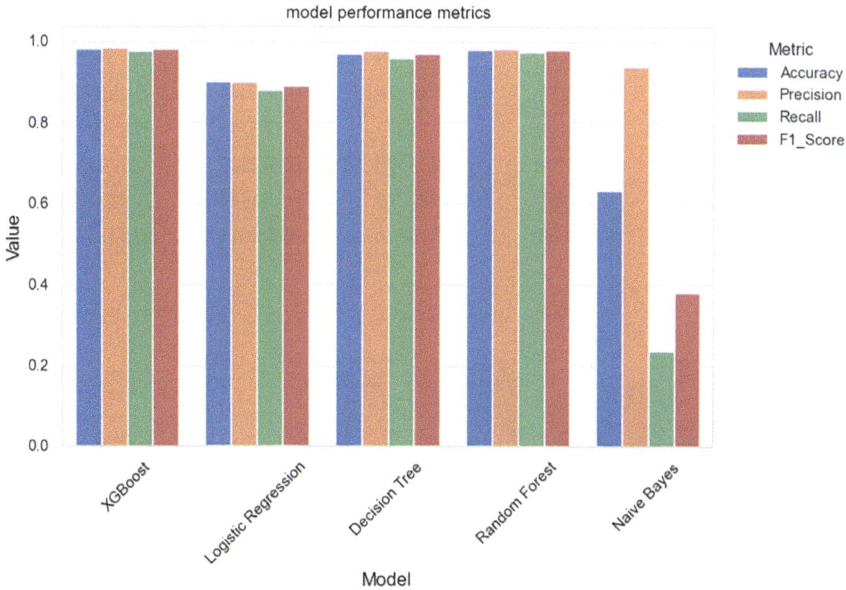

Fig. 1 Graphical representation of the algorithms with each evaluation metrics for dataset 1

Table 3 Boxplot statistics for dataset 1

	XGBoost	Logistic R	Decision T	Random F	Naïve B
Maximum value	0.985	0.901	0.977	0.982	0.938
3rd Quartile	0.983	0.899	0.972	0.980	0.709
Median	0.981	0.895	0.969	0.979	0.506
1st Quartile	0.980	0.889	0.966	0.978	0.343
Minimum value	0.976	0.881	0.959	0.975	0.237

4.2 Data Analysis for Dataset 2

Table 4 shows the result of the machine learning algorithm for Dataset 2. XGBoost and Random Forest both had the highest accuracy of 0.990016and 0.990432 respectively followed by Decision Tree with the accuracy of 0.982251then Logistics Regression with accuracy of 0.931385. The lowest is Naïve Bayes with an accuracy 0.414758. Naïve Bayes has high recall but low precision, resulting in low F1 scores and accuracy.

Figure 2 gives a clear graphical representation of the evaluation metrics and the algorithms using Dataset 2. The bar chart represents each of the machine learning algorithms and the results they produced across each of the evaluation metrics used (Table 5).

Table 4 Results obtained for dataset 2

Classifier	Accuracy	Precision	Recall	F1 Score
XGBoost	0.990016	0.986346	0.985041	0.985693
Logistic regression	0.931385	0.922873	0.876754	0.899222
Decision tree	0.982251	0.975590	0.973523	0.974556
Random forest	0.990432	0.987719	0.984842	0.986278
Naïve bayes	0.414758	0.373476	0.998014	0.543547

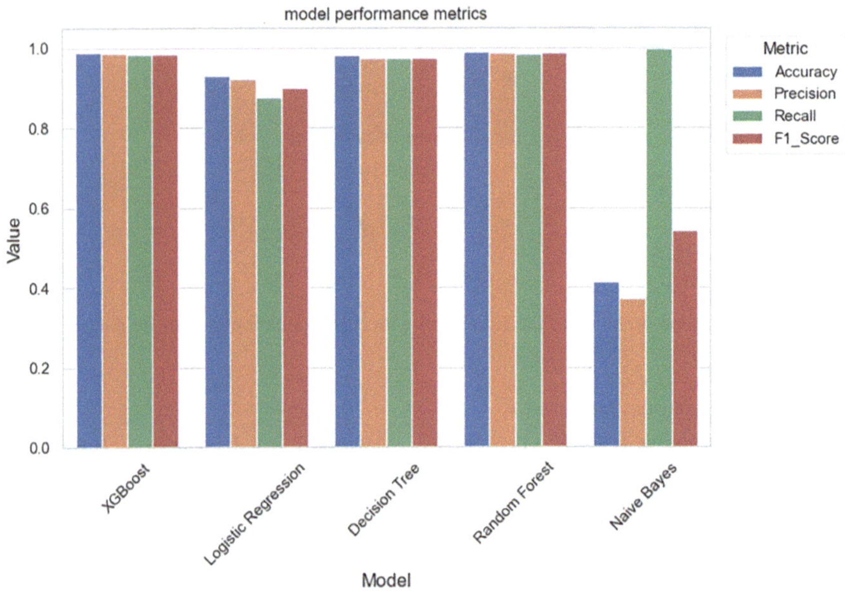

Fig. 2 Graphical representation of the algorithms with each evaluation metrics for dataset 2

Table 5 Boxplot statistics for dataset 2

	XGBoost	Logistic R	Decision T	Random F	Naïve B
Maximum value	0.990	0.931	0.982	0.990	0.990
3rd Quartile	0.987	0.925	0.977	0.988	0.657
Median	0.986	0.911	0.975	0.986	0.479
1st Quartile	0.985	0.893	0.974	0.985	0.404
Minimum value	0.985	0.876	0.973	0.984	0.373

4.3 Data Analysis for Dataset 3

Table 6 shows the result of the machine learning algorithm for Dataset 3. XGBoost, Random Forest and Decision Tree had the same accuracy of 1.0 meaning that its predictions are 100% accurate. Naïve Bayes has an accuracy of 0.621766 and Logistics Regression had the least accuracy of 0.5104.

Figure 3 gives a clear graphical representation of the evaluation metrics and the algorithms using Dataset 3. The bar chart represents each of the machine learning algorithms and the results they produced across each of the evaluation metrics used (Table 7).

Table 6 Results obtained for dataset 3

Classifier	Accuracy	Precision	Recall	F1 Score
XGBoost	1.0	1.0	1.0	1.0
Logistic regression	0.5104	0.506888	0.839531	0.632119
Decision tree	1.0	1.0	1.0	1.0
Random forest	1.0	1.0	1.0	1.0
Naïve bayes	0.621766	0.585586	0.838467	0.689574

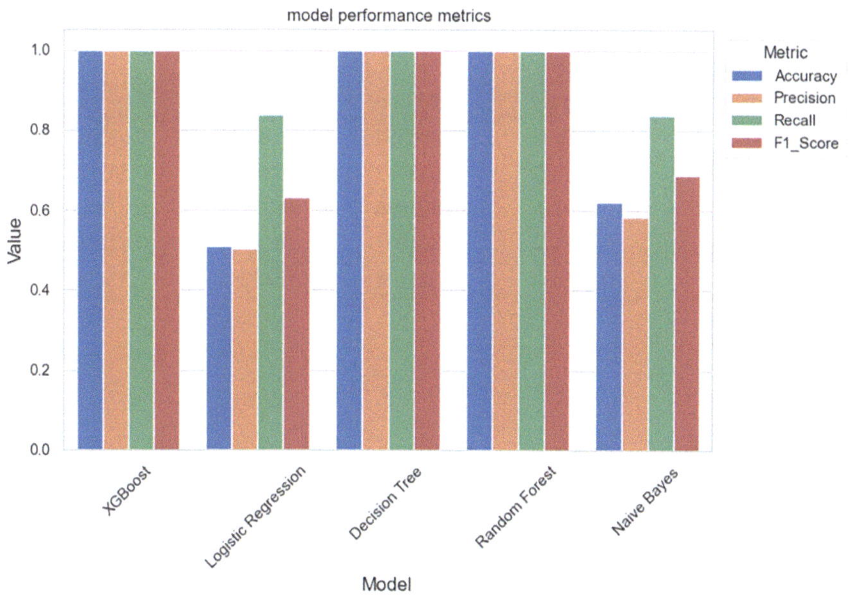

Fig. 3 Graphical representation of the algorithms with each evaluation metrics for dataset 3

Table 7 Boxplot statistics for dataset 3

	XGBoost	Logistic R	Decision T	Random F	Naïve B
Maximum value	1.0	0.839	1.0	1.0	0.838
3rd Quartile	1.0	0.683	1.0	1.0	0.726
Median	1.0	0.571	1.0	1.0	0.655
1st Quartile	1.0	0.509	1.0	1.0	0.612
Minimum value	1.0	0.506	1.0	1.0	0.585

4.4 Data Analysis for Dataset 4

Table 8 shows the result of the machine learning algorithm for Dataset 4. Random forest has the highest accuracy of 0.628474 followed by XGBoost with the second highest accuracy of 0.628135. Then Decision tree with an accuracy of 0.627118 followed by Logistic regression with accuracy of 0.584237. Naïve bayes with the least accuracy of 0.345084.

Figure 4 gives a clear graphical representation of the evaluation metrics and the algorithms using Dataset 4. The bar chart represents each of the machine learning algorithms and the results they produced across each of the evaluation metrics used (Table 9).

Table 8 Results obtained for dataset 4

Classifier	Accuracy	Precision	Recall	F1
XGBoost	0.628135	0.804354	0.628135	0.622012
Logistic regression	0.584237	0.638942	0.584237	0.565124
Decision tree	0.627118	0.803896	0.627118	0.621406
Random forest	0.628474	0.804091	0.628474	0.623264
Naïve bayes	0.345084	0.621104	0.345084	0.324384

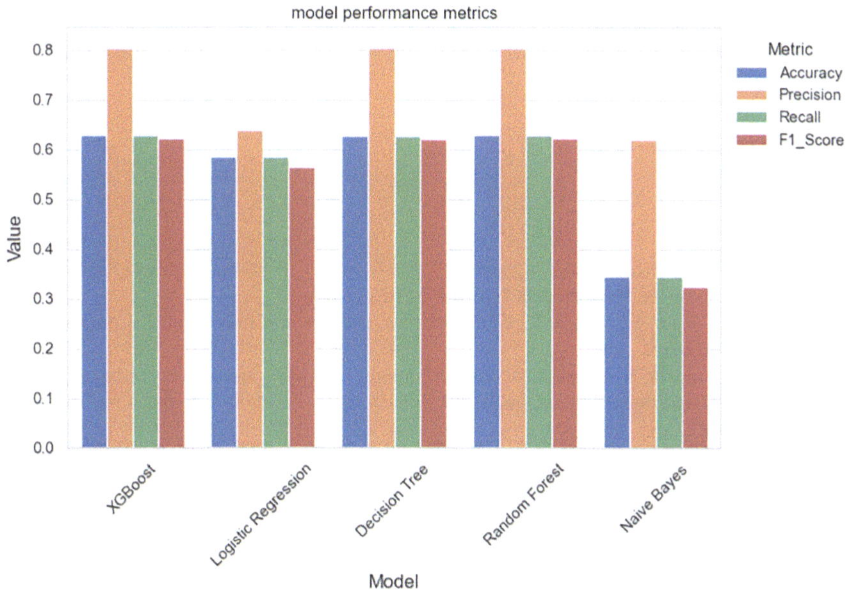

Fig. 4 Graphical representation of the algorithms with each evaluation metrics for dataset 4

Table 9 Boxplot statistics for dataset 4

	XGBoost	Logistic R	Decision T	Random F	Naïve B
Maximum value	0.804354	0.638942	0.803896	0.804092	0.621104
3rd Quartile	0.672190	0.597913	0.671313	0.672379	0.414090
Median	0.628136	0.584237	0.627119	0.628475	0.345085
1st Quartile	0.626605	0.579459	0.625691	0.627172	0.339910
Minimum value	0.622013	0.565125	0.621406	0.623265	0.324385

5 Conclusion

In this research, we explored the world of Malware detection with machine learning algorithms. We have been able to test the effectiveness of our techniques in distinguishing between normal system behavior and Malware activity. By training on different datasets containing both benign and malicious files, the models have demonstrated accuracy and precision in the detection of Malware attacks. As Malware attackers continue to evolve and adapt their strategies, we need to continually refine detection algorithms and improve defense. The results of this project demonstrate the importance of machine learning in enhancing Malware detection capabilities. Nevertheless, it should be stressed that machine learning should not be used in isolation but should be integrated into a comprehensive cybersecurity strategy in conjunction with other prevention and response measures for a comprehensive defense.

As we progress, the partnerships between researchers, cybersecurity professionals, and practitioners will be essential in keeping up with the ever-changing Malware landscape.

References

1. Alzahrani, A.I., Alshehri, M., Alghamdi, A.A.: A survey of malware: threats, attacks, and preventive measures. IEEE Access **7**, 82805–82825 (2019)
2. Sarma, K.K., Kaur, H., Ahuja, V.: Malware: past, present, and future. Int. J. Comput. Appl. **169**(2), 6–10 (2017)
3. Firdaus, A., Zakaria, A., Razak, N.A., Tuan Abdullah, T.A.: Malware: a survey on history, types, and detection and prevention techniques. J. Telecommun. Electron. Comput. Eng. **10**(1–13), 95–100 (2018)
4. Brik, D., Aviv, A.J., Russo, A.: A survey on Malware: Concepts, types, attacks, and mitigation techniques. Comput. Secur. **78**, 398–427 (2018)
5. Meidan, Y., Bohadana, M., Shabtai, A., Elovici, Y.: Detection and analysis of Malware: A survey. IEEE Secur. Priv. **16**(2), 68–76 (2018)
6. Alharbi, A., Manickam, S., Al-Muhtadi, J.: A survey on malware: threats, vulnerabilities, and prevention. J. Netw. Comput. Appl. **114**, 80–102 (2018)
7. Singh, A., Jain, P.: Malware: types, infection vectors, propagation techniques, detection, and prevention. Comput. Secur. **78**, 398–418 (2018)
8. Zohaib, M., Shahzad, W., Khattak, H.A.: Malware: a survey on critical factors, prevention and mitigation. J. Inf. Secur. Appl. **39**, 34–53 (2018)
9. Connolly, L., Borrion, H.: Your money or your business: decision-making processes in malware attacks. In: Proceedings of 2020 International Conference in Information Systems. Assoc. Inf. Syst. (2020)
10. Singh, N., Singh, N., Kumar, P.: Malware: threats, vulnerabilities, and prevention. J. Inf. Secur. Appl. **38**, 163–175 (2018)
11. Zawoad, S., Hasan, R., Ibne Motaleb, M., Ahmed, I.: Malware detection and mitigation: a survey. J. Netw. Comput. Appl. **130**, 79–111 (2019)
12. Kharraz, A., Robertson, W., Balzarotti, D., Bilge, L., Kirda, E.: A view of the cybersecurity threat landscape. Commun. ACM **61**(10), 43–48 (2018)
13. Hastie, T., Tibshirani, R., Friedman, J.: The elements of statistical learning: data mining, inference, and prediction. (2nd ed.) (2009)
14. Breiman, L.: Classification and regression trees. Chapman & Hall (1984)
15. Han, J., Kamber, M., Pei, J.: Data mining: concepts and techniques. Morgan Kaufmann (2011)
16. Hastie, T., Tibshirani, R., Friedman, J.: The elements of statistical learning: data mining, inference, and prediction (2009)
17. Mitchell, T.: Machine learning. McGraw-Hill (1997)
18. Rokach, L., Maimon, O.: Data mining with decision trees: theory and applications. World Scientific Publishing (2008)
19. Breiman, L.: Random forests. Mach. Learn. **45**(1), 5–32 (2001)
20. Liaw, A., Wiener, M.: Classification and regression by randomForest. R News **2**(3), 18–22 (2002)
21. Moore, C.: Detecting malware with honeypot techniques. In: Proceedings of the 2021 Cybersecurity and Cyberforensics Conference (CCC), pp. 77–81 (2021)
22. Masum, M., Faruk, J.H., Shahriar, H., Qian, K., Lo, D., Adnan, M.I.: Malware classification and detection with machine learning algorithms. In: IEEE Annual Computing and Communication Workshop and Conference (2022)

23. Azeez, N.A., Salaudeen, B.B., Misra, S., Damasevicius, R., Maskeliunas, R.: Identifying phishing attacks in communication networks using URL consistency features. Int. J. Electron. Secur. Digit. Forensics **12**, 200–213 (2020)
24. Azeez, N.A., Odufuwa, O.E., Misra, S., Oluranti, J., Damaševičius, R.: Windows PE malware detection using ensemble learning. Informatics **8**, 10 (2021). https://doi.org/10.3390/informatics8010010
25. Azeez, N.A., Der Vyver, C.V.: Digital education: assessment of e-learning and m-learning adoption in tertiary institutions in South Africa. In: Proceedings of the 2018 IEEE Conference on E-Learning, E-Management and E-Services (IC3e), Langkawi, Malaysia, pp. 23–28 (2018)
26. Azeez, N.A., Vyver, C.V.: Dynamic patient-regulated access control framework for electronic health information. In: 2017 International Conference on Computational Science and Computational Intelligence (CSCI), pp.1684–1690 (2017)

A Case Study of a Ransomware and Social Engineering Competition

Aunshul Rege and Rachel Bleiman

Abstract The average cost of a ransomware attack in 2022 was $4.49 million and the recovery time rose from 7.8 days in 2021 to 14.9 days in 2022. Organizations often bring in professional negotiators to represent them during these attacks to lower ransom demands, buy time for incident response, get evidence of decryptors, and determine who is behind the attack; these research and negotiation skills fall under the area of social engineering. This paper shares a case study of a real-time virtual Ransomware and Social Engineering Competition (RSEC), which occurred in summer 2022 and hosted 8 high school and 6 undergraduate teams. The paper demonstrates how students learned about using psychological persuasion in ransomware negotiations, leveraging OSINT and the ATT&CK framework for attack attribution, and interfacing with the clients to best represent their needs. The paper discusses how the skills and knowledge stemming from RSEC are mapped to the NICE framework.

1 Introduction

In 2022, there were two notable ransomware attacks. In March, the Toyota car manufacturer was impacted by a supply chain attack, which forced the company to suspend operations at 14 manufacturing plants in Japan, impacting the production of 13,000 vehicles [1]. The Costa Rican government was targeted by the Conti ransomware group in mid-April, disrupting several government agencies, such as the Finance and Labor ministries [1]. When the government refused to pay the ransom, the group published stolen information on the dark web as punishment [1].

Ransomware is a malware that denies users or organizations access to their systems; cybercriminals hold these locked systems hostage in exchange for a ransom [1]. The average cost of a ransomware attack (including the ransom amount) in

A. Rege (✉) · R. Bleiman
Temple University, Philadelphia, PA, USA
e-mail: Rege@temple.edu

R. Bleiman
e-mail: Rachel.bleiman@temple.edu

M. G. Jaatun et al. (eds.), *Proceedings of the International Conference on Cybersecurity, Situational Awareness and Social Media*, Springer Proceedings in Complexity,
https://doi.org/10.1007/978-981-96-0401-2_9

2022 was \$4.49 million, while for those that didn't pay the ransom, the cost was \$5.12 million [1]. Even though many organizations have strengthened their security measures, recovery time across various targeted sectors rose from 7.8 days in 2021 to 14.9 days in 2022 (a 91% increase) [2]. Adversaries remain adaptive and find ways to penetrate target systems using techniques such as social engineering [2].

Social engineering (SE) is a technique where psychological persuasion of humans is used to "conduct reconnaissance (identify systems operating at target facilities), obtain information intended to secure electronic systems (passwords), or to encourage targets to inadvertently provide access to electronic systems and information (downloading and executing malicious files that are disguised as familiar or benign)" [3]. The average organization is targeted by over 700 SE attacks each year [4]. The cost of an average breach using SE as the initial attack vector was estimated to be \$4.1 million in 2022 [5]. Forty three percent of IT workers and 80% of employees who don't have a financial or executive role were victims of SE attacks in 2020 [6]. Despite these statistics, only 27% of companies provide SE awareness training [6]. Given the relevance of SE-based attacks in general, and also specifically in the case of ransomware, more work needs to be done to educate the next generation workforce in both these spaces. Students need to understand how psychological persuasion is used in ransomware negotiations and how cybercriminals leverage OSINT in their attacks. Students also should learn how the ATT&CK framework can be useful for attack attribution. Finally, an important skill for students to develop is interfacing with the clients to best represent their needs.

This paper shares findings from a 2022 Ransomware and Social Engineering Competition (RSEC). The next two sections detail the competition, flags, techniques, participants, metrics, and ethics. The fourth section shares student experiences on their team dynamics, strategies, and opinions on various components of the competition. The fifth section discusses how RSEC allowed students to experience ransomware negotiations, understand the value of the ATT&CK framework, and attain knowledge and skills as listed in the NICE framework. The paper concludes by arguing for more SE exposure in cybersecurity education, as it offers a unique and much-needed set of skills to the future workforce.

2 Related Work

Numerous Capture the Flag (CTF) competitions are available to educate students about cybersecurity (Table 1). Although these competitions are crucial for training the next generation, they are highly specialized and designed primarily for technical STEM students. Additionally, they often focus on specific areas such as reverse engineering, hacking, cryptography, and exploitation.

To ensure effective cyber-defense among the general population and future cyberprofessionals, it is essential for students across all STEM disciplines to understand SE and its importance in cybersecurity. Integrating SE into existing competitions

Table 1 Current capture-the-flag (CTF) competitions for students (* for online CTFs)

Name	Host/Developer	When	Focus
PicoCTF*	Carnegie Mellon	All year	Reverse engineer, break, hack, decrypt
PlaidCTF*	Carnegie Mellon	Apr	Web hacking, binary reverse engineering, exploitation, forensics, cryptography
CSAW CTF	NYU-Poly	Nov	Reverse engineer, web, crypto, and forensics
UCSB iCTF*	UC Santa Barbara	Mar	Attack other teams while defending self
US Cyber Challenge	Center for Internet Security	Summer	Identify vulnerabilities, forensics analysis, packet capture analysis
Panoply	UT San Antonio	Sep	Network defense competition; control & operate critical services (SMTP, DNS, etc.)
CPTC	RIT	Oct	Pentest; attack & find vulnerabilities
CCDC	UT San Antonio	Apr	Detect and respond to outside threats, maintain availability of existing services, respond to business requests, and balance security needs vs. business needs
CyberPatriot*	Air Force Association	Nov-Feb	Find and fix security vulnerabilities in Windows and Linux operating systems
CyberAcademy*	MITRE	All year	Web hacking, binary reverse engineering, exploitation, forensics, cryptography

could dilute its significance, reducing it to a minor component of a larger, more technically focused event.

DefCon features a SE community where participants can attend conference-style talks and engage in a hands-on SE competition, attempting techniques like vishing, which is conducted over the phone [7]. The Layer8 conference is the first to focus exclusively on SE and intelligence gathering, offering attendees SE talks, hands-on workshops, and small-scale competitions [8]. Organizations like Social Engineer Inc. and the SANS Institute also provide specialized SE courses and training sessions. These events cover topics such as Open Source Intelligence (OSINT), communication styles, psychological manipulation techniques, reconnaissance, and phishing [9, 10]. Recently, video podcasts have become a popular form of SE awareness programs. Layer8 hosts a podcast series that delves into SE through the experiences and stories of professional social engineers [11]. Another project, OSINTCurio.us, offers various videos and podcasts that include case studies demonstrating the effective use of OSINT tools [12].

These SE programs and events provide participants with valuable knowledge and practical experiences in SE through podcasts, conferences, and training sessions. They served as an inspiration for the authors to develop their own SECTFs. In October 2020, the authors offered a pure Collegiate SECTF grounded in the social sciences. Students were provided with a target local to their geographic region whom they had to OSINT, phish, and vish. While they conducted OSINT on their target, the phish and vish flags were simulated to maintain proper ethics. Six professional social engineers were recruited to serve as judges for the event and pose as the target in the simulated phish and vish [3]. In the Summer of 2021, the authors organized a second SECTF by hosting a SE Pentesting Competition. This offered high school, undergraduate, and graduate students a simulated environment to experience the role of a penetration tester, someone whom companies hire to identify vulnerabilities in their defense systems. In the competition, student teams were "hired" by the authors' cybersecurity lab to perform a SE Penetration Test by conducting OSINT, phishing, and vishing members of the lab [13]. Following from their experiences hosting these two competitions, the next section describes the SE competition that the authors hosted in the Summer of 2022, which focused on the theme of ransomware. The authors are not aware of any ransomware related competitions in the public domain.

3 Competition Structure and Logistics

The premise of the competition was that the authors' cybersecurity lab was hit with a ransomware attack. Students experienced both offense (during the application process) and defense (during the live competition) aspects of ransomware attacks. The purpose of engaging students in an offensive point of view during the application process allowed them to get a feel for both sides of the situation. The competition was intended to focus on the defensive side, so the application process addressed the aim to give students a flavor of what the adversarial aspects would look like. The current section identifies the overall theme of the competition and how it was structured.

3.1 Competition Theme, Overview, Goals, and Objectives

The authors organized and implemented the Ransomware and Social Engineering Competition (RSEC) virtually in the summer of 2022 over three days.

Research has shown that ransomware groups conduct Open-Source Intelligence (OSINT) on their targets prior to launching their attacks to determine a ransom amount that the target can pay (e.g., a percentage of the victim's annual revenue) [14]. OSINT involves gathering information that can be "obtained legally and ethically from public sources" [15]. As such, the applications for the competition required students to do research on the lab to determine the ransom amount and develop a pretext to target the lab via a phishing email. Students also had to design a ransom note.

The defense aspects of ransomware and SE were baked into the live competition. Student teams posed as ransomware negotiators who the authors 'hired' to represent the lab during communication with the cybercriminal group.

The competition had four objectives: (i) determining attribution and stating which ransomware group was targeting the lab, (ii) representing the client's needs effectively, (iii) minimizing the loss experienced by the client, and (iv) demonstrating competency in professionally relaying outcomes and recommendations to the client.

Students had to complete these objectives via a set of predetermined flags (tasks), each of which demonstrated the effective use of SE. These included: (i) interfacing with the client (Day 1), to understand the situation, ask relevant questions about the lab's policy and procedures, and learn findings from the incident response team, (ii) negotiating with the ransomware group (Day 2) that represented the client's needs and abilities effectively, and (iii) relaying the outcome of the negotiations and making recommendations to the client via formal oral and written reports (Day 3).

The authors hosted an orientation session before the live competition where students learned about the event rules, instructions, and logistics. The authors also brought in subject matter experts to give talks on topics that students would need during the competition. These topics included: SE strategies and psychological persuasion techniques, the ATT&CK framework and tactics, techniques, and procedures, ransomware negotiation do's and don'ts. The MITRE ATT&CK framework is a globally accessible knowledge base maintained by the MITRE Corporation as a means of organizing and analyzing the tactics, techniques, and procedures (TTPs) used by real adversaries [16]. This framework is informed by real-world observations and enables practitioners to understand cyberadversary behaviors.

3.2 *Live Competition*

The 2022 RSEC was held virtually on June 3–5, 2022 (undergraduate track) and June 10–12, 2022 (high school track). It employed the zoom platform for real-time 'face-to-face' interaction. The organizers' email addresses were used to interact with the teams and/or address their inquiries during the live competition.

3.2.1 Day 1: Interfacing with the Client

The first day of the competition involved speaking with the client. Each team had three 8-min client meetings throughout the day.

In the first meeting, students were presented with the ransom note that the lab received.

привет.

As you can see, you cannot access any of your files or services at this time:-(

All your files, documents, photos, databases, and other important data are safely encrypted with the advanced algorithms RSA4096 and AES-256. If you try to use any recovery software your files will be deleted/removed and we won't be able to help you then:-/

All we want is some of the money you have. We do not ask for much. Just 17 BTC. But this can change if you take too long to write us back. Every day delay will add an extra + 0.5 BTC. We are business after all. Like you, we have bills to pay.

You have a FREE chance to test our service by choosing three encrypted files to unlock! If you would like this chance, write to emails (contacts are below) and attach the 3 encrypted files—but they cannot contain important information—Ha! Nice try! > :-).

You do not want to ignore us. We have also downloaded your files and will publish it if you do not respond. So contact us soon.

But we are nice people. You pay us, we unlock all files AND you get instructions on how to avoid this from happening again. It's like "buy one get one free".

Contact emails.

yourRWgroup@protonmail.com.

Students were given the opportunity to ask questions to obtain information, while demonstrating the application of tactical empathy. This required demonstrating to the client that they were "deeply listening to their words and, in effect, how thoughtfully [they were] considering" the client's position as they were experiencing an attack [17]. For instance, students could ask the lab if they had backups (the lab was not sure), what the policy was on paying ransoms (the lab had $50 k but needed time to gather more and get permission from higher ups), or if they had heard from the incident response team (the lab noted that the IR team was overwhelmed at that time).

Students had to use the information from the note, the client's answers, the ATT&CK framework, and other open-source information to determine which techniques were used, who might be behind the attack, and develop a game plan.

In the second meeting, students had to display tactical empathy and professionalism while sharing their findings. The lab provided students with information from the incident response team. For instance, the lab members told students "Our incident response team told us that portions of files were encoded to hide the plain-text strings that would have otherwise helped them discover the ransomware", which students could map on to ATT&CK technique T1027. Students were once again given the opportunity to ask questions with tactical empathy as the situation was depicted as continuously changing (which is reflective of reality when things are unpredictable). Students once again used this information and OSINT to work towards attribution, developing a game plan, and now also had to develop sample responses to the ransomware note.

In the third client meeting, students had to share their findings, and then were given some more information from the incident response team: "We have more updates from our IR team. They said that the ransomware group discovered details about our network configuration and settings. They also discovered details about the processes that were running on our system", which mapped to ATT&CK techniques T1016 and T1057. However, students were told that the lab was unable to secure permission and thus could not pay any ransom amount. The students then had to showcase the sample responses they developed and recommend the best one and why that should be sent to the ransomware group by the end of Day 1.

3.2.2 Day 2: Negotiating with the Ransomware Group

On Day 2, students first met with the client where they were given the response from the ransomware group. Students were provided with a time for a negotiation, which was simulated via zoom chat (no audio or video).

Each team had two rounds of negotiations with the ransomware group, each of which lasted for 15 min. The first of these negotiations was intentionally hard with students receiving pushbacks and no agreement with the ransomware group. The hurdles given to students were based on challenges experienced by our subject matter experts in real ransomware negotiations. For instance, typical client requests students brought into their conversations were to lower the ransom amount and/or buy more time; our subject matter experts who took on the role of the ransomware group would refuse and state "We have business to run, employees to pay."

In the second negotiation, the ransomware group gave in to some demands depending on the team's overall performance during the real-time exchange. During negotiations, students had to utilize the RASCLS (Reciprocity, Authority, Scarcity, Commitment and consistency, Likeness or commonality, and Social proof) psychological persuasion principles when negotiating with the ransomware group [18].

Reciprocity involves attackers tricking victims into divulging sensitive information in exchange for tempting potential offers or prizes. Authority is when an attacker poses as someone in a position of power or credibility, which makes the target more likely to comply. Attackers can tempt targets with limited time offers (scarcity) that require urgent action. Commitment and consistency refer to the fact that individuals are likely to act in ways consistent with the core values or beliefs they are committed to; attackers can leverage these to trick victims. In likeness or commonality, attackers exhibit shared traits or experiences (e.g., attending the same high school, shared birthdays) with their targets to gain instant rapport. Social proof leverages the fact that targets are more likely to engage in a task if someone they know or trust (e.g., friends, celebrities) has already completed it. Finally, attackers take advantage of natural inclination to help, as human beings typically help each other out.

Students then had to meet with the client to provide a quick summary about the current status, specifically focusing on how many of the client's requests were fulfilled, what the final ransom amount was, and an initial profile of the attacker. Students had to work towards their formal client reports and presentations for the next day.

3.2.3 Day 3: Formal Oral and Written Client Debrief

On the final day of the competition, students had to do a 10-min formal presentation where they provided a summary of what happened, the outcome (overall financial damage, whether the data had been decrypted, etc.), who was behind the attack, the point of entry, and how to prevent such attacks in the future all while being professional and demonstrating tactical empathy. They were then questioned by the subject matter experts, who posed as lab members, directors, and chief financial officers.

Students also had to provide a formal client report that expanded on the oral presentation and had additional components such as a detailed negotiations playbook that demonstrated the application of psychological persuasion principles and further reading on protecting against ransomware attacks.

4 Administrative Details of RSEC

4.1 Participants

The RSEC was open to high school, undergraduate, and graduate students across all disciplines. Students entered the competition with their team composition details, their designated mentor information, and the adversarial portion described in Sect. 2.1. (OSINT the lab to determine the ransom amount, develop a pretext to target the lab through a phishing campaign, and generate a ransom note).

A total of 14 teams (8 high school and 6 undergraduate) from the United States competed in the event. Team sizes ranged from 2 to 4 members. The RSEC had a good representation from females (31%), however the majority was still male (64%). In terms of competitors' racial composition, about 38% of students self-identified as White, 15% as Black or African American, 18% as Asian; and 30% chose not to disclose this information. Finally, just 8% of the students identified as Hispanic or Latino.

4.2 Industry, Government, and Non-Profit Engagement

For the 2022 RSEC, the authors had representatives from industry, government, and nonprofits engage with the students. These representatives posed as the bosses of the lab's director, chief financial officers, and the ransomware group during the negotiations. In total, five subject matter experts from these three sectors interacted with the students in rotation over the duration of the live competition.

4.3 *Ethics and Risk Management*

Like the previous competitions organized by the authors [3, 13], the RSEC included pre-and post-event surveys, which were reviewed and approved by the ethics board at the authors' home institution. The authors worked with the risk management unit at their home institution to design several waivers to ensure that teams would engage in ethical behavior during the competition. Each member of the selected teams had to complete three waivers to maintain participation eligibility. The waivers ensured that students would (i) not disclose any information found during the RSEC via any platform (ii) not cheat or use external/professional assistance, and (iii) allow the authors to use images, audio, text, and video generated during the competition for dissemination via conferences, publications, and podcasts by the authors and also for future event promotion.

4.4 *Measurement*

The organizers used pre and post event surveys, which had a combination of open- and close-ended questions. The pre-event survey asked students how they prepared for the event, what their expectations were, how their groups were formed, and what type of cybersecurity experience they had. The post-event survey asked students about their experiences with the overall RSEC and a summary of their strategies, divisions of labor, and how effective they thought their strategies were. Students also appeared as guests in the authors' podcast to share their thoughts about the event, which offered further insight into their experiences. All data shared in this paper are based on the experiences of the 14 teams who completed all parts of the competition.

5 Student Experiences

5.1 *Team Dynamics*

This subsection details students' team dynamics, with experiences regarding their group formation, preparation, and cohesion.

As reported in the pre-event survey, 83% of participants had never worked with their group in any previous cybersecurity exercise. Further, 81% only knew their teammates for less than one year before forming their team, with 35% knowing them for less than three months. Teams were formed through a variety of ways, including by random assignment, through school clubs, shared teachers/advisors, and in some instances, with existing friends.

Before the RSEC started, participants gave feedback on how well prepared they felt their team was. They later reflected on their preparation after the RSEC ended.

Before the event, only 2% of participants reported that they felt 'not prepared at all' and only 11% of participants felt 'completely prepared'. The rest of the participants were 'a little, 'fairly', or 'somewhat' prepared, with the majority of all participants (51%) reporting feeling 'somewhat prepared'. Reflecting after the event about how prepared they were, the 'completely prepared' option dropped down to 8%, while the 'not prepared at all' option increased to 5%, indicating that there were unexpected events that some teams could have prepared for more.

Group cohesion, including communication skills, was an important factor in each team's dynamic that was relevant throughout the duration of the competition, as the RSEC required a high level of teamwork. Some groups reported working well together and demonstrating good cooperation which improved the team's performance and efficiency. For instance, one team reported "we were all open to feedback from our group members [...] This definitely increased the quality of our work." However, not all groups worked well together. For example, one team that struggled to remain cohesive noted that "It felt like we were on different pages, especially as far as communication goes." Communication was a key factor in teams' successes. As the competition was hosted online, teams chose different ways to communicate with each other, with strategies that varied in platforms and frequency. Some teams maintained high levels of communication, such as one team who communicated and organized everything in their group folder through discord and set specific times for all members to join a call to discuss their findings and next steps.

Another aspect of group dynamics includes the team's division of labor. Sometimes the division of labor was based on each team member's skillset and strengths, so that, as one team noted, they could create "a plan that combined and utilized ideas and strategies from all [team members]." Some teams were multidisciplinary, influencing their division of labor to have the computer science students focus on the more technical information while other members led the negotiation strategy. Other teams similarly split tasks among their members, such as a team with some members who conducted OSINT while the others used the OSINT findings to work on their final report. This division of labor helped some teams to remain cohesive, as they had to rely on each other to fully succeed. Indeed, one team noted that "we moved forward as a team [even though] we were all working on different tasks at the same time." This style of division of labor was also seen during the client meetings, as some teams had designated members who would speak on behalf of the team, while all of the decision making was still discussed together in the background. However, not all teams took the approach of splitting tasks. Some teams took the approach of working individually on the same task in which they all worked on OSINT on their own that they later combined, so that they could gather as much information as possible.

5.2 Strategies and Adaptations

Teams developed strategies for how to approach conducting OSINT, negotiating with the ransomware group, and adapting to hurdles during the negotiations.

Strategies for conducting OSINT on the ransomware group commonly included using resources such as the ransom note, the incident response team's descriptions of the attack, and the ATT&CK website. From the ransom note, teams were able to gather information including encryption algorithms and techniques, tactics, and procedures (TTPs) that the client noted to them. After gathering evidence from the ransom note, referencing the ATT&CK framework was a strategy that many teams relied on. With this, a common strategy was mapping the ATT&CK TTPs to online articles that matched other details, such as the language used in the note. In regard to conducting OSINT to accurately profile the ransomware group, teams struggled with the lack of information immediately available to them. There was minimal information provided to the teams, and the ransomware group was not forthcoming during the negotiations, making the OSINT process difficult to begin. Teams had to rely on small details of the ransom note and the information they discovered during the client meetings to lead them in their OSINT process. Some teams even took to translating the note's Russian greeting to understand the formality that the attacker was using and if it matched a group's modus operandi. As more information was discovered, some teams struggled with organizing all of their findings in a way that would make it easy to piece together the whole story. Many teams were able to overcome this struggle, such as one team who adapted by making "an entire spreadsheet listing the groups and software that they used to match the techniques that [they] identified in the attack."

Students utilized RASCLS for their benefit in day 2. One team's approach to the negotiation can be seen in the following example of part of an exchange that took place between the ransomware group (played by an industry representative) and a student team.

Team: We can pay in a few days time. Need to setup a BTC wallet just for your payment, and figuring out what is available while we have limited access

RW group: I really don't understand why you asked to talk to us then

RW group: you know monty python?

Team: We just need a little more time. I thought for sure I would hear back from our finance guy by now

RW group: look

Team: Is it possible to get our files decrypted if we pay a small amount up front? We want to work with you

RW group: no

RW group: you pay everything and then you get a decryptor

As seen in this short exchange, teams demonstrated the use of RASCLS. For example, the team demonstrated the principle of scarcity by saying that they could "pay in a few days time." Further, reciprocity and committed were utilized by the team telling the RW group that they were going "to setup a BTC wallet just for your payment." Reciprocity was also expressed by their willingness to "pay a small amount up front" while the use of commitment was also demonstrated in their statement "we want to work with you." Meanwhile, the team utilized the principle of authority by shifting the focus to their "finance guy."

Teams varied in their approach for the negotiations with the ransomware group. Some groups prepared in advance with pre-written statements that they could copy-and-paste from another window. The teams that took this approach were more prepared for the responses they would receive back from the ransomware group. This preparation meant that they could adapt more quickly and not have to think of a response on the spot. This allowed teams to brainstorm and discuss their strategy for negotiation ahead of time, and to anticipate potential directions of pushback. Some teams also tried to conduct OSINT during the live negotiation but ended up finding it to be too difficult to juggle both OSINT and convincing the ransomware group to accept their offer. Meanwhile, other teams did not go into the meetings with much of a strategy. One team noted that it "could have done a better job at exploring ideas before jumping into" the negotiations. This made it more difficult for the team as their ideas were not thought through as thoroughly, which made it harder to adapt moment by moment. When asked about what the teams could have done to better prepare for the RSEC, teams several mentioned developing a better strategy or approach for negotiations.

5.3 Feedback on Five Main Components

During the post-event survey, students ranked each component of the competition from easiest to hardest as well as favorite to least favorite. The five components they ranked included OSINT, client meetings, ransomware negotiations, the formal debrief, and the report writing. Most students (62.5%) noted that the ransomware negotiations were the most difficult of the five tasks. In some instances, that perception of difficulty was due to frustration. As one competitor noted, "While negotiating, the attacker would always avoid the questions," making it harder for the team to learn any additional information. Others described the difficulty of the negotiations in relation to convincing the attackers to lower the ransom price. Using SE strategies, one competitor reported that their team "tried to convince them to lower the price as much as possible as well as get them to give up information about their group but whatever plans we had for them [...] wouldn't work because their only goal in mind was the money." Negotiating with the ransomware group gave students the chance to face hurdles in a live interaction with minimal time to decide how to proceed; as such, one team noted that "the way the roleplay played out was challenging to adapt to." Most teams had strategies prepared to engage with the ransomware gang, but the live interaction rarely went according to plan: "Every scenario changed so we always have to come up with another plan on the spot," because the ransomware group was "extremely persistent."

Most contestants (40%) were also in agreement that the client meetings were the second most difficult task. Similar to the ransomware negotiations, the client meetings required competitors to adapt on the spot with little time to prepare a response. Unlike the ransomware negotiations, the competitors also had to keep in mind their body language, appearance, and tone of voice during the client meetings, as they were face-to-face. Interestingly, after its ranking as the second most difficult task, the next most common ranking for the client meetings was as the second easiest task. Meanwhile, most competitors ranked the easiest tasks to be either creating/presenting the formal debrief (30%), report writing (25%), or conducting OSINT (22.5%).

While some competitors found the OSINT process to be on the easier side, some teams still struggled with conducting OSINT, with 25% of competitors ranking it in within the top two most difficult tasks. Some of this difficulty came from teams not having complete information. For instance, one team reported that "it was difficult to ... attribute the attack to a group/strain with 100% certainty, since we were unable to look at any encrypted files [or] other key features." Similarly, another team noted that "conducting OSINT on the ransomware group in this scenario was fairly difficult, as we were working with a limited amount of information," as they would have to do during a real ransomware attack. Nevertheless, the teams practiced utilizing SE to gain more information.

Despite most competitors ranking the ransomware negotiation as the most difficult task, it was also most ranked as the favorite task (37.5%). This indicates that some students were not put off by the challenge. After the ransomware negotiations, the overall next favorite task was OSINT. In fact, 75% of respondents ranked conducting OSINT as either their first or second favorite task. One student reported that "conducting OSINT for our group was incredible and really what made this competition such a learning experience." Another student noted, "there was just so information that made me feel passionate in a way I never would have predicted when I initially joined this competition."

Meanwhile, the overall least favorite task was report writing (37.5%). Much of the dislike of this task came from the amount of time, effort, and organization that was required. One team, who reported spending 7 h on it, said they "can't express how audibly tired we were in trying to have the report finished by 11:59 PM". While many competitors said the report writing was among their least favorite tasks, 20% of competitors said that it was among their top two favorites. This may be due to the lack of stress that they associated with the live interactions, or perhaps the teams had good organization, planning, and time management skills.

6 Training the Next Generation Workforce

6.1 Relevance of Social Engineering to Cybersecurity

Many students stated that they were used to the hard sciences and lacked people and SE skills. As one student noted: "Usually I deal with the technical aspects like machine learning or building your own network, but here, we're trying to engineer a person and I had not dealt with any of that." Another student echoed this statement: "This event allowed me to realize that along with the technical components to social engineering, the 'soft skills' (persuasion techniques, use of psychology) are also just as important, if not more important."

Students were interested in red teaming and penetration testing as future careers and realized that one aspect of this job included SE: "A part of red teaming is social engineering, and I'm only doing technical stuff—I stare at the computer screen the whole time, I don't really talk to anyone, so it's a bit hard if I want to red team—I can't talk two words to a person, so that's what drew me to this competition." Another student said that the RSEC "presented an environment in which I could practice my SE skills and find out which strategies work and which don't. This helped me a great deal in learning about being a social engineer, and I believe this has been a learning experience for all of us." Yet another student said that opportunities like RSEC "take you so much farther than a traditional classroom would."

Overall, students felt SE was relevant to a career in cybersecurity as shown in Fig. 1, and from a sentiment captured by this student's statement: "Many students used the competition as part of building their skill sets. One student said "I wanted to learn, I wanted to get experience, to put it on my resume. I wanted to learn and grow in cybersecurity."

Students also felt their confidence in being a social engineer improved after competing in the event, as shown in Fig. 2. As another student stated, "This event

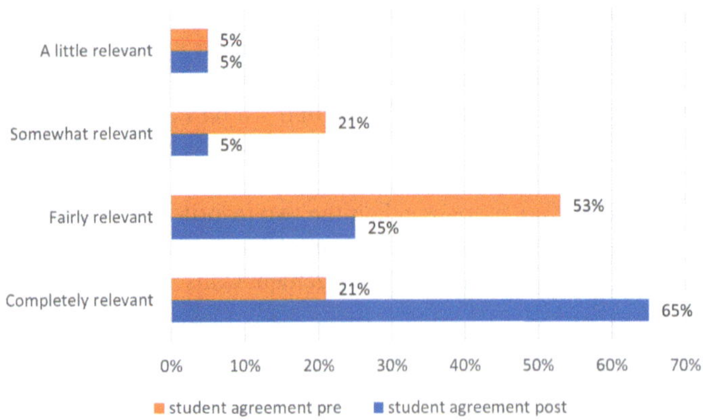

Fig. 1 Pre-post students' perspectives on SE relevance

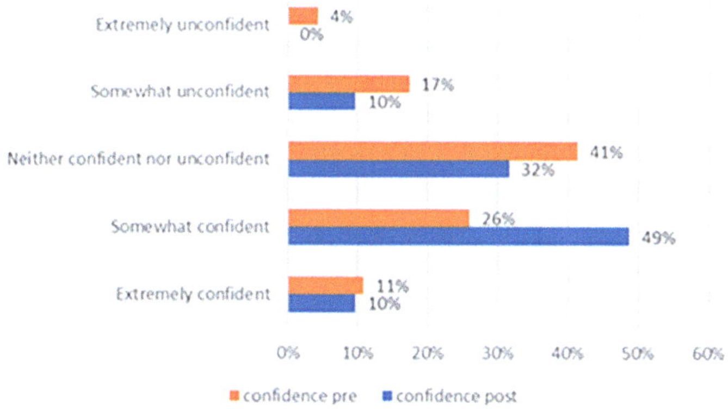

Fig.2 Pre-post students' confidence in being a social engineer

improved my understanding/ability/confidence in being an effective social engineer by teaching me how to be professional and how to come up with other plans on the spot." Students also felt that the event gave them a good overall experience of SE: "This event was awesome at showing me how it would be like if I was a social engineer. I feel that this event was a great first competition for me and provided me a solid introduction into the world of social engineering. I like how well the feedback was in providing me ways to improve and what to focus on."

6.2 *Experiencing the Art of Ransomware Negotiation*

Ransomware attacks have been increasing and ransom demands often reach seven figures [19]. Even if companies have backups to restore their encrypted data, they choose to pay the ransoms to prevent criminals from leaking it online [19]. This data extortion component, the need to delay payments, and the need to demonstrate that every possible approach was tried to protect data, has increased the need for ransomware negotiation services by companies, despite the U.S. Department of the Treasury's Office of Foreign Assets Control's (OFAC)'s advisory against making payments to entities on the U.S. sanctions list [19].

Professional negotiators have stated that the ability to communicate effectively with criminals is a "soft skill … that is not easily taught" [19, p. 3]. Negotiators need to get a feel for the threat actor and a good read on the tone of the conversation, which can be tricky, especially when criminals are nonnative English speakers [19]. They must establish trust with criminals by being upfront about who they are and what they are doing [19]. Furthermore, negotiators must assess the criminals and understand their characteristics and track record to develop a profile and determine whether they are 'credible' (i.e., provide decryptors and not leak data) and 'reasonable' (i.e., are open to negotiations) [19, 20].

Indeed, the RSEC exposed students to each of these components, giving them a chance to develop these 'soft skills', SE strategies, and psychological persuasion principles. As one student said, "the social aspect of ransomware and SE is underrepresented in today's society. I thought it was a good exposure to what real ransomware negotiations are going to be like." Specifically, students had to demonstrate these various abilities via:

1. Exhibiting professionalism and tactical empathy in speaking with the client, while simultaneously probing for information. Also important, was continuously updating clients effectively and tactfully as the real-time, stressful negotiations unfolded.
2. Developing a profile of the threat actor based on information provided by the incident response team, client responses, and OSINT.
3. Establishing trust with the criminals to contain the situation and not let it escalate (e.g., data extortion).
4. Representing the client's needs by attempting to (i) lower ransom demand, (ii) buy time for incident response, (iii) get evidence of decryptors, (iv) and determine who is behind the attack.

Thus, students had to have two conversations: one with their client and one with the ransomware group. One student captured this balance by stating "I learned how to negotiate, how to be curious and look into information that can lead into something, [and] how to make a deal that satisfies two parties as once."

6.3 Using ATT&CK for Attribution

The ATT&CK framework introduces individuals to the practice of modeling the cyber behaviors of threats, specifically for analyzing why and how adversaries make decisions and take corresponding actions [21, 22]. The ATT&CK framework allows students to understand tactics and techniques used by adversaries, which can be paired with defensive countermeasures to inform security decisions and operations [22].

Students competing in RSEC were given information from the incident response team, which was used primarily for attribution. For instance, both high school and undergraduate teams used ATT&CK technique IDs to narrow down which ransomware strains might have been used to infect the lab, which was then used to determine who might be behind the attack. Students cited some combination of T1005 (process injection), T1027 (obfuscated files or information), T1059 (command and scripting interpreter), and TA007 (discovery) as being used in strains such as Conti (S0575), Maze (S0449), WannaCry (S0366), REvil (S0496), and Ryuk (S0446). Students then used the ATT&CK framework to identify which groups were associated with those strains and narrowed down the criminal groups to: Wizard Spider (G0102), FIN6 (G0037), Lazarus Group (G0032), FIN7 (G0046), and Gold Southfield (G0115).

Students were thus able to understand how information provided by incident response could be mapped to the ATT&CK techniques, which helped identify strains, and thus helped identify a potential list of ransomware groups that may have targeted the lab. This helped students develop threat profiles and game plans for their negotiations.

Many students stated that they not only learned about how to use ATT&CK, but also enjoyed using it. One student noted: "I had heard of MITRE [ATT&CK] before but never used in a situation when you are finding information about the adversary. I had no idea how to do research on ransomware, but then I started digging into ATT&CK." While students who participated in RSEC may not take on the role of ransomware negotiators and use the ATT&CK framework for the specific purpose of attribution to develop threat profiles, comprehending the power and value of the ATT&CK framework and how to use it effectively are important skills that are required in the cybersecurity workforce.

6.4 Social Engineering and the NICE Framework

The NICE Cybersecurity Workforce Framework provides a comprehensive list of cybersecurity Knowledge and Skills to perform various cybersecurity tasks [23]. The framework provides educators with a "consistent way to develop learning across fields and helps in the development of credentials ... to consistently describe learner capabilities across fields and in roles that may pull from multiple domains" [24, p.1].

Knowledge is a "body of information applied directly to the performance of a function" [23, p. 5]. The RSEC can be mapped to the following Knowledge (K) IDs and descriptions:

K0110_1- Knowledge of adversarial tactics, tools, and techniques. During the orientation and the 3-day competition, students were exposed to, and had to apply, the ATT&CK framework. Students had to use the information they received about adversarial tactics and techniques throughout the event to identify ransomware groups.

K0162- Knowledge of cyberattack actor characteristics and *K0550- Knowledge of target characteristics.* The ATT&CK attribution helped students identify the various groups Wizard Spider (G0102), FIN6 (G0037), Lazarus Group (G0032), FIN7 (G0046), and Gold Southfield (G0115) who may have targeted the lab. Students could then research each of these to learn more about their structure and organizational dynamics.

K0177- Knowledge of cyberattack stages and *K0603- Knowledge of threat behavior*. When students played the adversarial role during the application process, they had to do reconnaissance (OSINT) on the lab to understand its operations and earnings, how to target the lab, and the amount of ransom to demand. When they took on the role of the negotiator during the live competition, they used the ATT&CK framework to identify how the various techniques shared by the incident response team mapped onto the different stages of the cyberattack. Thus, they developed adversarial and defender knowledge and understanding.

K0292- Knowledge of incident, event, and problem management policies and procedures. During Days 1 and 2 of the competition, students had to interact with the clients to understand what their incident management policies and procedures were around handling a ransomware attack.

K0426- Knowledge of dynamic/deliberate targeting principles and practices. During the orientation, students were provided real-world examples of ransomware and how they occurred. This knowledge helped the students learn about current patterns in ransomware and how these attacks were evolving.

In the pre- and post-assessment surveys, students felt that their knowledge about SE had improved. Participants who had never learned about or engaged in a formal SE exercise also gained the fundamental knowledge of SE, such as the psychological persuasion principles. The interactive aspect of learning about SE through RSEC gave students a deep understanding of ransomware and SE because, as this team describes, "this event [...] gave me a better understanding of what goes on in real life and not just what happens in a textbook scenario."

Skills involve the "application of tools, frameworks, processes, and controls that have an impact on the cybersecurity posture of an organization or individual" [23, p. 5]. The following cybersecurity skills were present in the RSEC and can be mapped to the following Skill (S) IDs and descriptions:

S0052- Skill in performing social engineering. Students used OSINT when playing the adversary and determining attack attribution. They also used various psychological persuasion principles when interacting with clients and negotiating with the ransomware group.

S0070- Skill in communicating effectively. Students had to first communicate effectively with the client by demonstrating tactical empathy, asking the right questions to understand the situation and lab policies, and relaying information from the negotiations back to the clients throughout the competition as the situation unfolded. Students also had to negotiate effectively and maintain a good balance between satiating the ransomware group while representing their clients.

S0244- Skill in managing client relationships. When students interacted with the client, they had to understand what their needs and capabilities were to represent them effectively during the ransomware negotiations. One student stated: "Dealing with the client interactions, no matter what position you are in cybersecurity, is such an important skill, and still one that I lack...if you're doing consulting, or [are] a cybersecurity engineer, you have to communicate with the rest of the company and all the C-level people, so being able to handle all that is incredibly important. I saw how valuable and difficult it was—this is a really [important] skill I need to build up on."

S0037- Skill in preparing reports, *S0249-Skill in preparing briefings*, and *S0344-Skill in presenting to an audience.* On the third day of the competition, all teams had to deliver a formal written report to the client and an oral report. Students had to defend their work, answer questions regarding the outcome of the negotiation, status, and recommendations to prevent ransomware attacks in the future.

Student experiences from the post survey and podcast interviews suggested that they felt their confidence in these skills improved. One student noted: "I learned some tools to effectively use OSINT and how to put it all together in a report. This definitely boosts both my ability and confidence when it comes to the tools and techniques applied for this competition. Also, I became a bit more comfortable with reporting to clients and negotiating with bad actors, although I hope I won't have to do the latter again (in the real world, that is)."

The RSEC's premise of ransomware negotiations on behalf of the lab provided a safe, fun, and ethical platform for students to gain a good knowledge base about adversaries and their behaviors, ransomware properties, the ATT&CK framework, and the target's policies and procedures. Students also developed an assortment of skills such as performing SE and communicating effectively in multiple settings (real-time cyberattack, formal debriefs, oral and written communication). While the RSEC was a simulation of a ransomware attack and negotiation, the knowledge and skills are portable as students can bring them in the workforce and apply them in different settings.

7 Conclusion

This paper shared findings from, and the relevance of, a ransomware and social engineering competition. It offered a safe and ethical platform for students to have a unique experience in the area of real-time ransomware negotiations, thereby developing important skills, such as speaking with clients, understanding their needs, negotiating with adversarial groups, using ATT&CK for attribution, writing formal reports, and presenting findings and making recommendations in professional settings. This competition thus showcased the many different aspects of, and skills needed in, the cybersecurity workforce as shown by mappings to the NICE Framework. By working with non-profit, government, and industry subject matter experts, more experiential learning events can be developed to enhance student learning about social engineering and train the next generation of cybersecurity professionals.

Acknowledgements This work was supported by the National Science Foundation Award # 2032292.

References

1. Security Boulevard (2023) Cyberattacks 2022: Phishing, ransomware & data breach statistics. https://securityboulevard.com/2023/04/cyberattacks-2022-phishing-ransomware-data-breach-statistics/
2. Brooks, A.: (2023) Cybersecurity trends & statistics; more sophisticated and persistent threats so far in 2023. https://www.forbes.com/sites/chuckbrooks/2023/05/05/cybersecurity-trends--statistics-more-sophisticated-and-persistent-threats-so-far-in-2023/

3. Rege, A., Bleiman, R.: (2021) Collegiate social engineering capture the flag competition. In 2021 APWG symposium on electronic crime research (eCrime). IEEE, pp. 1–11
4. Greig, J.: (2021) Average organization targeted by over 700 social engineering attacks each year: report. https://www.zdnet.com/article/average-organization-targeted-by-over-700-social-engineering-attacks-each-year-report/
5. Crane, C.: (2023) 12 Social engineering statistics that will make you question everything. https://www.thesslstore.com/blog/social-engineering-statistics/
6. Galov, N.: (2023) 17+ Sinister social engineering statistics for 2023. https://webtribunal.net/blog/social-engineering-statistics/
7. Social Engineer Community (2024) Social engineering community. https://www.se.community/
8. Layer8.com (2020) Layer8 conference: about us. https://layer8conference.com/about-us/"
9. Social-Engineer.org (2020) Social engineering training. https://www.social-engineer.com/socialengineering-training/
10. SANS (2020). SEC567: Social engineering for penetration testers. https://www.sans.org/course/social-engineeringforpenetration- testers
11. Layer8.com (2020) The layer 8 podcast. https://layer8conference.com/the-layer-8-podcast/
12. OSINTCurio.us (2020) OSINT videos and podcasts. https://osintcurio.us/osintvideosandpodcasts/
13. Rege, A., Bleiman, R., Williams, K.: (2024) The relevance of social engineering competitions in cybersecurity education. In: Proceedings from the IEEE Cyber Science Conference
14. Krebs, B.: (2022) Conti ransomware group diaries, Part III: Weaponry, 2022. https://krebsonsecurity.com/2022/03/conti-ransomware-group-diaries-part-iii-weaponry/
15. Lineberry, S.: The human element: the weakest link in information security. J. Account. **204**(5), 44 (2007)
16. Mitrecorp (2023) MITRE ATT&CK. https://attack.mitre.org/
17. Quito, A.: (2022). What is tactical empathy and how can it help in negotiations at work?. https://www.weforum.org/agenda/2022/01/tactical-empathy-key-workplace-negotiations-voss/
18. Cialdini, R.: (1984) Influence: the psychology of persuasion. Quill/William Morrow
19. Wright, R.: (2021) Ransomware negotiations: an inside look at the process. https://www.techtarget.com/searchsecurity/feature/Ransomware-negotiations-An-inside-look-at-the-process
20. Constantin, L.: (2021). How ransomware negotiations work. https://www.csoonline.com/article/3607689/how-ransomware-negotiations-work.html
21. Mitrecorp. (2018) MITRE ATT&CKcon 2018: ATT&CK as a teacher. https://youtu.be/4s3pZirFCPk
22. Rege, A., Williams, J., Bleiman, R., Williams, K.: (2022) Students' application of the MITRE ATT&CK® framework via a real-time cybersecurity exercise. In: Proceedings from the 22nd European Conference on Cyber Warfare and Security
23. National initiative for cybersecurity education (NICE) (2020). National initiative for cybersecurity education (NICE) cybersecurity workforce framework. https://nvlpubs.nist.gov/nistpubs/SpecialPublications/NIST.SP.800-181.pdf
24. NICE (2022) NICE framework knowledge and skill statement review: an introduction and summary of updates. https://www.nist.gov/system/files/documents/2022/04/18/NICEFramework_KSreview_intro11apr2022.pdf

Exploring DTrace as an Incident Response Tool for Unix Systems

Joe Duin, Sean McKeown, and Mwrwan Abubakar

Abstract Critical National Infrastructure (CNI) is often the target of sophisticated and sustained cyber attacks perpetrated by advanced threat actors with considerable resources. These attacks can lead to interruptions in core services such as energy and water supplies, transportation, healthcare, and telecommunications. The effective and swift remediation of such attacks is contingent on the respective Digital Forensics and Incident Response (DFIR) professionals possessing the appropriate tooling and resources for the target environments. However, the Unix systems which often run critical infrastructure are poorly accommodated in comparison to their Windows and Linux counterparts. This paper seeks to expand the options available to DFIR analysts on Unix systems by exploring the potential for DTrace to serve as an Incident Response utility. DTrace is included in many Unix operating systems by default, while also having support for Linux, Windows and macOS, making it a useful pre-packaged solution. We explore the utility of DTrace, and the visibility it provides into the OS and kernel, through a variety of proof-of-concept case studies based on tactics and techniques in the MITRE ATT&CK framework. We find that DTrace's functionality lends itself well to a real-time monitoring and probing solution for Unix systems, which could potentially form the basis of an Endpoint Detection and Response (EDR) solution to revolutionise Incident Response on such platforms.

J. Duin (✉) · S. McKeown · M. Abubakar
Edinburgh Napier University, Edinburgh, Scotland
e-mail: joeyduin@acm.org

S. McKeown
e-mail: s.mckeown@napier.ac.uk

M. Abubakar
e-mail: m.abubakar@napier.ac.uk

M. G. Jaatun et al. (eds.), *Proceedings of the International Conference on Cybersecurity, Situational Awareness and Social Media*, Springer Proceedings in Complexity,
https://doi.org/10.1007/978-981-96-0401-2_10

1 Introduction

Security breaches have seen a sharp increase in recent years, with ransomware attacks alone surging by 82% between 2020 and 2021 [12]. Attacks are also becoming more sophisticated [31], targeting companies of all sizes, with financial institutions being a popular target for both e-crime organisations and nation states alike [5]. However, attacks are not committed solely for financial gains, with the USA's Department of Homeland Security (DHS) [14] noting that nation state sponsored groups also target Critical National Infrastructure (CNI) to gain information and control important industrial control systems in the energy, water, aviation and manufacturing sectors, as well as the public telecommunications infrastructure. This is particularly concerning as skilled Advanced Persistent Threat (APT) groups are adept at avoiding detection mechanisms, and there is a wider trend towards malware-free approaches amongst adversaries more generally [12], meaning that endpoint security is less effective, and adversaries can go undetected for increased durations.

As breaches seem inevitable, it is of critical importance that Incident Response (IR) teams have the tools to rapidly identify the scope and nature of a given breach [11, 42]. Such IR teams rely on software to provide visibility into a victim's network in order to facilitate the investigation, however the most popular Endpoint Detection and Response (EDR) tools from Microsoft and Crowdstrike [16] do not offer uniform support across commonly deployed server operating systems, with functionality for Unix (e.g. BSD, Solaris), and Linux lagging behind that of Windows. This leaves defenders at a disadvantage, particularly given that Unix/Linux systems make up over 80% of Web servers [44], and there are examples of Unix systems being exploited in attacks against the telecommunications industry [22, 23, 35]. E-crime groups such as Lockbit have also recently shown more interest in attacking Unix systems, with ransomware binaries targeting FreeBSD systems [2]. While the introduction of the `extended Berkeley Packet Filter` (eBPF) [39] in Linux in recent years has improved Linux support somewhat, no such advances have been made for Unix.

In order to work towards better IR tooling for Unix, this paper explores the DTrace tool as a potential observation technology, which would allow for expanded tooling support on such platforms. DTrace is included in several Operating Systems (FreeBSD, macOS, Solaris), and with prior configuration it can be installed on Windows. We chose here to focus on an older version of Solaris (10) to represent the worst-case support, and a newer version of FreeBSD (12) for our best-case support configuration. The primary contributions of this work are: (i) An overview of DTrace's capabilities as an Incident Response tool; and (ii) An evaluation of DTrace with respect to both common Incident Response tasks [9], and the MITRE ATT&CK framework [33].

The remainder of this paper is structured as follows: Sect. 2 discusses background on Incident Response, related tooling, and particular considerations for Unix systems. Section 3 provides more information about the origins and capabilities of DTrace. To further understand these capabilities, proof of concept experiments are described in Sect. 4, with related findings and discussion in Sect. 5. Finally, conclusions and future work are presented in Sect. 6.

2 Background

2.1 Incident Response

While the field has evolved considerably since the inception of the original Computer Emergency Response Team (CERT) in 1988 [46], the overall approach to Incident Response remains largely the same: security incidents are investigated; intruders identified and evicted; and recommendations on security procedures are given to help prevent future attacks. However, each investigation is unique and different archetypes may involve different strategies within each of these stages. For example, after spending days to observe the network and gain elevated privileges, ransomware such as Ryuk [4] begins data encryption within minutes post-deployment, such that early detection in the initial exploration phase is critical to avoid large-scale damage. However, an entrenched Advanced Persistent Threat may be operating covertly over a long timeframe and may require more in-depth assessment of compromise before taking action. In either case, triage and EDR tools are essential for providing visibility across a network in order to detect and assess the levels of damage and intrusion [7, 13].

Applications, servers, and security tools all generate large quantities of information, necessitating the centralisation and organisation of events and logs generated by these solutions. A Security Information and Event Management (SIEM) [26] is often used to provide a single, queryable, interface to this data, which includes logs from EDR and triage tooling, facilitating a wider spatial and temporal view of a network. That is, while EDR and triage tooling provide vertical depth of analysis, a SIEM provides horizontal width [27], with overviews and dashboards allowing for the rapid detection of Indicators Of Compromise (IOCs) for known threats which would be difficult without this higher-level data management. A wide and longitudinal view of networked resources also facilitates active threat hunting [25], which is particularly useful when adversaries evade detection by EDR systems.

2.2 Related Tools for Unix-Like Systems

GRR Rapid Response (GRR) is a Python-based Incident Response framework, created by Google, which can be used to conduct live forensics on remote systems. GRR uses a client-server architecture where agents are installed on endpoints and managed from a central server. The agent supports Windows, Linux and macOS but notably lacks support for other Unix flavours such as Solaris AIX, HP-UX, and the BSDs [17]. GRR primarily focuses on the aggregation of host-based forensic artefacts such as logs and makes them available through a Web-based interface. While analysts can investigate individual machines, GRR's power is in the ability to hunt across many systems.

Osqery [1] provides OS system instrumentation for Linux, macOS, and Windows, but does not support Unix systems. It provides an SQL interface for a wide range of real-time artefacts, but does not possess analysis functionality. Existing work has used Osquery as the foundation for Linux EDR solutions [3, 15], in one case combining both Osquery with GRR [40]. In this latter work, the solution was evaluated for its coverage against the MITRE ATT&CK framework [33], finding that platform specific queries to Osquery yielded better results.

The Unix-like Artifacts Collector (UAC) is a tool for Incident Response which utilises native applications in order to collect live response triage data. Unlike GRR and Osquery, UAC is a run-once application that creates an archive with the collected live response data for upload and ingestion into further tooling for analysis. UAC aims to support macOS, the BSDs, and Linux, as well as major Unix platforms such as Solaris and AIX [29]. As an example, when run on a Solaris system, UAC will capture the artefacts from common commands such as `mount`, `ifconfig`, `ptree`, `dmesg` etc. While UAC is a useful tool for collecting live response data, this kind of information tends to be surface level, rather than in-depth. Together with its run-once behaviour, UAC is better geared towards triage of specific devices, and is inappropriate for a real-time monitoring system. However, UAC appears to be one of the few examples of security systems providing Unix support, though Linux support of some kind appears to be typical for leading commercial EDR solutions, such as Crowdstrike Falcon, Microsoft Defender and SentinelOne.

3 DTrace

Like UAC, DTrace is a cross-platform observation technology, providing instrumentation points across the OS, orchestrated by the D programming language. In addition to Windows/Linux/macOS, strong support is provided on FreeBSD and Solaris/Illumos, besting the Unix support of other technologies. DTrace also allows for the aggregation of underlying data and the ability to perform speculative tracing, making it flexible and useful for facilitating data-based decision making. By default, it is safe to use in production [10], though some explicit unsafe operations are permitted.

DTrace has had relatively little usage in the security space despite having significant potential. Some prior work has utilised DTrace to gather system logs [28, 32], but does not comment on its functionality in detail. Other work has demonstrated its use for CPU-level integrity checking [45], and for use as a Host-based Intrusion Detection System (HIDS) on Mac OS X [6]. It is clear from these latter works that DTrace is capable of providing very detailed system introspection, such that it is prudent to further explore its potential here.

3.1 DTrace Usage and Visibility

DTrace takes inspiration primarily from the Awk and C programming languages. DTrace can be invoked either using one-liners on the command line or by executing a D script from a file. The structure of a DTrace command line is shown in Listing 10.1 and D script structure is shown in Listing 10.2. The probe naming convention has the format `provider:module:function:name`, and components between colons may be omitted and can contain wildcards such as `*` to match patterns. The predicate is an optional expression which evaluates to either true or false such as `/uid == 101/`, when true, the subsequently specified actions are executed. Actions are statements followed by semicolons, for example `{ x++; printf("x is %d", x); }` [21].

```
dtrace -n 'probe /predicate/ { actions }'
```

Listing 10.1 DTrace command line example—shows the structure of a DTrace command [21].

```
#!/usr/sbin/dtrace -s
probe
/predicate/
{
  actions /* comment */
}
```

Listing 10.2 D script example—shows the structure of a D script [21].

DTrace has the concept of built-in variables, as shown in Table 1, which expose much of what is happening on the system. For example, which processes are running on the CPU, their Process IDs (PIDs), parent PIDs, timestamps, etc.

Arguments vary between probes, but take on three general forms: (i) Built-in D types, such as `uint64_t`; (ii) DTrace specific structs, such as the IP metadata used by the `ip` provider; and (iii) Original arguments used by related kernel functions, as seen with the `fbt` provider which observes kernel functions in the `entry` probe, and returns memory offsets in the `return` probe. As kernel functions can change over time, or be deprecated, this third form is considered to be 'unstable' in long term use. Therefore, where possible, stable alternatives which do not rely on kernel arguments directly should be used by the DTrace user.

In addition to the built-in variables, DTrace has functions for interrogating the system such as `trace` which takes an argument and prints its value, or `printf` which can print any number of formatted values. For more hands-on observation, `tracemem` can be used to print a region of memory, and stack tracebacks can be dumped by the functions `stack`, `ustack` and `jstack`.

Aggregations are a variable type used for summarising data. Aggregating functions are predominantly simple arithmetic operations: `count` tracks the number of times it is called, `sum` is the total value, `avg` the average, `min` and `max` can be used to find the smallest and largest value, while `stddev`, the standard deviation. Visualisations can be created with the functions `quantize` to create a power-of-

Table 1 DTrace built-in variables [21]

Variable name	Type	Description
`arg0...arg9`	uint64_t	Probe arguments; content is provider-specific
`args[]`	*	Typed probe arguments; content is provider-specific
`cpu`	processorid_t	CPU ID of the current CPU
`curpsinfo`	psinfo_t	Process state info for the current thread
`curthread`	kthread_t	Operating system internal kernel thread structure for the current thread
`errno`	int	Error value from the last system call
`execname`	string	The name of the current process
`pid`	pid_t	Process ID for current process
`ppid`	pid_t	Parent process ID of the current process
`probeprov`	string	Provider name of the current probe
`probemod`	string	Module name of the current probe
`probefunc`	string	Function name of the current probe
`probename`	string	Name of the current probe
`stackdepth`	uint_t	Current thread's stack frame depth
`tid`	id_t	Thread ID of the current thread
`timestamp`	uint64_t	Elapsed time since boot in nanoseconds
`uid`	uid_t	Real user ID of the current process
`uregs[]`	uint64_t	The current thread's saved user-mode register values
`vtimestamp`	uint64_t	Current thread's on-CPU time in nanoseconds
`walltimestamp`	uint64_t	Nanoseconds since epoch (January 1, 1970)

two distribution and `lquantize` which populates a linear frequency distribution. Aggregations can be printed using the `printa` function while `normalize` can be used to divide an aggregation by a value which is a useful operation for converting values to, for example, kilobytes. Additionally, the `trunc` and `clear` functions can truncate an aggregation to a certain number of keys specified or clear the values, leaving only the keys.

We identify specific DTrace instrumentation areas that initially appear interesting from a DFIR perspective, and discuss the alignment with the MITRE ATT&CK Framework in Sect. 4.3.

4 Approach

In order for DTrace to be considered for use in Incident Response engagements, the scope of its capabilities was investigated. DTrace's ability to gain unprecedented visibility into the system, particularly on Unix devices, makes it likely that novel sources of forensically valuable information are discovered. A full DTrace Incident Response approach should contain tooling to take advantage of DTrace's unique capabilities, both in terms of new functionality and that already present in existing frameworks. What we explore here is essentially a proof of concept of its utility and to demonstrate the value of DTrace to defenders, rather than enumerating every anticipated Incident Response capability of DTrace.

4.1 Scoping: Hardware and Software Considerations

While DTrace is available in multiple Operating Systems, we chose to focus on Unix system support here to better understand deployment in the CNI space. To reflect a mixed range of DTrace probe support, Solaris 10 and FreeBSD 12 were chosen to represent older software with poor support and newer and more supportive OS, respectively. Some CNI systems may be running legacy software, such that understanding the minimal use case is important. To this end, virtual machines were utilised to run each operating system to facilitate snapshot generation for convenient repetition. Solaris 10 was run on VMware Workstation on Windows 10, while FreeBSD ran on Virtualbox on Linux. Future work should look to explore how well DTrace functions as an Incident Response utility on other platforms.

4.2 Investigative Procedure

The general approach taken for this work is described below:

1. DTrace Providers of interest were identified by categorising each provider against tactics in the MITRE ATT&CK framework. This involved manually inspecting documentation and creating initial preliminary scripts to observe provider outputs.
2. DTrace was instrumented to collect data from probes which are likely to contain relevant information.
3. Tooling was evaluated against a success criteria defined for the respective MITRE ATT&CK tactic.
4. DTrace output from each experiment was compared against other tooling on Solaris such as UAC or tooling available for IR on other platforms and the tradeoffs or novelty discussed.

DTrace scripts were developed in the D language, with the popular MITRE ATT&CK framework (discussed below) being used to ground the exploration of relevant case studies/use cases and the associated providers in DTrace. This focused approach reduced the list of potential probes from tens of thousands to a selection of 16 providers which are relevant for DFIR tasks in Solaris 10 and FreeBSD 12. This approach in referencing MITRE ATT&CK framework to ground exploration and findings is similar to the one used by Park et al. [40] to evaluate their Osquery/GRR solution.

4.3 MITRE ATT&CK Alignment and Instrument Area Selection

MITRE ATT&CK [34] is a widely recognised knowledge base of tactics and techniques used by threat actors, which was derived from empirical observations from real cyber incidents. All 14 tactics in the framework are depicted in Table 2, with the first two (Reconnaissance and Resource Development) not creating host-level footprints which could feasibly be observed by DTrace, and are therefore not in scope for this work.

A list of providers available in our Solaris 10 and FreeBSD 12 versions, and their alignment with the MITRE ATT&CK framework, is provided in Table 3, though not all of them are explored experimentally. This table is not exhaustive but highlights the most interesting providers from an IR perspective. It should be noted that coverage of the providers is not uniform, and while FreeBSD 12 has better support overall than Solaris 10, it still misses some providers that were present in 10. SPARC Platform-

Table 2 The 14 MITRE ATT&CK techniques [34], of which only the latter twelve are observable to DTrace, the two unsupported tactics are highlighted in grey

Techniques	Description
Reconnaissance	The adversary gathers information that can be used to plan future operations
Resource development	The adversary establishes resources which can be leverated in future operations
Initial access	The adversary gains access to the network
Execution	The adversary runs malicious code
Persistence	The adversary establishes long term access
Privilege escalation	The adversary gains higher-level permissions
Defense evasion	The adversary takes steps to evade being detected
Credential access	The adversary accesses account names and passwords
Discovery	The adversary maps out the environment
Lateral movement	The adversary gains access to additional computers
Collection	The adversary gathers data of interest to their goal
Command and control	The adversary communicates with compromised systems to control them
Exfiltration	The adversary moves sensitive data to a remote location
Impact	The adversary manipulates, interrupts, or destroys systems and data.

Table 3 DTrace providers on Solaris 10 and FreeBSD 12 and their anticipated alignment to MITRE ATT&CK tactics

Provider	Solaris 10	FreeBSD 12	ATT&CK tactics
`dtrace`	✓	✓	N/A
`profile`	✓	✓	Execution, impact
`fbt`	✓	✓	Privilege escalation, credential access, impact
`syscall`	✓	✓	Privilege escalation, credential access, impact
`sdt`	✓	✓	*
`mib`	✓		Exfiltration
`pid`	✓	✓	*
`sysinfo`	✓		Execution, impact
`vminfo`	✓		Execution, impact
`proc`	✓	✓	Execution
`sched`	✓	✓	*
`io`	✓	✓	Collection, exfiltration, impact
`ip`		✓	Discovery, lateral movement, C2, exfiltration
`tcp`		✓	Discovery, C2, exfiltration
`udp`		✓	Discovery, C2, exfiltration
`sctp`		✓	Discovery, C2, exfiltration

specific providers, such as `fpuinfo`, are not in scope of this paper due to the lack of available hardware.

Five common attacker scenarios have been identified to serve as proof of concept case studies, with the general areas, and their mapping to MITRE ATT&CK, depicted in Table 4. **Process monitoring** is particularly important for exposing running malware or misbehaving applications. This can be achieved by monitoring system calls (which is somewhat brittle to changes in kernel implementation) or via higher-level behaviours. We discuss Process monitoring in Sect. 5.1. **RAM introspection** is often avoided in IR due to difficulties in dumping memory, but DTrace allows for highly targeted memory interrogation, while also possessing the ability to manipulate data in real-time, as seen in Sect. 5.2. **Network monitoring** is essential, and while this will typically happen at a centralised point for a network, inspection at the host-level may prove valuable, as explored in Sect. 5.3. Many Incident Response engagements have

Table 4 Experimental instrumentation areas and their alignment with MITRE tactics and techniques

Instrumentation area	MITRE ATT&CK alignment (Tactic → Technique)
Process monitoring	Execution → Command and Scripting
Reading and writing memory	Persistence, credential access → Credential dumping
Network monitoring	Persistence (Generic)
Login and authentication monitoring	Credential access → Modify authentication process
Disk IO monitoring	Impact → Data destruction, data wipe

an element of credential theft, so monitoring **Login and Authentication** via DTrace should allow for the detection of compromised accounts, as in Sect. 5.4. Finally, we explore **Disk IO Monitoring** in Sect. 5.5, as it is a key component in detecting ransomware, data exfiltration, and other data focused attacks. We use two main scenarios to explore there various instrumentation areas: i) monitoring ssh / reverse shell processes, and ii) monitoring specific functions or system calls for malicious activity, such as disk wiping.

5 Implementation and Findings

Here we present the implementation and experimental findings for each of the instrumentation areas, each within its own subsection.

5.1 Process Monitoring

Process execution can be monitored in real-time, or logged via security and audit logging mechanisms for later analysis. Shell histories are also useful, however there are means of preventing, obfuscating, or deleting such logs. Audit logging on Solaris is provided by the 'Basic Security module' (BSM) while Linux and FreeBSD use Auditd. However, these mechanisms are not enabled by default and require configuration. DTrace can provide real-time monitoring of command execution which differs from the snapshot view provided by tools such as `ps`, `ptree` and UAC, and is therefore more comparable to the command execution logging of an audit log. DTrace is more flexible than typical audit logs, however, and has a consistent experience between platforms. The performance characteristics of DTrace have not been compared to Auditd or BSM, but prior work [39] suggests that eBPF is more efficient than Auditd.

In order to assess DTrace's process monitoring capabilities, shell commands and DTrace were executed in two opposing configurations on the Solaris 10 VM: (i) DTrace via SSH and user commands via the local console, and (ii) DTrace via the local console and user commands via SSH. The corresponding D script, **pexec.d** (Listing 10.3) is capable of capturing process execution in real time and associating it with a timestamp, its PID, PPID and UID, as provided by the `proc:::exec-success` probe. The script was left running in the background in an attempt to also capture non-interactive processes.

```
#!/usr/sbin/dtrace -s

#pragma D option quiet
#pragma D option switchrate=10hz

dtrace:::BEGIN
{
    printf("%-20s %6s %6s %6s  %s\n", "ENDTIME",
        "UID", "PPID", "PID", "PROCESS");
}

proc:::exec-success
{
    printf("%-20Y %6d %6d %6d  %s\n", walltimestamp,
        uid, ppid, pid, curpsinfo->pr_psargs);
}
```

Listing 10.3 pexec.d—Observe process execution. A version of this script first appeared in the DTrace book [21]. However, this version has been edited to switch out the `execname` for the full command, with arguments.

Output and Limitations

Example output from pexec.d is provided in Listing 10.4, and we note that while it provides clear command and argument details, some items are duplicated (identical PID and timestamp). One benefit of DTrace here is that it provides visibility into commands run via shell scripts, which, for example, would only show as './myscript.sh' in the shell history, rather than individual commands. However, ideally, a combination of the two approaches is useful, as the name of the script itself does not appear in the pexec output. Additionally, using this approach, DTrace does not log commands built-in to the shell, such that results differ between shell environments (e.g. `pwd` is built-in to `sh` but not `csh`).

```
ENDTIME              UID  PPID  PID  PROCESS
2022 Nov  3 21:32:13   0   919  978  ls
2022 Nov  3 21:32:17   0   919  979  cat log.txt
2022 Nov  3 21:36:07   0   982  983  svcs
2022 Nov  3 21:36:07   0   919  982  grep ssh
2022 Nov  3 21:36:14   0   919  984  format
2022 Nov  3 21:41:14   0   919  985  cat log.txt
2022 Nov  3 21:47:15   0   746  986  /usr/openwin//lib/xscreensaver/bin/xscreensaver-lock
2022 Nov  3 21:49:40   0   822  989  /usr/bin/ps -f -u root
2022 Nov  3 21:49:40   0   822  989  /usr/bin/ps -f -u root
```

```
2022 Nov  3 21:49:40   0  993   995 basename /usr/sbin/pprosetup
2022 Nov  3 21:49:40   0  993   996 dirname /usr/sbin
2022 Nov  3 21:49:40   0  993   998 dirname /usr/sbin
2022 Nov  3 21:49:40   0  822   990 /usr/lib/cc-ccr/bin/ccr -g cns.assetid
2022 Nov  3 21:49:40   0  822   991 /usr/bin/hostid
2022 Nov  3 21:49:40   0  822   992 /usr/sbin/smpatch get patchpro.patchset
2022 Nov  3 21:49:40   0  992   993 /sbin/sh /usr/sbin/pprosetup -@ patchpro.patchset
2022 Nov  3 21:49:40   0  993   994 dirname /usr/sbin/pprosetup
2022 Nov  3 21:49:40   0  993   997 dirname /usr/sbin
2022 Nov  3 21:49:40   0  993   999 sed s/:/ /g
2022 Nov  3 21:49:40   0  993  1001 /usr/jdk/latest/bin/java -version:1.5+ com.sun.patchpro
      .cli.PatchProSetup -@ pa
```

Listing 10.4 Output from pexec.d—The output from `pexec.d` while observing command execution via a terminal emulator and non-interactive processes.

DTrace introduced some modifications to the input and recorded commands. In particular, when utilising the `printf` command quotation marks are stripped in the DTrace output, which could be a relevant omission in some cases. Pipe characters ('|') also appear to be stripped in favour of presenting sequences of piped commands as individual items, such that some commands do not look complete in isolation. However, it may be possible to reconstruct pipelines by mapping PIDs to Parent Process Identifiers (PPIDs). The command '`grep ssh`' would spawn grep interactively however, it actually received its input from the `svcs` command via a pipeline. This means that the original command from Listing 10.4 is then '`svcs | grep ssh`'. Additionally, the script does not capture commands which failed to execute, which is achieved using the `proc:::exec-failure` probe. Failed commands may sometimes be useful to identify attacker intent, as they may try to execute commands they lack the permissions for, or attempt to probe available options when living off the land.

Command source is important as some processes, such as a netcat reverse shell, spawn a sister process which executes commands on its behalf. This means that simply tracing the parent process, of netcat in this case, would not make the adversary's shell commands visible. We return to this problem in Sect. 5.3.

Finally, in our experiments it was noted that system commands, such as `xscreensaver`, were captured by the probes. This kind of command execution would not be present in shell histories, though it would be logged in auditing systems, such as auditd and BSM, if configured. As DTrace is native to the environments in our tests, this gives it a slight advantage, though the script needs to be running at the time of execution, as it does not log historical commands.

5.2 *Reading and Writing Memory*

Many advanced adversaries utilise defense evasion techniques which enable them to minimise their disk footprint and remain entirely in memory [35]. Currently, memory analysis is a cumbersome task involving capturing memory dumps and performing forensics on them with specialised tooling, a task further complicated on Solaris and other Unix systems due to a lack of support in common tooling.

DTrace is capable of reading and writing memory, which has previously been used to alter the Unix `uname` command to display false system details [24]. Therefore we aim to experimentally identify whether this capability can be employed by incident responders for threat hunting and triage in order to bridge the gap between full memory forensics and current methods. Additionally, its memory manipulation capabilities may be useful in preventing attacks.

To represent the detection of a potential backdoor or Command and Control (C2), a netcat reverse shell, running in memory, was deployed on a FreeBSD 12 VM. Two primary objectives were explored here: (i) To observe netcat's `read/write` functions using the `pid` provider, and (ii) to intercept and modify a respective syscall to modify command behaviour in real-time. Both of these functionalities are provided in the **mem.d** (Listing 10.5) script.

The main steps of this small experiment are as follows:

1. Set up a netcat reverse shell.
2. Use the `pid` provider to intercept netcat's `read/write` functions.
3. Connect to the reverse shell and run `cat /etc/passwd`.
4. Attempt to edit this in memory to print an alternate file instead.

While DTrace can alter the contents of memory on the fly [20], there are risks involved in overwriting data in memory as it can lead to a program crashing or silent data corruption. In order to protect users from damaging the system, the `copyout` function (used for writing memory) is only available when the `#pragma D option destructive` is used.

mem.d

```
#!/usr/sbin/dtrace -s

/*#pragma D option quiet*/
#pragma D option destructive

dtrace:::BEGIN { printf("Tracing read() on PID %d...\n", $target); }

pid$target::read:entry
{
   tracemem(copyin(arg1, 256), 256);
}

pid$target::write:entry
/stringof(copyinstr(arg1)) == "cat /etc/passwd\n"/
{
   this->defang = (char *)alloca(sizeof (char)*11);
   this->defang[0] = '/';
   this->defang[1] = 't';
   this->defang[2] = 'm';
   this->defang[3] = 'p';
   this->defang[4] = '/';
```

```
    this->defang[5]  = 'd';
    this->defang[6]  = 'e';
    this->defang[7]  = 'f';
    this->defang[8]  = 'a';
    this->defang[9]  = 'n';
    this->defang[10] = 'g';

    copyout(this->defang, arg1+4, sizeof (char)*11);
    tracemem(copyin(arg1, 256), 256);
}
```

Listing 10.5 mem.d— Observe the `read` and `write` syscalls by a defined process and rewrite the command `cat /etc/passwd` to `cat /tmp/defang` within the `write` syscall.

The `mem.d` script uses three probes. Firstly, the `dtrace:::BEGIN` probe prints some information about the script and shows the PID being tracked. Secondly, the `pid$target::read:entry` probe traces the entry point of the `read` syscall to allow tracing of function memory, which will be used again in Sect. 5.3. The second argument to the `read` syscall, a pointer to memory, is traced to dump 256 bytes in hex to provide proof of concept visibility. Finally, `pid$target::write:entry` traces the entry point of the `write` syscall, it also takes a predicate. The predicate causes the probe to only trigger when the pointer value contains `"cat /etc/passwd\n"`. It then allocates and fills a buffer with the string `"/tmp/defang"`. This effectively rewrites requests for the *passwd* file to point at the *defang* file which contains dummy data, avoiding possible credential dumping. before returning its contents.

In this limited experiment, the command '`cat /etc/passwd`' is executed via the reverse shell. Figure 2 demonstrates the **mem.d** script predicate triggering on the *passwd* path, rewriting the path to *defang*, and then dumping the first 256 bytes of memory. The result from the perspective of the remote attacker is depicted in Fig. 1. The reverse shell runs normally when any command is run, except when it interacts with the *passwd* file path. This successfully shows that memory interaction can be used both to inspect active processes, but also to manipulate them. This could be used to limit damage during an ongoing breach, or to monitor particular syscalls as part of a logging or honeytoken setup [41].

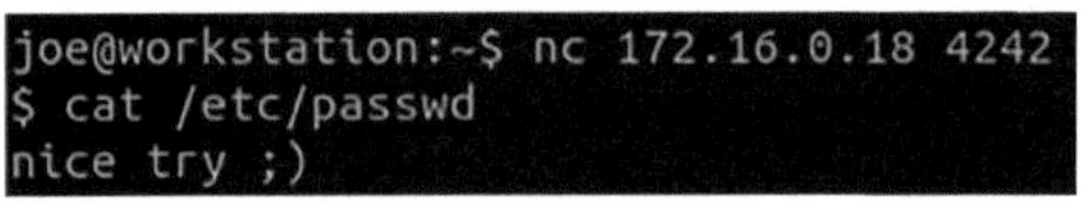

Fig. 1 Demonstration of the remote attacker connecting to the netcat shell and attempting to read `/etc/passwd`, but being redirected to `/tmp/defang`

```
0  64794                          write:entry
          0  1  2  3  4  5  6  7  8  9  a  b  c  d  e  f  0123456789abcdef
      0: 63 61 74 20 2f 74 6d 70 2f 64 65 66 61 6e 67 0a  cat /tmp/defang.
     10: 00 00 00 00 00 00 00 00 00 00 00 00 00 00 00 00  ................
     20: 00 00 00 00 00 00 00 00 00 00 00 00 00 00 00 00  ................
```

Fig. 2 mem.d output (256 bytes) when rewriting the syscall accessing `/etc/passwd`

5.3 Network Monitoring

In order to accomplish their mission, a threat actor must traverse the network, whether for Command and Control, data exfiltration, or lateral movement, such that network activity is key in IR. While much of this data is provided by logs from network appliances, host-based logs from utilities such as `netstat` are often incorporated. DTrace has the potential to gather data similar to a host-based packet filter log, while also providing visibility into processes and process memory, as seen above. We use the same `netcat` reverse shell from Sect. 5.2 to simulate a threat actor persistence mechanism on the FreeBSD 12 VM. Three providers were used in this experiment: (i) `proc` for command execution visibility, and (ii) `pid` to intercept netcat's functions. The `ip` provider was briefly tested for its ability to trace and dump IP content, however `tracemem` only appears to accept a constant as its argument, while packets are variable in length, causing crashing. Additionally, we expect such communications to be encrypted.

Tracking Parent Processes With proc

As noted previously, some programs spawn additional processes to handle command execution. In order to accommodate this, DTrace needs to listen for the creation of these sister processes so they can be mapped back to the parent process. In this way, netcat reverse shell commands can be monitored.

The `trace_parent.d` script (Listing 10.6) depicts an implementation of this approach, with output in Fig. 3. The script uses `proc:::exec`, which triggers before the new process is loaded [38], allowing for the capture of the parent `execname` which is stored in a thread-local variable to make it available in the `proc:::exec-success` probe. A predicate is used on this probe to ensure the parent `execname` is defined, the parent `execname` is then printed, followed by executed command, its arguments, and a timestamp.

```
#!/usr/sbin/dtrace −s

#pragma D option quiet
#pragma D option switchrate=10hz

dtrace:::BEGIN
{
  printf("%−20s %6s %6s %6s  %s\n", "ENDTIME",
    "UID", "PPID", "PID", "PROCESS");
}
```

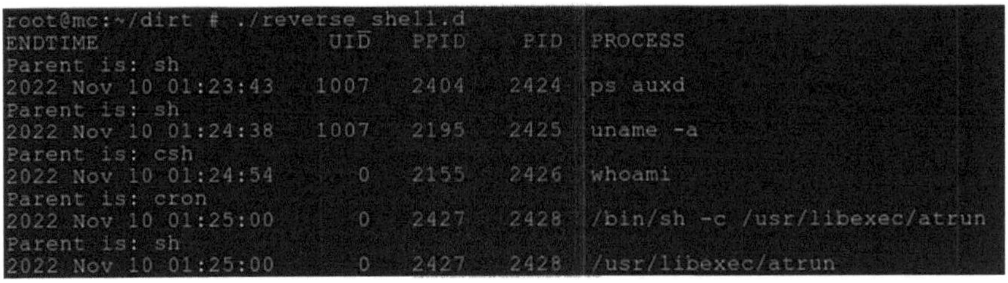

```
root@mc:~/dirt # ./reverse_shell.d
ENDTIME                    UID   PPID    PID  PROCESS
Parent is: sh
2022 Nov 10 01:23:43      1007   2404   2424  ps auxd
Parent is: sh
2022 Nov 10 01:24:38      1007   2195   2425  uname -a
Parent is: csh
2022 Nov 10 01:24:54         0   2155   2426  whoami
Parent is: cron
2022 Nov 10 01:25:00         0   2427   2428  /bin/sh -c /usr/libexec/atrun
Parent is: sh
2022 Nov 10 01:25:00         0   2427   2428  /usr/libexec/atrun
```

Fig. 3 **trace_parent.d output** depicting parent processes

```
proc:::exec
{
  self->parent = execname;
}

proc:::exec-success
/self->parent != NULL/
{
  printf("Parent is: %s\n", self->parent);
  printf("%-20Y %6d %6d %6d  %s\n", walltimestamp,
     uid, ppid, pid, curpsinfo->pr_psargs);
}
```

Listing 10.6 **trace_parent.d**—Observe process execution and trace parent process. Attempted detection of `netcat` reverse shells.

The pid Provider

As discussed in Sect. 5.1, the `pid` provider can be used to observe processes. To do this we must pass the target process PID using the `-p` argument, which is passed to the probe by the `$target` variable in the code examples to follow. It is important to note that this approach is somewhat brittle, and stopping the trace resulted in the reverse shell process being killed. Similarly, if the trace is started after the remote machine connects to a shell via netcat, the process is also killed. In order to avoid this, the following process was used:

1. Launch reverse shell.
2. Find netcat PID with `ps aux | grep nc`.
3. Run DTrace with the target PID.
4. Connect to the shell with netcat from another machine.
5. Run commands in reverse shell.

The `pid` provider has 3072 probes, which can be listed using the command in Listing 10.7. To narrow down the list, it is possible to check which probes are actually triggering for this process during execution, using the command in Listing 10.8. Three functions were used repeatedly by netcat in the operation of the reverse shell: `pid$target::read:entry`, `pid$target::write:entry` and `pid$target::poll:entry`. We focus on exploration of the read function, below.

```
dtrace -ln 'pid$target:::entry' -p 2385
```

Listing 10.7 Finding Entry Probes—a DTrace one-liner to discover entry probes in the `pid` provider [18].

```
dtrace -n 'pid$target:::entry { trace(probename); }' -p 2519
```

Listing 10.8 Finding In-use Probes—a DTrace one-liner to discover which entry probes are firing in the `pid` provider.

To further understand the `read` function, stack backtraces were printed using the command in Listing 10.9. These showed the list of linked libraries called by netcat and lists the `nc` command itself. The information is limited, but would be useful for determining if a process is behaving unusually by linking to unexpected libraries, or for understanding the structure of live malware.

```
dtrace -n 'pid$target::read:entry { ustack(); }' -p 2566
```

Listing 10.9 PID Stack Backtrace—a DTrace one-liner to print user stack backtraces in the `pid` provider.

Further exploration yielded (via Listing 10.10), that the second argument to the function provides pointers to the memory space of the target function. In this experiment, the netcat process runs in userspace whereas DTrace runs in the kernel, necessitating the use of the DTrace functions `copyin` and `copyinstr` to access memory data directly [20, 37]. The DTrace command in Listing 10.11 was then used to dump 256 bytes of memory to the console, with Fig. 4 depicting clear evidence that the `ps` command was run in the reverse shell. The output can be interpreted as strings for better readability by using `trace(copyinstr(arg1));`.

Based on the above, DTrace can be used to provide insights into running networked programs using observations of both network traffic and process memory/execution, together with associated metadata.

```
dtrace -n 'pid$target::read:entry { printf("%d\t%d\t%u\n", arg0, arg1, arg2); }' -p 2598
```

Listing 10.10 Printing All read Function Arguments—a DTrace one-liner to print all of the `read` function arguments.

```
dtrace -n 'pid$target::read:entry { tracemem(copyin(arg1, 256), 256); }' -p 2943
```

Listing 10.11 Tracing Memory for the read Function—a DTrace one-liner to trace the memory of the `read` function arguments.

5.4 Login and Authentication Monitoring

Credential theft is a significant issue and common occurrence in attack chains, with Pluggable Authentication Module (PAM) attacks recently being exploited explicitly on Solaris systems to facilitate backdoors and credential stealing [35].

```
0  64794                           read:entry
             0  1  2  3  4  5  6  7  8  9  a  b  c  d  e  f  0123456789abcdef
         0: 20 50 49 44 20 54 54 20 20 53 54 41 54 20 20 20   PID TT  STAT
        10: 20 54 49 4d 45 20 43 4f 4d 4d 41 4e 44 0a 32 32   TIME COMMAND.22
        20: 39 34 20 76 30 20 20 49 20 20 20 20 30 3a 30 30  94 v0  I    0:00
        30: 2e 30 31 20 2d 73 75 20 28 73 68 29 0a 32 32 39  .01 -su (sh).229
        40: 38 20 76 30 20 20 49 2b 20 20 20 30 3a 30 30 2e  8 v0  I+   0:00.
        50: 30 31 20 2f 62 69 6e 2f 73 68 20 2f 75 73 72 2f  01 /bin/sh /usr/
        60: 6c 6f 63 61 6c 2f 62 69 6e 2f 73 74 61 72 74 78  local/bin/startx
        70: 0a 32 33 31 31 20 76 30 20 20 49 2b 20 20 20 30  .2311 v0  I+   0
        80: 3a 30 30 2e 30 30 20 78 69 6e 69 74 20 2f 68 6f  :00.00 xinit /ho
        90: 6d 65 2f 6b 6f 64 69 2f 2e 78 69 6e 69 74 72 63  me/kodi/.xinitrc
        a0: 20 2d 2d 20 2f 75 73 72 2f 6c 6f 63 61 6c 2f 62   -- /usr/local/b
        b0: 69 6e 2f 58 20 3a 30 20 2d 61 75 74 68 20 2f 68  in/X :0 -auth /h
        c0: 6f 6d 65 2f 6b 6f 64 69 2f 2e 73 65 72 76 65 72  ome/kodi/.server
        d0: 61 75 74 68 2e 32 32 39 38 0a 32 33 31 34 20 76  auth.2298.2314 v
        e0: 30 20 20 49 20 20 20 20 30 3a 30 30 2e 33 37 20  0  I    0:00.37
        f0: 78 66 63 65 34 2d 73 65 73 73 69 6f 6e 0a 32 33  xfce4-session.23
```

Fig. 4 256 bytes dumped from a read function pointer in netcat when the ps command is run in the reverse shell

```
|-- /usr/sbin/sshd
| |-- sshd: root@pts/0 (sshd)
| | `-- -csh (csh)
| |   `-- ps auxd
| `-- sshd: root [priv] (sshd)
|   |-- sshd: root [net] (sshd)
|   `-- sshd: root [pam] (sshd)
```

Fig. 5 Cropped screenshot of ps auxd output for the sshd process

Unix authentication evidence is primarily found in logs such as the plain-text auth.log which contains various authentication information. Login information is captured in the binary files: /var/run/utmp for current logins, /var/log/wtmp for current and previous logins, and /var/log/btmp for bad logins. However, these logs do not facilitate an in-depth view of the authentication process, which may be required to identify backdoors. Logins can also be queried on a live system using the commands users, who and w, providing increasingly more information.

DTrace is capable of drilling down into the authentication process, and the extent to which this can be utilised for Incident Response was investigated. The limited proof of concept experiment was conducted as follows:

1. Use FreeBSD 12 VM with sshd configured with password authentication.
2. Use the pid provider to investigate potentially valuable probes.
3. Trace the authentication process of the PAM module used by sshd.
4. Attempt to find contextual information that provides deeper insight into authentication.

PAM is used by many different applications and modules exist to authenticate using many methods. We focus here on SSH due to its common usage as a vehicle for lateral movement on Unix-based systems, with OpenSSH being chosen in particular due to its open-source nature. The sshd process tree is depicted in Fig. 5, which shows a related PAM process.

Finding Probes

There was again a need to scope the investigation from a large number of probes. The DTrace command shown in Listing 10.12 was used to discover which functions are called by the SSH PAM process in order to find candidates for tracing, with output in Listing 10.14. The command essentially counts the number of times the process invokes a function (combining `probefunc` and `count`) by observing the `entry` probe. This process is similar to the the one used in Listing 10.8 for `pid`, except in this case it is functions that are being explored.

```
dtrace -n 'pid$target:::entry { @[probefunc] = count(); }' -p 3113
```

Listing 10.12 Finding sshd Probes—a DTrace one-liner to find useful probes in `sshd`.

```
dtrace: description 'pid$target:::entry ' matched 13321 probes
dtrace: pid 3113 has exited

[...]
  __malloc                                  37
  malloc                                    37
  strcasecmp                                47
  strcasecmp_l                              47
  _openpam_log                              57
  __free                                    78
  free                                      78
  fgets                                     79
  memchr                                    80
  strlen                                    80
  __pthread_cleanup_pop_imp                 81
  __pthread_cleanup_push_imp                81
  __tls_get_addr                           137
  __runes_for_locale                       387
  __SHA512_Final                          5005
  __SHA512_Init                           5005
  explicit_bzero                          5013
  memset                                  5062
  __SHA512_Update                        17657
  memcpy                                 22812
```

Listing 10.13 SSH PAM Functions—Example output of Listing 10.12, cutting all but the top 20 most called functions by the SSH PAM process.

Tracing SSH PAM Authentication

The PAM functions were inspected and `pam_sm_acct_mgmt` (only called once, so not present in Listing 1.13) was considered to be of interest as it verifies if the user has authenticated successfully [30]. The DTrace one-liner shown in Listing 10.14 was created to trace the return value of this target function, which is the second argument passed to the `return` probe [19]. In this particular case, the function returns 0 (`PAM_SUCCESS`) upon successful authentication [43], which can be used to track successful SSH logins.

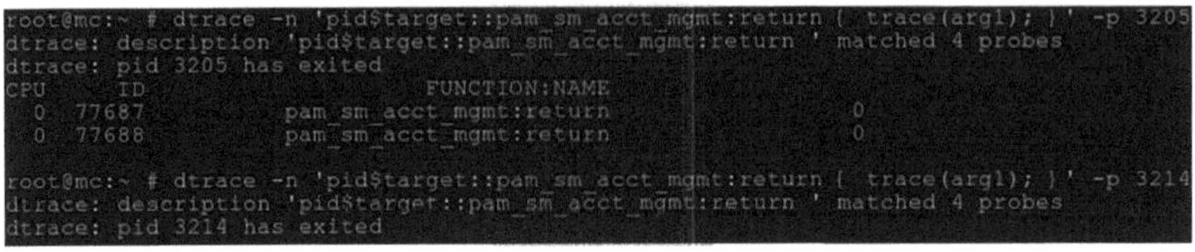

Fig. 6 Screenshot showing the DTrace one-liner tracing a successful and failed authentication attempt

```
dtrace −n 'pid$target::pam_sm_acct_mgmt:return { trace(arg1); }' −p 3205
```

Listing 10.14 Tracing sshd Authentication—a DTrace one-liner to trace successful authentication in the PAM module used by `sshd`.

The output of the DTrace one-liner is shown in Fig. 6, which shows the difference between successful authentication (invocation 1), and failed authentication (invocation 2). It is worth noting that the function returns `0` twice during a successful authentication and does not return during a failed authentication attempt, likely because the function is not called. This visibility could potentially help to identify a backdoor or rootkit. To be useful, this data should be correlated with other information, such as the host's IP address.

5.5 *Disk IO Monitoring*

Disk IO is an important component of a system's response time and therefore many tools monitor it for performance reasons. Cloud providers often make metrics available and myriad performance tools, such as `iostat`, `atop` and `sar` are used to monitor performance on a host. Abnormal disk IO activity can tip off an administrator to malfunctioning software or configuration issues, but are also indicative of attacks such as Denial of Service, ransomware, wipers, or data exfiltration. DTrace facilitates a deep understanding of the causes of disk IO activity, which can generate critical insights when detection and response is time critical, as is the case for ransomware and wiper attacks in particular.

A pair of experiments, defined below, was used as a proof of concept to test DTrace's ability to detect wiper malware on Solaris and FreeBSD. This involved tracing the `unlink` syscall to detect which program is responsible for deletion. Monitoring this call is fairly limited, as is it not used to delete directories or overwrite files, and it would typically be called frequently on high-traffic storage. Rootkits which do not use system calls to modify files would also be missed, but for basic test purposes it should suffice, here.

Two main scenarios were tested: active wiper deletion using shell commands (Listing 1.15) and a perl script (Listing 1.16). The latter perl script was chosen as prior testing indicated that shell scripts themselves do not always get captured directly by DTrace probes (Sect. 5.1), but the hypothesis was that other scripts from outside core shell functionality would be captured. It should also be noted that the system's typical usage patterns for `unlink` system calls needs to be taken in to account here when determining if behaviour is anomalous.

The process used was as follows:

1. Use the Solaris 10 VM.
2. Create simulated wiper programs using shell and perl.
3. Trace the `unlink` syscall using the DTrace `syscall` provider.
   (a) While running DTrace in the background:
   (b) Create test files in a test directory.
   (c) Run the wiper shell/perl script to delete the test files.

```
#!/bin/sh

rm /Documents/test/*
```

Listing 10.15 wiper.sh—Shell script which deletes the contents of the `/Documents/test` directory.

```
#!/usr/bin/perl

use strict;
use warnings;

for (glob '/Documents/test/*') {
    unlink;
}
```

Listing 10.16 wiper.pl—Perl program which deletes the contents of the `/Documents/test` directory.

unlink System Call Tracking

The DTrace command shown in Listing 10.17 uses the `syscall` provider [36] to trace the program which is executing when the `unlink` syscall is called. The count and executable name are stored in an aggregation and printed at the end of execution. It should be noted that multi-core CPUs were not tested here, such that additional steps may need to be taken to avoid the conflation of results across multiple execution cores.

```
dtrace -n 'syscall::unlink:entry { @f[execname] = count(); }'
```

Listing 10.17 unlink Syscall One-liner—a DTrace one-liner to trace the number of times an executable calls the `unlink` syscall.

While monitoring, each wiper script was run in succession to delete files in the specified test directory. In the case of the shell script we can only verify that `rm` is executed, while the source perl script is present in the latter example, as depicted in Fig. 7. Attribution becomes easier with external scripts, then, but a baseline of

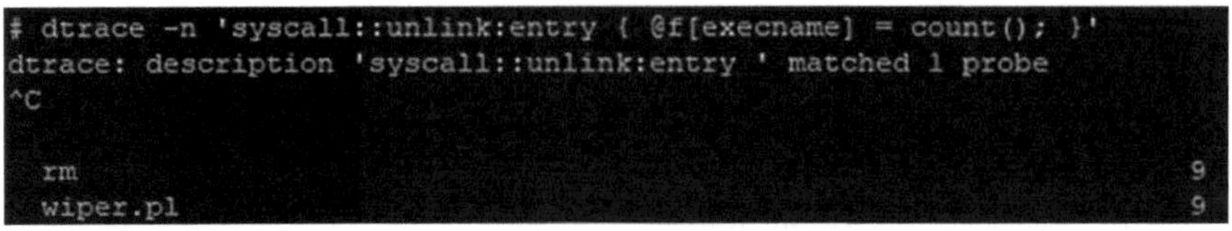

Fig. 7 Screenshot showing the DTrace one-liner used to trace the `unlink` syscall when running the perl wiper. Clearly depicts the `rm` command and the wiper.pl script

activity could still enable the output to trigger a warning that many files are being deleted, even when a shell script is being executed. In light testing there were not a large number of calls to unlink happening in the background of a clean system which was not running distinct servers.

6 Conclusions

This paper introduced the concept of using DTrace to gather telemetry from legacy Solaris and modern FreeBSD systems in the context of tracing security events. We found that the real-time system visibility provided by DTrace enables: (i) System calls to be monitored, including for spawned processes via their PPID, and (i) Process execution to be observed and combined with in-depth examination of the processes, allowing the detection of sensitive information in memory, as well as providing visibility into network metadata and content.

However DTrace does not lend itself to the acquisition of forensic artefacts that enable the analysis of historical events, and is better used as a real-time monitoring/probing utility. We also found that while DTrace has the technical capability to write to a systems memory, the loss of the safety that DTrace otherwise provides makes it unlikely that this feature will be of significant value in the field, particularly when used on critical infrastructure. This positions it differently to UAC, which is better served to create snapshot views, or perform high-level triage. Similarly, GRR is positioned as a reactive solution, enabling a network to be triaged for threats and remediation actions to be taken, as opposed to proactive monitoring and analysis.

The capabilities of DTrace are therefore best suited for integration into an EDR platform. As such, DTrace may help fill the current real-time monitoring and EDR gap on Unix systems, as this has been largely overlooked by existing tooling. In practice, building an EDR solution around DTrace would allow Unix systems to come close to the state-of-the-art Linux support offered by many vendors, helping bridge the Critical National Infrastructure gap as telecommunications and other industries become increasingly targeted.

6.1 Future Work

Further work is required to investigate the full potential of the use of DTrace for Incident Response purposes. One of the most promising areas is the integration of DTrace into an EDR product to enable comprehensive monitoring of supported systems. Because this is a large undertaking, a valuable stepping stone towards this goal would be the development of standalone tools, which could be consolidated into a framework, that allow an operator to monitor a system and detect attacks. The robustness and effectiveness of this framework should be tested and evaluated. Furthermore, the frequent use of containers in modern computing environments makes it valuable to investigate DTrace's ability to snoop on the activity within a container in order to perform security monitoring. Of particular interest are Solaris Zones, FreeBSD jails and the popular Docker platform.

While this paper has focused on the use of DTrace for defensive purposes, the extent to which DTrace can be used for offensive operations should be explored in future work, as advanced attackers are already leveraging eBPF for malicious purposes [8].

Acknowledgements The authors would like to express their sincere gratitude to Jamie Harries, Charles J Bevacqua Dylan Watson, and the anonymous reviewers for their constructive comments.

References

1. osquery/osquery (2024). https://github.com/osquery/osquery. Original-date: 2014-07-29T20:27:33Z
2. Abrams, L.: Lockbit ransomware encryptors found targeting Mac devices (2023). https://www.bleepingcomputer.com/news/security/lockbit-ransomware-encryptors-found-targeting-mac-devices/. Accessed 22 Sept 2023
3. Agarwal, S., Sable, A., Sawant, D., Kahalekar, S., Hanawal, M.K.: Threat detection and response in linux endpoints. In: 2022 14th International Conference on COmmunication Systems & NETworkS (COMSNETS), pp. 447–449. IEEE, Bangalore, India (2022)
4. Agence Nationale de la Sécurité des Systèmes d'Information: Ryuk ransomware. Technical Report, Paris France (2021)
5. BBC: The Lazarus heist: how North Korea almost pulled off a billion-dollar hack (2021). https://www.bbc.co.uk/news/stories-57520169. Accessed 28 Feb 2022
6. Beauchamp, T., Weston, D.: Dtrace: the reverse engineer's unexpected swiss army knife. Blackhat USA 2008 (2008). https://www.blackhat.com/presentations/bh-usa-08/Beauchamp_Weston/BH_US_08_Beauchamp-Weston_DTrace.pdf
7. Bhardwaj, A.: Cybersecurity Incident Response Against Advanced Persistent Threats (APTs), pp. 189–209. Springer International Publishing, Cham (2021)
8. Bogle, D.: eBPF: a new frontier for malware (2023). https://redcanary.com/blog/ebpf-malware/
9. Brozek, M.: 6 questions you must ask for a successful incident response (2016). https://www.paloaltonetworks.com/blog/security-operations/6-questions-you-must-ask-for-a-successful-incident-response/. Accessed 12 Nov 2022
10. Cantrill, B.M., Shapiro, M.W., Leventhal, A.H.: Dynamic instrumentation of production systems. In: 2004 USENIX Annual Technical Conference (USENIX ATC 04). USENIX Association, Boston, MA (2004)

11. CrowdStrike: Crowdstrike on dark reading: Why "breakout time" is critical to your security strategy (2018). https://www.crowdstrike.com/blog/crowdstrike-discusses-breakout-time-in-an-article-on-dark-reading/. Accessed 1 Mar 2022
12. CrowdStrike: Crowdstrike 2022 global threat report (2022). https://www.crowdstrike.com/resources/reports/global-threat-report/
13. De Wiest, R., The BlackBerry Incident Response Team: 13 deadly sins when dealing with apt incidents—part 2 (2022). https://blogs.blackberry.com/en/2022/09/13-deadly-sins-when-dealing-with-apt-incidents-part-2. Accessed 17 Oct 2022
14. Department of Homeland Security: Secure cyberspace and critical infrastructure (2022). https://www.dhs.gov/secure-cyberspace-and-critical-infrastructure. Accessed 19 Oct 2022
15. Divya, P.J., George, R.S., Madhusudhan, G., Padmasree, S.: Organization-wide IOC monitoring and security compliance in endpoints using open source tools. In: 2022 IEEE 3rd Global Conference for Advancement in Technology (GCAT), pp. 1–6 (2022)
16. Gartner: Gartner magic quadrant for EPP 2021 (2021). https://www.crowdstrike.com/resources/reports/gartner-magic-quadrant-endpoint-protection-platforms-2021/. Accessed 15 Mar 2022
17. Google: Grr rapid response: remote live forensics for incident response (2022). https://github.com/google/grr. Accessed 16 Sept 2022
18. Gregg, B.: DTrace pid provider (2011). http://dtrace.org/blogs/brendan/2011/02/09/dtrace-pid-provider/. Accessed 10 Nov 2022
19. Gregg, B.: Dtrace pid provider return (2011). http://dtrace.org/blogs/brendan/2011/02/14/dtrace-pid-provider-return/. Accessed 20 Nov 2022
20. Gregg, B.: Tweaking memory on the fly (2011). http://dtrace.org/blogs/brendan/2011/06/30/tweaking-memory-on-the-fly/. Accessed 10 Nov 2022
21. Gregg, B., Mauro, J.: DTrace: Dynamic Tracing in Oracle Solaris, 1st edn. Mac OS X and FreeBSD. Prentice Hall Press, USA (2011)
22. Harries, J.: Hunting a Global Telecommunications Threat: Decisive Architect and Its Custom Implant Justforfun (2022). https://www.crowdstrike.com/blog/how-to-hunt-for-decisivearchitect-and-justforfun-implant/. Accessed 10 June 2022
23. Harries, J., Mayer, D.: Lightbasin: A Roaming Threat to Telecommunications Companies (2021). https://www.crowdstrike.com/blog/an-analysis-of-lightbasin-telecommunications-attacks/. Accessed 15 Feb 2022
24. Haslam, J.: uname lies ... (jon haslam's weblog) (2005). https://web.archive.org/web/20111102063631/, http://blogs.oracle.com/jonh/entry/it_s_been_a_long. Accessed 29 Oct 2022
25. IBM: What is threat hunting (2022). https://www.ibm.com/uk-en/topics/threat-hunting. Accessed 26 May 2022
26. Johnson, L.A., Dempsey, K.L., Ross, R.S., Gupta, S., Bailey, D.: Sp 800-128. guide for security-focused configuration management of information systems. Technical Report, Gaithersburg, MD, USA (2011). https://nvlpubs.nist.gov/nistpubs/SpecialPublications/NIST.SP.800-128.pdf
27. Kipling, L., Petito, A.: Changing security incident response by utilizing the power of the cloud-dart tools, techniques, and procedures: part 1 (2019). https://www.microsoft.com/security/blog/2019/11/14/security-incident-response-utilizing-cloud-dart-tools-techniques-procedures-part-1/. Accessed 26 May 2022
28. Kwon, Y., Wang, F., Wang, W., Lee, K.H., Lee, W.C., Ma, S., Zhang, X., Xu, D., Jha, S., Ciocarlie, G., Gehani, A., Yegneswaran, V.: MCI: modeling-based causality inference in audit logging for attack investigation. In: Proceedings 2018 Network and Distributed System Security Symposium. Internet Society, San Diego, CA (2018)
29. Lahr, T.C.: Unix-like artifacts collector (2022). https://github.com/tclahr/uac. Accessed 15 Sept 2022
30. Linux PAM Project: pam_sm_acct_mgmt(3)—linux manual page (2016). https://man7.org/linux/man-pages/man3/pam_sm_acct_mgmt.3.html. Accessed 20 Nov 2022

31. Mandiant: Highly evasive attacker leverages solarwinds supply chain to compromise multiple global victims with sunburst backdoor (2020). https://www.mandiant.com/resources/evasive-attacker-leverages-solarwinds-supply-chain-compromises-with-sunburst-backdoor. Accessed 27 Feb 2022
32. Milajerdi, S.M., Gjomemo, R., Eshete, B., Sekar, R., Venkatakrishnan, V.N.: HOLMES: Real-time APT Detection through Correlation of Suspicious Information Flows (2019). arXiv:1810.01594 [cs]
33. MITRE Corporation: Mitre att&ck (2022). https://attack.mitre.org/. Accessed 17 Oct 2022
34. MITRE Corporation: Mitre att&ck enterprise tactics (2022). https://attack.mitre.org/tactics/enterprise/. Accessed 28 Oct 2022
35. Moore, J., Ledzion, W., Rocha, L., Pisarczyk, A., Caban, D., Rincon, S., Susin, D., Monaca, A.: Live off the land? how about bringing your own island? an overview of unc1945 (2020). https://www.mandiant.com/resources/live-off-the-land-an-overview-of-unc1945. Accessed 15 Feb 2022
36. Oracle: syscall provider (2006). https://docs.oracle.com/cd/E18752_01/html/819-5488/gcgkk.html#gcgld. Accessed 18 Nov 2022
37. Oracle: copyin() and copyinstr() subroutines (2010). https://docs.oracle.com/cd/E19253-01/817-6223/chp-user-1/index.html. Accessed 10 Nov 2022
38. Oracle: Proc provider (2012). https://docs.oracle.com/cd/E23824_01/html/E22973/proc-provider.html#scrolltoc. Accessed 12 Nov 2022
39. Ortiz, R.: eBPF for security: a beginner's guide (2022). https://redcanary.com/blog/ebpf-for-security/
40. Park, S.H., Yun, S.W., Jeon, S.E., Park, N.E., Shim, H.Y., Lee, Y.R., Lee, S.J., Park, T.R., Shin, N.Y., Kang, M.J., Lee, I.G.: Performance evaluation of open-source endpoint detection and response combining google rapid response and osquery for threat detection. IEEE Access **10**, 20259–20269 (2022)
41. Petrunić, A.R.: Honeytokens as active defense. In: 2015 38th International Convention on Information and Communication Technology, Electronics and Microelectronics (MIPRO), pp. 1313–1317. IEEE (2015)
42. Pool, R.: 1-10-60: measuring the speed of incident response. Cyber. Secur. A Peer-Rev. J. **3**(4), 308–314 (2020)
43. The FreeBSD Project: freebsd-legacy pam_unix.c (2022). https://github.com/freebsd/freebsd-legacy/blob/de1aa3dab23c06fec962a14da3e7b4755c5880cf/lib/libpam/modules/pam_unix/pam_unix.c. Accessed 20 Nov 2022
44. w3techs: Usage Statistics and Market Share of Unix for Websites (2023). https://w3techs.com/technologies/details/os-unix
45. Wang, X., Huang, F., Chen, H.: DTrace: fine-grained and efficient data integrity checking with hardware instruction tracing. Cybersecurity **2**(1), 1 (2019). https://doi.org/10.1186/s42400-018-0018-3, https://cybersecurity.springeropen.com/articles/10.1186/s42400-018-0018-3
46. West-Brown, M.J., Stikvoort, D., Kossakowski, K.P., Killcrece, G., Ruefle, R., Zajicek, M.: Handbook for Computer Security Incident Response Teams (CSIRTs). Carnegie Mellon University, Software Engineering Institute (1998)

Utilizing YARA: An Effective Method for Phishing Attack Response

Ferenc Leitold

Abstract Phishing attacks continue to be a major threat to cybersecurity, leveraging deception to extract sensitive information from unsuspecting victims. This paper explores the use of YARA (Yet Another Recursive Acronym), a powerful tool for malware identification and classification, in enhancing the detection and response to phishing attacks. We delve into several phishing techniques such as spear phishing, whaling, and clone phishing, outlining their evolution and the increasing sophistication of these threats. The paper discusses the creation and application of YARA rules to identify phishing indicators within emails, including suspicious language, generic greetings, and malicious links or attachments. By integrating YARA into cybersecurity operations, organizations can create customizable and efficient detection frameworks that adapt to emerging threats. Through practical examples and case studies, we demonstrate the effectiveness of YARA in identifying both traditional and obfuscated phishing attacks, thereby improving incident response and mitigating the impact of these pervasive threats to individuals and businesses alike. This research highlights YARA's versatility and its significant role in bolstering cybersecurity defenses against phishing.

1 Introduction

In today's interconnected world, phishing attacks have become one of the most prevalent and damaging forms of cyber threats [1–3]. These attacks involve malicious actors impersonating legitimate entities to deceive individuals into divulging sensitive information such as usernames, passwords, and credit card numbers. Despite advancements in security technologies, phishing remains a significant challenge due to its evolving nature and the sophistication of the techniques used by attackers.

F. Leitold (✉)
Óbuda University, Budapest, Hungary
e-mail: leitold.ferenc@nik.uni-obuda.hu

M. G. Jaatun et al. (eds.), *Proceedings of the International Conference on Cybersecurity, Situational Awareness and Social Media*, Springer Proceedings in Complexity,
https://doi.org/10.1007/978-981-96-0401-2_11

1.1 Understanding Phishing Techniques

Phishing attacks are typically executed through deceptive emails, messages, or websites that appear to come from trustworthy sources. There are several forms of phishing [4–7], each with unique characteristics and methods of execution:

1. Spear Phishing: This highly targeted form of phishing involves personalized messages crafted using detailed information about the victim. Attackers might gather this information from social media profiles or other sources to create convincing emails that appear to come from trusted colleagues or associates.
2. Whaling: Targeting high-profile individuals such as executives or senior management, whaling attacks often aim to access sensitive corporate information or financial data. These attacks typically feature official-looking and urgent messages, often posing as legal or financial requests.
3. Clone Phishing: In this method, attackers create nearly identical copies of legitimate emails that the victim has previously received. The cloned email, which appears to be a follow-up or related communication, contains malicious links or attachments designed to deceive the recipient.

1.2 Evolution of Phishing Attacks

Phishing attacks have a history dating back to the mid-1990s, coinciding with the rise of the internet and email usage. Early phishing attacks, such as the "AOL phishing" in the mid-1990s, involved hackers posing as AOL staff to trick users into revealing their passwords. Over the years, phishing tactics have become more sophisticated and widespread, targeting online payment systems like PayPal and eBay in the early 2000s.

A significant development in phishing occurred in 2003 with the emergence of the "Rock Phish" group, which used automated and scalable methods to redirect legitimate website traffic to fraudulent sites. This shift marked the transition from individual email scams to organized and professional phishing operations.

2 YARA

YARA, an acronym for "Yet Another Recursive Acronym," is a powerful tool used in cybersecurity to identify and classify malware [8]. Created by Victor Alvarez of VirusTotal, YARA is widely used by security researchers and analysts for detecting and describing malware patterns. While it is possible to detect phishing threats without YARA [9–16], but YARA is much more efficient. Its effectiveness lies in its ability to create rules based on textual or binary patterns, making it a versatile tool for identifying various types of malware and other security threats.

2.1 Key Features of YARA

YARA has couple of features that aids its use in threat detection:

1. Pattern Matching: YARA uses pattern matching to identify strings of text or binary sequences within files. These patterns can be simple strings or complex regular expressions, providing flexibility in detection.
2. Rule-Based Approach: Users write YARA rules that define the patterns to search for and the conditions under which these patterns should be flagged. Each rule consists of a name, metadata, strings to search for, and conditions to be met.
3. Modularity and Reusability: YARA rules can be reused and combined in different ways, allowing for modular and scalable detection capabilities. This modularity enables the creation of comprehensive detection frameworks that can evolve over time.
4. Integration Capabilities: YARA can be integrated with other security tools and systems such as antivirus software, intrusion detection systems (IDS), and forensic analysis tools, enhancing the overall effectiveness of security operations.

2.2 The Structure of YARA Rules

YARA operates through a straightforward process involving the creation of rules, scanning files or data against these rules, and identifying matches. A YARA rule includes the following parts:

- Rule Name: Each YARA rule begins with a name that describes its purpose.
- Meta Section: This section includes metadata such as the rule's author, description, and date of creation. Metadata helps in managing and understanding the rules over time.
- Strings Section: This section defines the specific strings or patterns that the rule will search for. Strings can be plain text, hexadecimal values, or regular expressions.
- Condition Section: The condition section specifies the logic that determines when the rule should trigger. This logic can involve one or more of the defined strings and may include Boolean operators and other logical constructs.

Example of a simple YARA Rule:

```
rule Example_Rule {
    meta:
        author = "Security Analyst"
        description = "Detects sample malware"
        date = "2024-05-17"

    strings:
        $pattern1 = "malicious_string"
        $pattern2 = { 6D 61 6C 69 63 69 6F 75 73 }
        $pattern3 = /malware[0-9]{3}/

    condition:
        any of ($pattern1, $pattern2, $pattern3)
}
```

Once the rules are written, YARA can scan files or data streams against these rules. The YARA engine processes the files, searching for the defined patterns. If a match is found, YARA flags the file and provides information about which rule triggered the match. This information includes the specific strings that matched and their locations within the file.

After scanning, the results are analyzed to determine the next steps. This might involve further investigation, quarantine of the identified files, or automated responses such as alerting security personnel. Detailed logging and reporting features in YARA help analysts track and understand the nature of detected threats, aiding in the development of more effective security measures. YARA's versatility makes it applicable in several scenarios within cybersecurity:

- Malware Analysis: YARA is extensively used in malware research to identify and classify malware samples. Researchers write rules to detect specific malware families or characteristics, aiding in the rapid identification of known threats.
- Threat Hunting: Security teams use YARA to proactively search for signs of compromise within their networks. By deploying YARA rules across endpoints and network traffic, they can detect malicious activity early and respond quickly.
- Incident Response: During incident response, YARA helps in identifying the presence of malware and other indicators of compromise (IOCs). It enables quick and accurate detection, supporting efficient incident containment and remediation.
- File and Memory Scanning: YARA can be used to scan files on disk or the contents of system memory, providing a comprehensive tool for detecting malware regardless of where it resides.
- Integration with Security Platforms: Many security platforms and tools, such as SIEMs (Security Information and Event Management systems) and EDRs (Endpoint Detection and Response), integrate YARA for enhanced detection capabilities.

The adoption of YARA in cybersecurity practices offers several significant benefits:

- Customizability: YARA rules can be tailored to detect specific threats relevant to an organization, providing a high degree of customization.
- Efficiency: YARA's pattern matching is efficient and scalable, making it suitable for large-scale deployments across numerous systems.
- Community and Collaboration: The cybersecurity community actively shares YARA rules, creating a rich repository of detection rules that can be leveraged by organizations worldwide.
- Continuous Improvement: As new threats emerge, YARA rules can be updated and refined, ensuring that detection capabilities remain current and effective.

3 Identifying Phishing Attacks with YARA

Phishing attacks often have distinct characteristics that can help identify them. These indicators include:

- Suspicious Email Addresses: Phishing emails frequently come from addresses that mimic legitimate ones, with slight variations that may go unnoticed by the untrained eye.
- Urgent or Threatening Language: Phishing messages often create a sense of urgency or fear, compelling the recipient to act quickly without verifying the legitimacy of the request.
- Generic Greetings and Poor Grammar: Unlike legitimate emails that are usually personalized, phishing emails often use generic greetings and may contain grammatical errors or awkward phrasing.
- Unusual Requests for Personal Information: Legitimate organizations rarely ask for sensitive information via email. Requests for passwords, credit card details, or other personal information should be treated with suspicion.
- Malicious Links and Attachments: Phishing emails often include links or attachments that, when clicked or downloaded, install malware or lead to fraudulent websites designed to steal information.

Example of a Basic YARA Rule for Phishing Detection:

```
rule Phishing_Email_Detector {
    meta:
        author = "Cybersecurity Team"
        description = "Detects common indicators"
        date = "2024-05-17"

    strings:
        $phish1 = "urgent action required"
        $phish2 = "verify your account"
        $phish3 = "update your password"
        $phish4 = "click here to login"

    condition:
        any of ($phish1, $phish2, $phish3, $phish4)
}
```

For advanced rule customization the following features can be used:

- Regex Patterns: Utilize regular expressions to match more complex patterns within phishing emails, such as email addresses or URLs that fit certain criteria.
- Combination of Strings: Create conditions that combine multiple strings or patterns, increasing the accuracy of phishing detection by reducing false positives.
- File Attributes: Include file attributes in the rules, such as file type or size, to further refine the detection criteria.

Example of an advanced YARA rule:

```
rule Advanced_Phishing_Detector {
    meta:
        author = "Cybersecurity Team"
        description = "Advanced detection"
        date = "2024-05-17"

    strings:
        $phish1 = "urgent action required"
        $phish2 = /verify.*account/i
        $phish3 = /update.*password/i
        $phish4 = "click here to login"
        $url_pattern = /https?:\/\/[a-zA-Z0-9.-]+\.[a-zA-
Z]{2,6}(\/[a-zA-Z0-9.-]*)?/

    condition:
        (any of ($phish1, $phish2, $phish3, $phish4)) and
$url_pattern
}
```

Real-world case studies illustrate the effectiveness of YARA in detecting phishing attacks. By analyzing past incidents, we can understand how YARA rules are applied and fine-tuned to catch phishing attempts.

Case Study 1: Financial Sector Phishing Campaign

- Scenario: A financial institution experienced a wave of phishing emails targeting its customers, asking them to verify their account details.
- YARA Implementation: The cybersecurity team developed YARA rules based on the common phrases and patterns observed in the phishing emails. The rules included keywords like "verify your account" and "click here to login," along with regex patterns for detecting suspicious URLs.
- Outcome: The YARA rules successfully identified and flagged incoming phishing emails, allowing the institution to warn customers and mitigate the impact of the attack.

Case Study 2: Corporate Environment Spear Phishing Attack

- Scenario: A large corporation faced a spear-phishing attack where emails appeared to come from senior executives, requesting employees to transfer funds.
- YARA Implementation: YARA rules were created to detect the specific language and email patterns used in the spear-phishing emails. The rules included the names of executives and phrases typically used in fraudulent requests.
- Outcome: The customized YARA rules detected the spear-phishing attempts, preventing financial losses and prompting the company to enhance its email security measures.

By leveraging YARA, organizations can create precise and effective rules to detect phishing attacks, significantly improving their incident response capabilities. Through continuous refinement and adaptation, YARA remains a vital tool in the fight against phishing attacks as well.

4 Detecting Obfuscated Phishing Attacks

Obfuscated phishing attacks represent a sophisticated evolution in phishing tactics, where malicious actors use techniques to hide or disguise their malicious intent. These techniques make it challenging to detect phishing attempts using traditional methods. However, provides a powerful framework to create detection rules that can uncover these hidden threats. This section explores the application of YARA in identifying obfuscated phishing attacks through practical examples.

4.1 Understanding Obfuscated Phishing

Obfuscated phishing involves various techniques to mask the true nature of the phishing attempt. These techniques can include:

- URL obfuscation: Using encoded or shortened URLs to disguise malicious links.
- Content obfuscation: Employing HTML/CSS tricks to hide phishing content within legitimate-looking emails.
- Attachment obfuscation: Embedding malicious scripts within seemingly benign attachments.

The goal of these techniques is to evade detection by traditional security measures and trick users into interacting with malicious content.

4.2 Applying YARA to Detect Obfuscated Phishing

YARA's rule-based approach allows security analysts to define specific patterns and conditions to identify obfuscated phishing attempts. Below are examples of YARA rules designed to detect different types of obfuscation in phishing emails.

Example 1: Detecting URL obfuscation

URL obfuscation often involves encoding URLs in ways that are not immediately recognizable as malicious. This YARA rule detects common obfuscation patterns in URLs within email content.

```
rule Obfuscated_URL_Phishing {
meta:
   author = "Cybersecurity Analyst"
   description = "Detects obfuscated URLs"
   date = "2024-05-17"

strings:
   $url_encoded = /https?:\/\/(%[0-9A-Fa-f]{2}){1,}/
   $url_hex = /0x[a-fA-F0-9]{2,}/
$url_ip = /http:\/\/\d{1,3}\.\d{1,3}\.\d{1,3}\.\d{1,3}/
$url_short =
/https?:\/\/(bit\.ly|tinyurl\.com|goo\.gl|t\.co)\//

condition:
   any of ($url*)
}
```

Example 2: Detecting content obfuscation

Phishers may use HTML and CSS tricks to hide malicious content. This rule identifies suspicious HTML tags and encoding techniques used to obscure phishing content.

```
rule Obfuscated_Content_Phishing {
    meta:
        author = "Cybersecurity Analyst"
        description = "Detects obfuscated content"
        date = "2024-05-17"

    strings:
        $html_hex = /&#x[a-fA-F0-9]{2,};/
        $html_css_hide = /display\s*:\s*none/
        $html_base64 = /data:text\/html;base64,/
        $html_unescape = /unescape\(\'/

    condition:
        any of ($html*)
}
```

Example 3: Detecting attachment obfuscation

Attachments in phishing emails can be obfuscated to bypass security filters. This rule detects common obfuscation techniques within attachments.

```
rule Obfuscated_Attachment_Phishing {
    meta:
        author = "Cybersecurity Analyst"
        description = "Detects obfuscated scripts"
        date = "2024-05-17"

    strings:
        $script_encoded = /eval\(\s*atob\(/i
        $script_hex = /\\x[a-fA-F0-9]{2}/
$script_var = /var\s+.*?=\s*function\s*\(.*?\)\s*{.*?}/

    condition:
        any of them
}
```

As can be seen, YARA provides a robust framework for detecting obfuscated phishing attacks. By defining precise patterns and conditions, security analysts can uncover hidden threats that evade traditional detection methods. Continuous refinement and customization of YARA rules enable organizations to stay ahead of evolving phishing techniques, enhancing their overall cybersecurity posture.

5 Conclusion

This article delves into the persistent and evolving threat of phishing attacks, highlighting their various forms, historical development, and the sophisticated techniques used by attackers. Phishing, a cyber-attack method where malicious actors impersonate legitimate entities, aims to deceive individuals into divulging sensitive information. The article outlines several types of phishing, including spear phishing, whaling, and clone phishing, each with unique tactics and targets.

The application of YARA in phishing detection is a focal point, demonstrating how customized YARA rules can identify phishing indicators in emails, such as suspicious language, generic greetings, and malicious links or attachments. Practical examples of YARA rules are provided, showcasing their effectiveness in detecting both traditional and obfuscated phishing attacks.

Through real-world case studies, the article illustrates the successful implementation of YARA rules in several scenarios as well. For instance, a financial institution's response to a phishing campaign and a large corporation's defense against spear-phishing attacks highlight how YARA can significantly enhance incident response and mitigate the impact of phishing threats.

The article concludes by emphasizing YARA's versatility and its critical role in strengthening cybersecurity defenses against phishing attacks. By continuously refining and updating YARA rules, organizations can stay ahead of evolving phishing techniques, ensuring robust protection and improved security operations. The experiences and insights shared in this article demonstrate the practical value and effectiveness of YARA in combating phishing, making it an indispensable tool in the modern cybersecurity arsenal.

References

1. Anti-Phishing Working Group (APWG): Phishing Activity Trends Report. APWG. https://apwg.org/trendsreports/ (2022)
2. Mandiant: phishing trends and intelligence report. FireEye. https://www.fireeye.com/blog/threat-research/2019/04/phishing-trends-and-intelligence-report.html (2019)
3. Ponemon institute: cost of phishing: why phishing attacks are successful. Ponemon Institute. https://www.ponemon.org/cost-of-phishing.html (2021)
4. Bisson, D.: The evolution of phishing attacks: why it's a threat to everyone. Tripwire. https://www.tripwire.com/state-of-security/security-awareness/the-evolution-of-phishing-attacks/ (2020)
5. Google: protecting against phishing: the state of phishing attacks. Google Safety Center. https://safety.google/stories/protecting-against-phishing/ (2021)
6. Krebs, B.: Target: names of customers Stolen. Krebs on security. https://krebsonsecurity.com/2014/01/target-names-of-customers-stolen/ (2014)
7. SANS institute: phishing attacks: common types & how to protect against them. SANS Institute. https://www.sans.org/white-papers/phishing-attacks-common-types-how-to-protect-against-them/ (2022)
8. Alvarez, Victor: YARA: Documentation. VirusTotal (2023). https://virustotal.github.io/yara/

9. Abu-Nimeh, S., Nappa, D., Wang, X., & Nair, S.: A comparison of machine learning techniques for phishing detection. In: Proceedings of the Anti-Phishing Working Groups 2nd Annual eCrime Researchers Summit. (2007). https://dl.acm.org/doi/https://doi.org/10.1145/1299015.1299019
10. Bergholz, A., De Beer, J., Glahn, S., et al.: A real-life study of the applicability of a Naive Bayes classifier and a decision tree to phishing email detection. In: Proceedings of the 23rd ACM Symposium on Applied Computing (SAC '10). (2010). https://dl.acm.org/doi/https://doi.org/10.1145/1774088.1774091
11. Basnet, R., Mukkamala, S., & Sung, A.H.: Detection of phishing attacks: a machine learning approach. Soft Computing Applications in Industry. Studies in Fuzziness and Soft Computing, vol 226. (2008). https://link.springer.com/chapter/https://doi.org/10.1007/978-3-540-77465-5_12
12. Fette, I., Sadeh, N., & Tomasic, A.: Learning to detect phishing emails. In: Proceedings of the 16th International Conference on World Wide Web. (2007). https://dl.acm.org/doi/https://doi.org/10.1145/1242572.1242660
13. Garera, S., Provos, N., Chew, M., & Rubin, A.D.: A framework for detection and measurement of phishing attacks. In: Proceedings of the 2007 ACM Workshop on Recurring Malcode. (2007). https://dl.acm.org/doi/https://doi.org/10.1145/1314389.1314391
14. Ma, J., Saul, L. K., Savage, S., & Voelker, G. M.: Beyond blacklists: learning to detect malicious web sites from suspicious URLs. In: Proceedings of the 15th ACM SIGKDD International Conference on Knowledge Discovery and Data Mining. (2009). https://dl.acm.org/doi/https://doi.org/10.1145/1557019.1557036
15. Afroz, S., & Greenstadt, R.: PhishZoo: detecting phishing websites by looking at them. In: Proceedings of the 5th IEEE International Conference on Semantic Computing. https://ieeexplore.ieee.org/document/6069444 (2011)
16. Aburrous, M., Hossain, M. A., Dahal, K., & Thabtah, F.: Intelligent phishing detection system for e-banking using fuzzy data mining. Expert. Syst. Appl. **37**(12), 7913–7921. https://www.sciencedirect.com/science/article/abs/pii/S095741741000346X (2010)

Threat Intelligence and Emerging Techniques for Fraud Detection and Prevention

TIPS: Threat Sharing Information Platform for Enhanced Security

Lakshmi Rama Kiran Pasumarthy, Hisham Ali, William J. Buchanan, Jawad Ahmad, Audun Josang, Vasileios Mavroeidis, and Mouad Lemoudden

Abstract There is an increasing need to share threat information for the prevention of widespread cyber-attacks. While threat-related information sharing can be conducted through traditional information exchange methods, such as email communications etc., these methods are often weak in terms of their trustworthiness and privacy. Additionally, the absence of a trust infrastructure between different information-sharing domains also poses significant challenges. These challenges include redactment of information, the Right-to-be-forgotten, and access control to the information-sharing elements. These access issues could be related to time bounds, the trusted deletion of data, and the location of accesses. This paper presents an abstraction of a trusted information-sharing process which integrates Attribute-Based Encryption (ABE), Homomorphic Encryption (HE) and Zero Knowledge Proof (ZKP) integrated into a permissioned ledger, specifically Hyperledger Fabric (HLF). It then provides a protocol exchange between two threat-sharing agents that share encrypted messages through a trusted channel. This trusted channel can only be accessed by those trusted in the sharing and could be enabled for each data-sharing element or set up for long-term sharing.

L. R. K. Pasumarthy (✉) · H. Ali · W. J. Buchanan · J. Ahmad · M. Lemoudden
Blockpass ID Lab, Edinburgh Napier University, Edinburgh, UK
e-mail: lakshmiramakiran.pasumarthy@napier.ac.uk

H. Ali
e-mail: h.ali@napier.ac.uk

W. J. Buchanan
e-mail: b.buchanan@napier.ac.uk

J. Ahmad
e-mail: j.ahmad@napier.ac.uk

A. Josang · V. Mavroeidis
University of Oslo, Oslo, Norway
e-mail: osang@ifi.uio.no

V. Mavroeidis
e-mail: vasileim@ifi.uio.no

M. G. Jaatun et al. (eds.), *Proceedings of the International Conference on Cybersecurity, Situational Awareness and Social Media*, Springer Proceedings in Complexity,
https://doi.org/10.1007/978-981-96-0401-2_12

Keywords Distributed ledger technology · Cyber threat intelligence · Encryption · GDPR

1 Introduction

With increasing threats in cybersecurity, the requirement to share threat information becomes ever more important. Traditional transport mechanisms such as electronic email often lack any real form of privacy and trust, and so we need improved methods that can identify the sender and receiver of threat-sharing information. We also need to integrate threat-sharing methods that integrate privacy-aware techniques, especially to comply with data regulations such as the General Data Protection Regulation (GDPR) etc. This should support enhanced features such as the redaction of information, the Right-to-be-forgotten (RTBF), and location/time-based access controls. And, so, end-to-end encryption using messaging applications, such as WhatsApp and Signal, can be used for information sharing; the trust infrastructure is generally hosted on external systems, which limits the usage of digital wallets and in the auditing of the threat-sharing exchange. Risks can also exist in external parties *ghosting* onto communications [1].

This proposed TIPS platform tackles these issues by combining cutting-edge cryptographic methods like Attribute-Based Encryption (ABE), Homomorphic Encryption (HE), and Zero Knowledge Proofs (ZKP) within a permissioned blockchain framework (Hyperledger Fabric). It supports compliance and privacy laws like the General Data Protection Regulation (GDPR) by guaranteeing that only authorised entities can access shared threat information. It then builds on this to provide a secure and trusted channel for exchanging information on threats using a permissioned blockchain (Hyperledger Fabric). The main contribution of this paper is to provide a protocol that will allow threat agencies to pass threat information in an encrypted format within a trusted communication channel. A permission blockchain is used as this allows for the integration of trusted threat-sharing agencies, where each threat-sharing agent within a trusted agency can prove their identity and share their public keys in a trusted way. The enactment of this is achieved through the use of chain code.

Maasimo et al. [2] presented the state-of-the-art for Threat information sharing standards and platforms. Several standards have been introduced to facilitate the sharing of threat information, such as Structured Threat Information CybereXpression (STIX), Cyber Observable eXpression (CybOX), Incident Object Description Exchange Format (IODEF), and Trusted Automated eXchange of Indicator Information (TAXII). Some platforms based on these standards are Malware Information Sharing Platform, Collaborative Research Threats (MITRE CRIT), Collective Intelligence Framework (CIF), EclecticIQ Platform and LookingGlass Cyber.

Overall, the paper outlines the creation of a trusted communication channel between threat agents where we use digital wallets to identify them and thus digitally sign for trusted information flows. In the case of Alice sending threat-sharing information to Bob, Bob generates his public key pair on his computer—(E_{priv}, E_{pub}). Bob and Alice then set up a trusted Hyperledger channel that cannot be accessed by other threat agents. It refers to the establishment of a private, secure communication channel that is only accessible by the parties that are trusted, such as Alice and Bob, within the Hyperledger Fabric network. Cryptographic techniques are used to protect this channel and guarantee the integrity and secrecy of the transferred data.

Bob and Alice created a secured private Hyperledger channel. The configuration of this channel restricts access to people who possess the appropriate permissions. This makes sure that other threat agents are unable to see in on the conversation without the necessary authorization.

Bob then places his public key in that channel. Alice can then receive this and encrypt a symmetric key with Bob's public key (E_{pub}). She can then place the encrypted threat data and the encrypted key in the channel for Bob to receive. Bob then decrypts this on his computer using his private key. An important advantage of this is that we can audit and log each part of this process so that Alice knows when Bob reads the encrypted message and when Alice posts it to the secure channel. We can thus make the channel a long-term location for information sharing, or create for each transfer. In this case, RSA public key encryption is used to support the generation of the key pair.

1.1 Permissioned Blockchain

A permissioned blockchain, alternatively called a private blockchain, is another type of blockchain network in which accessibility is limited to specific user segments. Hyperledger Fabric is the best example of this kind of blockchain network, whereas Ethereum is an example of a public blockchain. Several industries, such as healthcare, defence, and investigation agencies, can be applications of permissioned blockchains [3]. Hyperledger Fabric, an open-source framework, is pivotal for building distributed ledgers using blockchain technology, encompassing applications, tools, and libraries [4]. One significant component is the *Fabric Channel*, providing secure communication among network participants for transactions within a subnet. It ensures trusted data sharing and demonstrates high transaction performance compared to traditional blockchain processes [5].

1.2 GDPR and RTBF

The General Data Protection Regulation (GDPR), instituted by the European Union in May 2018 [6], safeguards the privacy of personal data exchanges among EU member states and affiliated entities globally. Compliance with GDPR is obligatory for all entities engaging in data transactions with EU members, necessitating prior consent. Moreover, GDPR presides over pivotal facets of data authorisation and authentication, encompassing rights such as data access, modification, deletion, and combination, ensuring comprehensive oversight of data-sharing practices.

The Right to be Forgotten (RTBF) pertains to GDPR mandates concerning data deletion [7]. However, disseminating this right across the network proves impractical and unfeasible. Consequently, the significance of discovering covert protocols arises to facilitate content deletion upon request. Furthermore, this directive may be enforced by content proprietors or their designated representatives.

1.3 Cyber Threat Intelligence (CTI))

Cyber Threat Intelligence (CTI) is evidence-based knowledge that plays an important role in cybersecurity. The CTI lifecycle involves the process of data collection, analysis and dissemination among trusted users. Threat intelligence sharing has become crucial, enhancing organisations' security posture against emerging threats and keeping them vigilant to repel potential attacks. To plan for, respond to, and mitigate these risks effectively, the main objective of CTI is to assist organisations in understanding the threats they face, including the tactics, methods, and procedures (TTPs) employed by cyber attackers. Considering threat actors' background, intents, and capabilities is equally important to CTI in comprehending cyber threats' technical components. However, the most significant challenges threat intelligence agents encounter are sharing threats across different organisational boundaries, navigating privacy regulations, and implementing suitable technological solutions. We introduce a novel solution to these issues using a unified threat language (STIX) and privacy-preserving methods [8].

1.4 Structured Threat Information eXpression (STIX)

Structured Threat Information Expression (STIX) is a standardised language for describing cyber threat intelligence in a structured manner [8]. It provides a common framework for organisations to share, analyse, and respond to cyber threats effectively. STIX enables the exchange of threat information in a machine-readable format, facilitating automation and interoperability among security systems through different geographical locations. Using STIX, organisations can better understand and mitigate cyber threats, improving their overall cybersecurity posture.

2 Related Work

This section provides insights into recent advances in privacy regulations, blockchain technology, and threat information sharing.

2.1 Privacy-Preserving Data Sharing

Coppolino et al. [9] proposed an IDS architecture that offers customers the benefits of managed security services without giving third-party entities full access to their sensitive information. The growing number of cyber threats has led organisations and companies to seek managed security services (MSS) to address their information security concerns. However, privacy concerns regarding MSS outsourcing have resulted in a conflict between privacy and security monitoring solutions. This paper proposes an intrusion detection system (IDS) architecture for privacy-preserving security monitoring in MSS using homomorphic Encryption (HE) technology. Homomorphic Encryption enables computations on encrypted data without direct access to decryption, allowing secure processing, storage, and exchange of confidential information.

José et al. [10] describe privacy-enhancing Cybersecurity Information Sharing (CIS), which plays a vital role in combating cyber threats and incidents by facilitating improved situational awareness among its members. In the surge of sophisticated cyber attacks, partnerships via CIS are critical to read and respond to future threats. As firms use the cloud as a tool for transmitting data, the risk of data privacy is increasing. It proposed a protocol called PRACIS, a system for CIS networks that ensures the confidentiality of data as it is forwarded and aggregated. This system uses the established STIX standard data format. PRACIS uses traditional format-preserving techniques and homomorphic encryption methods to achieve its goals. Additionally, after conducting tests on a prototype for a portion of STIX, the outcome displays that up to 689 incidents per minute can be reported, much higher than the typical estimation of 81. Furthermore, the consolidation of 104 incidents can be completed within just 2.1 s, and the data transmission overhead is minimal at 13.5 kbps. Overall, these findings suggest that the cost of implementing PRACIS is reasonable and practical for practical use.

David et al. [11] proposed a collaborative model to address these problems in sharing threat information. Sharing Cyber Threat Information (CTI) can significantly aid in forecasting, preventing, and mitigating cyber-attacks. The person or organisation that owns the Data can choose a suitable level of trust and method for sanitising Cyber Threat Information (CTI) data. This can involve using plain text, anonymisation or pseudo-anonymisation strategies, or homomorphic encryption. Before the CTI data is distributed for analysis, it is processed and secured using these techniques. This paper proceeds with associated research on cloud-edge computing, CTI data sharing, and data security concerns in cloud-edge computing. The proposed trust model is then described, and it goes in-depth into the data-sharing infrastructure and explores the diverse deployment models available. Four real-world projects serving as test cases for the data-sharing infrastructure are explained, including their chosen levels of trust and deployment models to meet their unique trust requirements. An overview of the progress made thus far, the testing methodology was implemented, and information on validation work completed to date was provided. It concludes by discussing the current limitations and identifying areas requiring additional research.

2.2 Architecture of C3ISP

Cyber threat intelligence has been proposed to achieve cyber threat information sharing between organisations. Various frameworks, such as C3ISP, have been proposed. C3ISP (Collaborative and Confidential Information Sharing and Analysis for Cyber Protection) is funded by the European Commission. The main goal of C3ISP is to create a framework in which entities can share cyber threat information securely and confidentially. This is designed to allow more collaboration between organisations, industries, and countries to better respond to and prevent cyber threats [12].

3 TIPS Infrastructure and Methodology

This section defines a privacy-aware threat-sharing architecture, which will be built by analysing the use case, which will be implemented, and the experimental work will provide insights into the overall architecture. These will address the key gaps within the existing research work and practice. This work uses Hyperledger Fabric, encryption and STIX Protocol [13], along with using TAXII (Trusted Automated eXchange of Indicator Information) [14]. With this platform, the research will integrate smart contracts and Hyperledger Fabric channels to provide trusted identity checks and private areas for information sharing. To control access to threat-sharing messages, attribute-based encryption (ABE), an access control mechanism for cloud storage, will be used. This provides secure and adaptive sharing encryption-based policies and authentication. Two core attributes include time and location, where a receiver must give a signed attestation of the current time and place for these to be included in the policy.

Trusted information sharing would happen through a trusted ledger, which will integrate a digital wallet, where Bob and Alice's public keys are stored on the shared ledger. Figure 1 outlines the communication between Alice and Bob using the ABE and in supporting interoperability. In ABE, the data is decrypted based on different attributes and policies.

Based on the literature review and background study, Hyperledger Fabric can be one of the best frameworks for the interoperability and implementation of this model-the design aspects for a sharing data element between Bob and Alice. Alice, who represents a cyber threat agency, is a sender who sends a cyphertext encrypted by the receiver's public key (Bob); and then Bob retrieves the message from the trusted cyber threat agency.

In this scenario, the communication happens within a private channel in a ledger, as explained in the steps shown in Fig. 1:

1. MSP manages the lifecycle of user identities, including registration, renewal, and revocation, to maintain network security and integrity. Managing user identities in Hyperledger Fabric involves key pair generation, certificate issuance, enroll-

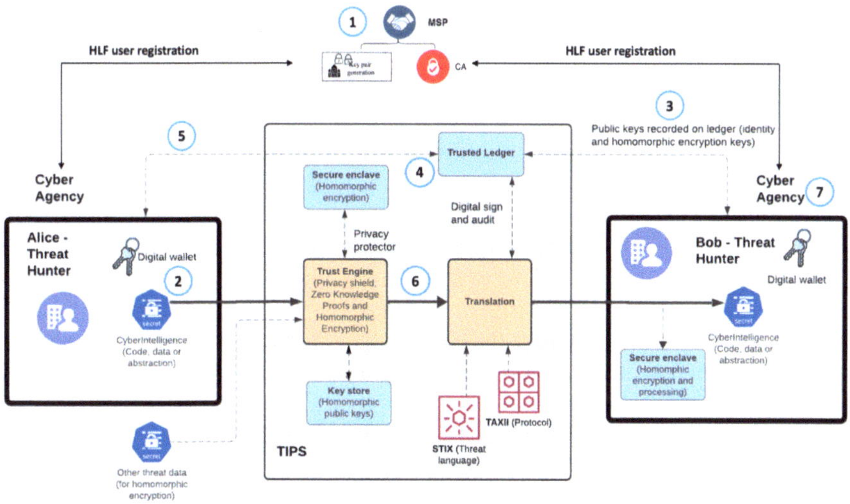

Fig. 1 Threat information sharing model with ABE, Homomorphic encryption and zero knowledge proof

ment, MSP configuration, and ongoing identity management to ensure secure and authenticated interactions within the network.
2. Use Hyperleger fabric to create two wallets.
3. The ledger stores users' public keys, identifies Bob and Alice and supports digital signing. They may be able to blind their identity using a privacy-aware digital signature. The ledger is used to audit shared information.
4. Setup a trusted channel on Hyperledger.
5. Get the other person's public key (generated off-chain).
6. Alice passes the encrypted data (STIX file) to the channel using Bob's public key.
7. ABE defines the attributed attributes to read the threat-sharing information. Bob uses its private key to decrypt STIX files that contain threat information. ZKPs are used to verify things without revealing the data source, such as for someone's proof of identity (such as proof of the ownership of a hardware token). Homomorphic encryption is used to aggregate data together from multiple threat-sharing instances.

A well-designed threat-sharing platform has several key attributes which increase its trustworthiness [15–17]. A firm basis to enhance data security is provided by the permissioned blockchain system Hyperledger Fabric, particularly when it involves threat intelligence information exchange between threat agencies. This is especially important for data security, as timely and safe threat intelligence sharing can greatly strengthen defences against collective threats. Here are a few ways that Hyperledger Fabric can help with this.

To enable secure communications, Bob and Alice set up a Hyperledger channel for their communication, either for a one-time session or with a long-term session. Bob will generate a key pair, and where the private key is:

$$k_d \in \mathcal{K}_d \tag{1}$$

and a public key of:

$$k_e = \mathcal{E}(k_d) \in \mathcal{K}_e \tag{2}$$

This public key is then passed to the Hyperledger channel, and where Alice picks it up. She then generates a random symmetric key of k_m and her threat-sharing message (m) of:

$$m \in \mathcal{M} \tag{3}$$

Alice will then encrypt the threat-sharing message with:

$$c = e_{k_m}(m) \tag{4}$$

and encrypt the symmetric key with:

$$k_s = e_{k_e}(k_m) \tag{5}$$

She will then pass c and k_s into the Hyperledger channel. Bob will pick these up, and then discover the symmetric key using:

$$k_m = d_{k_d}(k_s) \tag{6}$$

and then the message with:

$$m = d_{k_m}(c) \tag{7}$$

4 Implementation and Results

Sharing threat information is vital in cybersecurity battles, yet several key challenges persist. Firstly, there are no clear guidelines for sharing among organisations. Secondly, privacy and legal concerns arise. Thirdly, interoperability issues hinder understanding of different systems' threat data. Lastly, building trust across organisations proves difficult. This work aims to tackle these challenges by proposing strategies to facilitate seamless and trusted threat information sharing, ultimately enhancing online security for all.

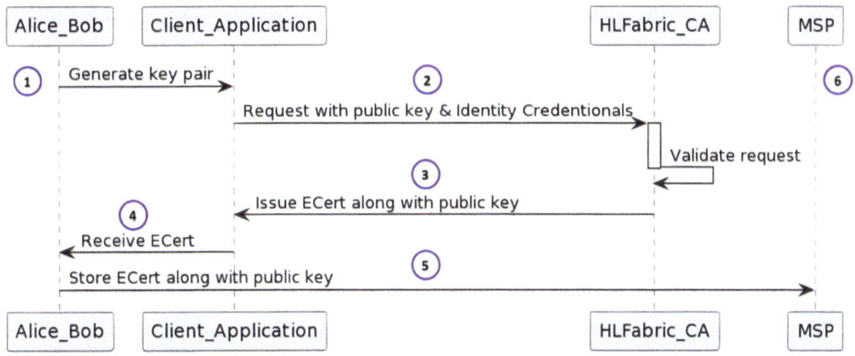

Fig. 2 Sequence diagram of Public key management and storage in Hyperledger Fabric

This section will provide a practical aspect using Hyperledger Fabric and robust encryption techniques to address these challenges. Specifically, we will utilise a specific environment to install a Hyperledger Fabric network and adapt the encryption techniques.

4.1 Users Registration on MSP, and Encryption Approach

The Membership Service Provider (MSP) registration and encryption process is pivotal in achieving the objectives. MSP registration involves authenticating network participants and assigning them cryptographic identities, while encryption ensures the confidentiality of data exchanged within the network. Understanding the intricacies of MSP registration and encryption processes is essential for effectively managing access control and safeguarding sensitive information within Hyperledger Fabric networks. This section delves into the MSP registration and encryption process, elucidating the underlying mechanisms and their significance in bolstering the security posture of Hyperledger Fabric-based blockchain networks.

Here's an overview of the process for managing user identities in HLF, starting from the generation of key pairs as shown in Fig. 2:

1. Key Pair Generation: Users generate a key pair consisting of a public key and a private key using RSA cryptographic algorithms.
2. Certificate Signing Request (CSR): Users create a Certificate Signing Request (CSR) containing their public key and identity information (e.g., Common Name, Organisation). The CSR is sent to the Certificate Authority (CA) for certificate issuance.
3. Certificate Issuance: The CA receives the CSR and verifies the user's identity. If the identity is valid, the CA issues an X.509 digital certificate containing the user's

public key and identity information. The CA's private key signs the certificate, making it tamper-proof and verifiable.

4. Membership Service Provider (MSP) Configuration: The HLF network defines an MSP for managing identities. MSP configuration specifies the trusted CA(s), certificate revocation lists (CRLs), and access control policies.
5. Enrollment: Users enrol in the HLF network by presenting their digital certificate to the Membership Service Provider (MSP).
   MSP verifies the certificate's authenticity and validity against the trusted CA(s). Upon successful enrollment, users are granted membership status within the network.
6. Identity Management: MSP manages user identities, including registration, revocation, and attribute-based access control. Users are assigned roles and permissions based on their identity attributes (e.g., organisation, role).

Overall, managing user identities in Hyperledger Fabric involves key pair generation, certificate issuance, enrollment, MSP configuration, and ongoing identity management to ensure secure and authenticated interactions within the network.

4.2 Messages Exchange Phases

Message transaction processes through HLF, enabling secure and transparent data exchange among network participants. This process includes transaction proposal, endorsement, ordering, validation, and commitment, ensuring the integrity of transactions on the ledger. Understanding this process is essential for grasping the mechanisms governing blockchain operations in HLF. This section explores the message transaction processing in HLF, outlining its key components and workflow for executing and validating transactions within a Hyperledger Fabric-based blockchain network as shown in Fig. 3 explained in the following sequences.

- Peer and Client Configuration: Peers and clients in the HLF network are configured with MSP settings, including the trusted CA(s) and identity information. MSP ensures that only authorised entities can participate in network activities.
- Transaction Endorsement and Verification: During transaction endorsement, peers verify the identity and signature of the transaction proposer using their digital certificate. MSP enforces access control policies to ensure that only authorised users can endorse transactions.
- Consensus and Commitment: Peers reach a consensus on transaction validity based on endorsement policies and transaction content. MSP verifies the integrity of transaction endorsements and ensures that only valid transactions are committed to the ledger.
- Identity Revocation: If a user's privileges need to be revoked (e.g., due to departure from the network or policy violation), MSP updates the Certificate Revocation List (CRL) to invalidate the user's certificate.

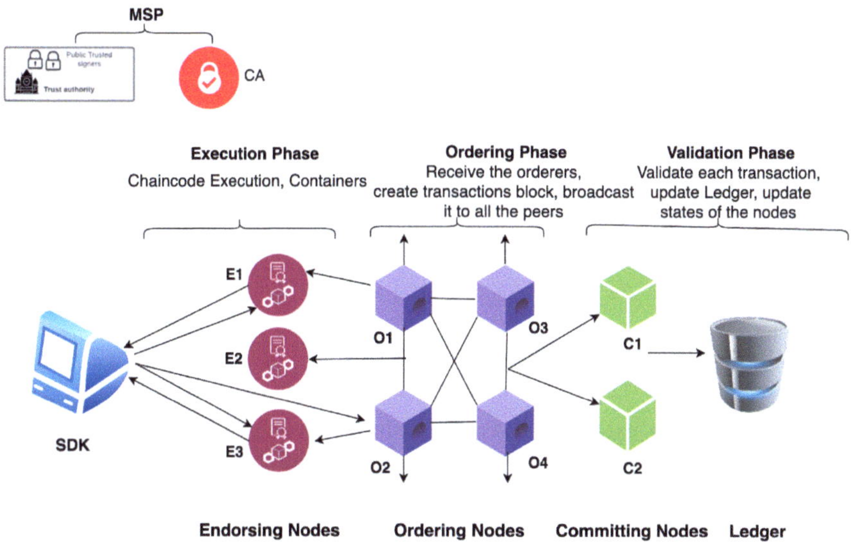

Fig. 3 Sequence diagram of Message Transaction within Hyperledger Fabric [18]

- Identity Lifecycle Management: MSP manages the lifecycle of user identities, including registration, renewal, and revocation, to maintain network security and integrity.

4.3 Environment Setup and Test Configurations

This section includes network performance and functionality tests. Initially, fabric binaries are installed via a command line interface to establish a two-organisation network, with each organisation having two peers linked through a private channel. Subsequently, the benchmark engine engages with chain code to deploy, execute, analyse, and produce network performance reports, depicted in Fig. 5. The setup of the Hyperledger Fabric network requires specific prerequisites, detailed in Table 1 includes the installation of Hyperledger Fabric binaries package version 2.2.0; Ubuntu version 20.04; Docker version 20.10.7; and Docker Compose version 1.25.0 for generating and managing network entities. Hyperledger Fabric offers three optional languages: Golang, Java, and JavaScript. Here, we utilised JavaScript (Node.js version 10.19.0) to build chaincode (smart contract).

Table 1 Hyperledger fabric environment setup

System and tools	Version
Operating system	Ubuntu 20.04
Hyperledger fabric	2.2.0
Docker	20.10.7
Docker-compose	1.25.0
Node.js	10.19.0

4.4 Interaction System Implementation

The implementation extends the asset-transferbasic/chaincode-javascript using a Solo Ordering Service provided by Hyperledger Fabric and a test network. Performance testing of the smart contract on a Fabric network is conducted using Caliper. The basic workflow involves implementing smart contracts using Hyperledger Fabric, deploying them with JavaScript, and testing by installation, approval, commitment to the channel, and invocation of chain code [19].

The chain code handles various data queries and addresses constraints of threat information sharing by using hash keys for lightweight implementation on endorsing peers. The system's chaincode operations include storing data, querying checksums, retrieving objects, extracting versions, retrieving lineage, and more.

Performance testing encompasses system throughput, latency, scalability, and resource usage, evaluated with different transaction loads and asset batch sizes. Benchmarks involve 'getAssetsFromBatch' transactions, with evaluations conducted using Caliper benchmarking tools. The assessment includes real-time data reporting and resource consumption statistics for Hyperledger Fabric V2.2.0. The subsequent steps below illustrate different functions within Hyperledger infrastructure setup and network performance assessment:

- Start the test network and establish the channel.
- Package and deploy the smart contract.
- Approve the chaincode definition.
- Commit the chaincode definition on the channel.
- Execute the chaincode, as depicted in Fig. 4.
- Integrates benchmarking tools to assess network performance, monitoring network latency, send rate, and throughput, as demonstrated in Fig. 5.

4.5 Results

Performance benchmarking was conducted using the Hyperledger Caliper Benchmarking Tool, focusing on fundamental performance indicators during the message

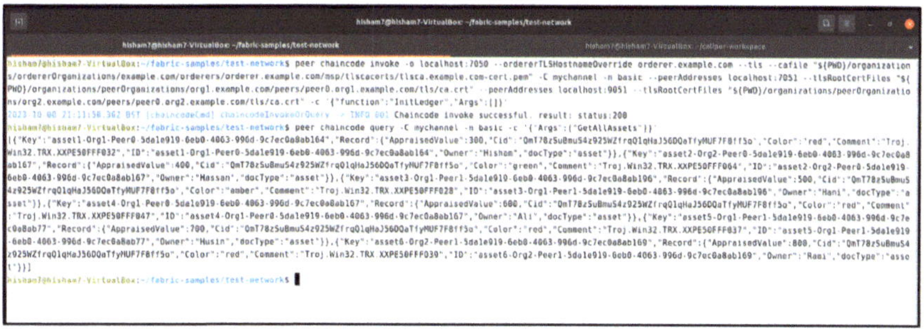

Fig. 4 Chaincode invocation with majority endorsement peers

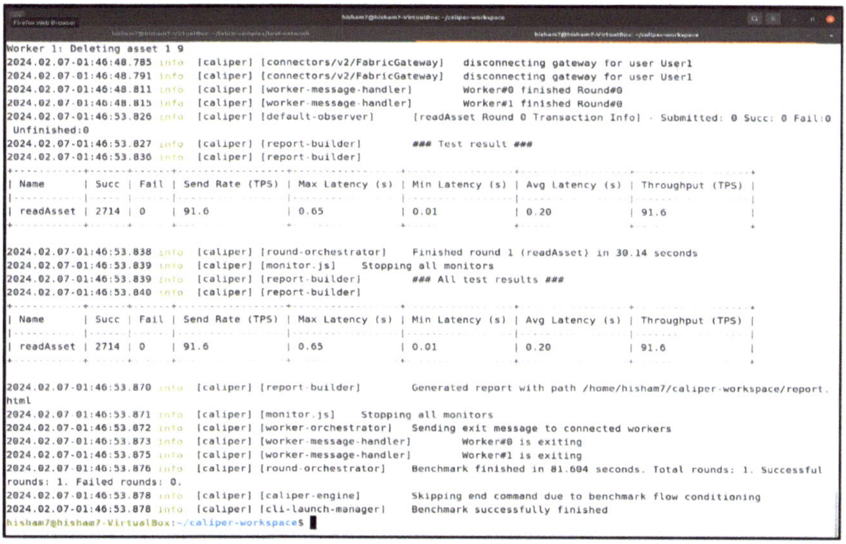

Fig. 5 Running Caliper benchmark and getting the network performance report

exchange between Alice and Bob. Our results show reading measurements, achieving approximately 91.6 transactions per second. During extensive testing, we observed a maximum latency of 0.65 s, a minimum latency of 0.01 s, and an average latency of 0.20 s. The latency is the time for the ciphertext and encrypted session key to be committed to the ledger and then to be read by the recipient.

These scalability, throughput and latency results are derived from Hyperledger Fabric's Execute-Order-Validate method, which separates transaction execution and ordering. This separation enhances scalability and performance and reduces node workload. Unlike traditional blockchain architectures, Fabric's approach enables parallel transaction processing, mitigating smart contract non-determinism and yielding higher throughput with lower latency. Consequently, it fosters an efficient and high-performance blockchain ecosystem for trusted sharing.

5 Conclusions and Future Work

Although this study describes TIPS's theoretical foundation and initial implementation, future research will primarily focus on empirical performance evaluations. These assessments will include,

- Scalability Testing: Use tools like Hyperledger Calliper to monitor throughput, latency, and scalability under different loads.
- User Experience studies: Through user studies, possible adoption roadblocks and the platform's usability are investigated. Testing for interoperability and ensuring smooth interaction with current threat intelligence platforms, such as MISP, are important aspects of interoperability testing.

By addressing these pointers, we aim to validate TIPS's theoretical advantages and offer a direct route to implementation and use in real-world scenarios. Based on the performance benchmarking results and an in-depth understanding of message transaction processes through HLF, this research highlights the importance of privacy-preserving threat information sharing. Using RSA encryption to pass the threat-sharing information provides a mechanism to allow the recipient to pass their public key within a trusted channel and for the private key to be stored in a secure enclave or on the recipient's computer. This channel can be set up either for each session or for a long-term session. While acceptable when a secure enclave is used, it has weaknesses related to *forward secrecy* in that the recipient's private key is breached, especially in creating a long-term channel. For this, an enhancement would look towards PAKE (Password-Authenticated Key Exchange) methods, such as OPAQUE [20], to allow for a symmetric key to be exchanged within an asynchronous hand-shake of the key.

In this paper, the TIPS platform is introduced, a new threat-sharing information platform that combines cutting-edge cryptography methods inside a permissioned blockchain. Our method tackles important issues in exchanging threat information in a secure and privacy-aware manner. The goal is to offer the cyber security community a reliable and useful solution by addressing user adoption and usability, combining empirical performance evaluations, and more. TIPS is a significant contribution to the world of cyber security community because of its novel elements of combining blockchain and privacy-focused proposal.

References

1. Baraniuk, C.: Ghosts in the machines. New Sci. **227**(3028), 38–41 (2015)
2. Guarascio, M., Cassavia, N., Pisani, F.S., Manco, G.: Boosting cyber-threat intelligence via collaborative intrusion detection. Future Gener. Comput. Syst. **135**, 30–43 (2022)
3. Polge, J., Robert, J., Le Traon, Y.: Permissioned blockchain frameworks in the industry: a comparison. ICT Express **7**(2), 229–233 (2021). [Online]. Available: https://www.sciencedirect.com/science/article/pii/S2405959520301909
4. Elrom, E., Elrom, E.: Hyperledger. The Blockchain Developer: A Practical Guide for Designing, Implementing, Publishing, Testing, and Securing Distributed Blockchain-based Projects, pp. 299–348 (2019)
5. A.-A. C. services: Blockchain Technologies (2022)
6. Grundstrom, C., Väyrynen, K., Iivari, N., Isomursu, M.: Making sense of the general data protection regulation-four categories of personal data access challenges (2019)
7. Havelange, A., Dumontier, M., Wouters, B., Linde, J., Townend, D., Riedl, A., Urovi, Luce: a blockchain solution for monitoring data license accountability and compliance (2019). arXiv preprint arXiv:1908.02287
8. Ali, H., Ahmad, J., Jaroucheh, Z., Papadopoulos, P., Pitropakis, N., Lo, O., Abramson, W., Buchanan, W.J.: Trusted threat intelligence sharing in practice and performance benchmarking through the hyperledger fabric platform. Entropy **24**(10), 1379 (2022)
9. Sgaglione, L., Coppolino, L., D'Antonio, S., Mazzeo, G., Romano, L., Cotroneo, D., Scognamiglio, A.: Privacy Preserving Intrusion Detection via Homomorphic Encryption, pp. 321–326 (2019)
10. de Fuentes, J.M., González-Manzano, L., Tapiador, J., Peris-Lopez, P.: Pracis: privacy-preserving and aggregatable cybersecurity information sharing. Comput. Secur. **69**, 127–141 (2017). Security Data Science and Cyber Threat Management. [Online]. Available: https://www.sciencedirect.com/science/article/pii/S0167404816301821
11. Chadwick, D.W., Fan, W., Costantino, G., de Lemos, R., Di Cerbo, F., Herwono, I., Manea, M., Mori, P., Sajjad, A., Wang, X.-S.: A cloud-edge based data security architecture for sharing and analysing cyber threat information. Future Gener. Comput. Syst. **102**, 710–722 (2020). [Online]. Available: https://www.sciencedirect.com/science/article/pii/S0167739X19300895
12. Ec c3isp project (2022). [Online]. Available: https://c3isp.eu
13. Rantos, K., Spyros, A., Papanikolaou, A., Kritsas, A., Ilioudis, C., Katos, V.: Interoperability challenges in the cybersecurity information sharing ecosystem. Computers **9**(1), 18 (2020)
14. Connolly, J., Davidson, M., Schmidt, C.: The Trusted Automated Exchange of Indicator Information (taxii). The MITRE Corporation, pp. 1–20 (2014)
15. Burger, E.W., Goodman, M.D., Kampanakis, P., Zhu, K.A.: Taxonomy Model for Cyber Threat Intelligence Information Exchange Technologies, pp. 51–60 (2014)
16. Wagner, T.D., Palomar, E., Mahbub, K., Abdallah, A.E.: A novel trust taxonomy for shared cyber threat intelligence. Secur. Commun. Netw **2018** (2018)
17. Mavroeidis, V., Bromander, S.: Cyber Threat Intelligence Model: An Evaluation of Taxonomies, Sharing Standards, and Ontologies Within Cyber Threat Intelligence, pp. 91–98 (2017)
18. Ali, H., Papadopoulos, P., Ahmad, J., Pitropakis, N., Jaroucheh, Z., Buchanan, W.J.: Privacy-preserving and trusted threat intelligence sharing using distributed ledgers. In: 2021 14th International Conference on Security of Information and Networks (SIN), vol. 1, pp. 1–6. IEEE (2021)
19. Ali, H., Abubakar, M., Ahmad, J., Buchanan, W.J., Jaroucheh, Z.: Passion: permissioned access control for segmented devices and identity for IoT networks (2023). arXiv preprint arXiv:2310.05032
20. Jarecki, S., Krawczyk, H., Xu, J.: Opaque: an asymmetric pake protocol secure against pre-computation attacks. In: Advances in Cryptology—EUROCRYPT 2018: 37th Annual International Conference on the Theory and Applications of Cryptographic Techniques, Tel Aviv, Israel, April 29–May 3, 2018 Proceedings, Part III 37, pp. 456–486. Springer (2018)

TRIST: Towards a Container-Based ICS Testbed for Cyber Threat Simulation and Anomaly Detection

Carol Lo, Jack Christie, Thu Yein Win, Zeinab Rezaeifar, Zaheer Khan, and Phil Legg

Abstract Cyber-attacks on Industrial Control Systems (ICS), as exemplified by the incidents at the Maroochy water treatment plant and the Ukraine's electric power grid, have demonstrated that cyber threats can inflict significant physical impacts. These incidents caused widespread service disruptions and substantial economic losses, underscoring the urgent need for an in-depth understanding of cyber threats in industrial environments. Industrial security research is usually conducted on physical testbeds to avoid safety issues, production interruptions and other operational constraints in industrial processes. Nevertheless, security defenders often encounter obstacles in developing or accessing physical testbeds due to associated costs and complexities. These factors hinder research progress to devise early detection mechanisms for cyber threats—essential for effective incident response. To overcome these obstacles, this paper presents a container-based virtual testbed. Its lightweight architecture enables replicable and efficient deployment of testbeds at low cost for simulating cyber threats on Cyber-Physical Systems (CPS)—the cornerstone of industrial automation and control systems. Also, the container-based virtual testbed provides a cost-effective option for producing datasets for training, testing and optimization of unsupervised anomaly detection models. Besides, an evaluation on resource consumption is conducted. The paper also discusses the benefits and limitations of proposed container-based ICS testbeds and suggests future research areas.

C. Lo (✉) · J. Christie (✉) · Z. Rezaeifar · Z. Khan · P. Legg
University of the West of England, Bristol BS16 1QY, UK
e-mail: carol.lo@uwe.ac.uk

J. Christie
e-mail: jack2.christie@live.uwe.ac.uk

T. Y. Win
University of Sunderland, Sunderland SR1 3SD, UK

M. G. Jaatun et al. (eds.), *Proceedings of the International Conference on Cybersecurity, Situational Awareness and Social Media*, Springer Proceedings in Complexity,
https://doi.org/10.1007/978-981-96-0401-2_13

1 Introduction

1.1 Security of Industrial Control Systems in Industry 4.0

Industrial Control Systems (ICS), also referred to as Operational Technology (OT), are complex Cyber Physical Systems (CPS) that use hardware and computational components to automate and control physical processes of critical infrastructure and manufacturing systems. Traditionally, protection of these systems can be achieved through security by obscurity or air-gapped strategies [1]. However, convergence of IT and OT in Industry 4.0 has brought new challenges in securing these interconnected systems. For instance, expanded attack surface allowing threats to move laterally from IT network to critical OT systems that have high availability and safety requirements, and increasing prevalence of Advanced Persistent Threats (APT) that targeting ICS through various application of Cyber Kill Chain (CKC).

Cyber-attacks on ICS can bring devastating real-world impacts, as exemplified by the cyber incidents at the Maroochy water treatment plant in 2000, which resulted in the release of 800,000 liters of raw sewage into the local community. Another cyber incident has demonstrated adversary's capabilities to trigger hours of power outages by attacking Ukraine's electric power grid [2]. These incidents highlight the urgent need for early threat detection for effective incident response.

1.2 Obstacles in ICS Security Research

Although penetration test on production systems is useful for identifying actual risks against systems [33], it is an uncommon practice for testing ICS due to safety concerns and the requirement for uninterrupted operational processes. Typically, research communities often choose to build testbeds to investigate threats. Nonetheless, the traditional practices can hinder collaborative security research and overall research progress with three main reasons:

1. **Confidentiality or sensitivity concerns**: Data captured from production systems are undisclosable due to confidentiality or sensitivity concerns.
2. **Cost and time**: Physical testbeds that mimic production environments are costly and time-consuming to set up, operate, maintain, and dismantle.
3. **Replication difficulties**: Replicating production environment or physical testbeds requires substantial hardware investment and specialist knowledge to setup.

These obstacles often render cybersecurity research unrepeatable and hard to verify. Recognizing the growing need for alternative solutions, recent work by [3] has stressed the importance of building virtual or hybrid testbeds for security research. In alignment with these observations, our work explores the practical design of virtual testbed, for which non-physical nature of the CPS could be viewed as a Digital

Twin without direct interaction with its physical components for simulating and analyzing CPS data for developing threat detection model, making minimal impact to availability and functionalities of the underlying physical parts and processes.

1.3 Research Questions and Proposed Approach

To overcome these obstacles and explore alternatives to physical testbeds, this paper aims to address the following two research questions:

Research question 1 (RQ1): To what extent are the results of threat simulation in a virtual testbed relevant to corresponding physical testbeds?

Research question 2 (RQ2): How can a threat detection model and its training data simulated in virtual testbed be effectively used for optimization of threat detection?

To answer the above questions, Threat Research and Intelligence Sharing Testbed (TRIST) is proposed. TRIST is a virtual testbed leveraging containerization technology for simulation and detection of threats on CPS. Comparing with setting up physical testbeds, two major benefits could be obtained from the proposed approach:

- Scalability and Replicability: Virtual testbeds can be sharable and repeatedly built with minimum effort and be portable to different operating systems and computing settings, thereby facilitating research collaboration and verification.
- Accessibility: Virtual testbeds remove geographical barriers to physical testbeds, and eliminate the complexity associated with the acquisition, installation and ongoing cost and maintenance of physical hardware components.

As this paper does not utilize physical testbed, to ensure fidelity of TRIST, Secure Water Treatment (SWaT) sample case published on MiniCPS [4] is used as a basis to simulate key process in water treatment plant. Utilizing the SWaT case study enables us to address RQ1 by examining whether the consequences of selected attacks observed in the SWaT physical testbed can be replicated in a virtualized environment, ensuring simulation is realistic (Sect. 4.1). Also, simulated data collected from TRIST is used to address RQ2, showcasing how machine learning-based threat detection systems could be optimized to detect targeted attacks (Sect. 4.2).

1.4 Contribution of this Paper

Our research addresses two fundamental research questions through the development and application of a container-based virtual testbed. The key contributions include:

1. **Design, development and validation of a virtual testbed**: We designed and developed a virtual testbed with Docker containers that accurately simulates raw water supply and storage processes in a water treatment plant. We also simulated

a selected set of attacks by manipulating control logic on CPS. Based on empirical analysis, we validated that virtual testbed achieved the same attack consequence for selected attacks, as if they were conducted on SWaT physical testbeds.

2. **Machine learning model development and validation**: We developed a processing pipeline in Python to automatically collect data from virtual testbed for development of unsupervised time series anomaly detection model. Its detection capability is evaluated by using datasets containing simulated attacks. The experiment showed that our virtual CPS testbed facilitated better threat detection.
3. **Resource consumption and scalability analysis**: We evaluated the CPU and memory consumption of the container-based testbed and assessed the system performance if the workload of the testbed is doubled. This analysis provides an understanding of the potential impact if the testbed was to be deployed in computing environment with resource limitations.
4. **Open accessibility and replicability**: To foster collaborative research, we have made all the relevant scripts and machine learning models publicly available on a GitHub repository [34]. The docker-compose functionality allows any level of expertise to easily replicate our testbed for research result verifications.

The remainder of this paper is as follows: we explain the common purposes of creating and using ICS testbeds (Sect. 2.1) and analyze the advantages and disadvantages of different ICS testbed types (Sect. 2.2). In the third section, we explain the design requirements for virtual testbeds, and reasons for choosing containers for developing proposed testbed (Sect. 3.1). We develop TRIST as a proof-of-concept and describe its system architecture and key components (Sect. 3.2). Preliminary evaluations on the container-based testbed are explained in Sect. 4, including threat simulation (Sect. 4.1), threat detection (Sect. 4.2), and resources consumption (Sect. 4.3). The paper discusses the benefits and limitations of TRIST (Sect. 5). We conclude with proposed future research areas (Sect. 6).

2 Related Work

This section introduces different types of ICS testbeds and analyses their advantages and disadvantages in cybersecurity research.

2.1 Purposes of Creating and Using ICS Testbeds

Testbeds are scaled-down versions of real-world industrial counterparts, typically developed in high-risk industries, especially when system failures could profoundly impact public health, safety, and the environment. These sectors include energy (e.g., wind farms [5], smart grids [6], gas pipelines [7]), utilities (e.g., 5G telecommunications [8], water treatment [9], and water distribution [10], railway [11]), and

Table 1 Key purposes of ICS testbeds

No	Purpose of testbed
1	**Testing security controls**. Evaluate the effectiveness of security controls within a test system before implementation in production
2	**Identifying system vulnerabilities**. Expose system vulnerabilities and assess the impact by threat simulation to reveal bugs, latency, or other issues
3	**Creating datasets**. Collect data for developing machine learning models while avoiding collection of confidential or sensitive information
4	**Training and educating users**. Provide users knowledge of the mechanics, operations, networks and system components in safe and controlled settings

safety–critical manufacturing [12, 13]. For cybersecurity purposes, Table 1 outlines four key motivations behind the creation of ICS testbeds.

Currently, there are no specific regulations imposing the creation of ICS testbeds and their specifications for cybersecurity research. Indeed, establishment of testbeds is influenced by context-specific regulations [14], and best practices [15, 16].

2.2 *Advantages and Disadvantages of Different ICS Testbed Types*

Based on the implementation of system and network components, ICS testbed could be broadly categorized into three types: physical, virtual, and hybrid. Table 2 summarized pros and cons of different ICS testbed types.

Physical Testbeds. Physical testbeds blend actual ICS hardware and software, such as [5–13] are typical examples. Physical testbeds offer high realism but demand considerable investment cost and time for setup and maintenance [17]. Thus, it is challenging to replicate.

Virtual Testbeds. Virtual testbeds use software simulators, such as Simulink [18] for process simulation, and MiniNet [19] for network simulation. Virtual testbeds

Table 2 Comparisons between physical, virtual and hybrid testbed

Factors	Physical	Virtual	Hybrid
Fidelity	High	Low	Moderate
Cost	High	Low	Moderate
Time	High	Low to moderate	Moderate
Hardware skills	High	Low	High
Accessibility	Low	High	Low to moderate
Reproducibility	Low	High	Low to moderate
Scalability	Low	High	Low to moderate

such as GRFICS [20] are cost-effective and scalable, but face criticism over their realism, as they rely solely on 3D visualization tools and software like SCADA Sim [21] and MiniCPS [4] to simulate Programmable Logic Controllers (PLC) or Human–Machine Interface (HMI). In recent years, various researchers [7, 22–24] demonstrated successful use of virtual testbeds for cyber security research.

Hybrid Testbeds. Hybrid testbeds, with fewer ICS hardware components than physical testbeds, blend real or simulated ICS software to mimic ICS processes. Hybrid testbeds like [3, 11] strike a balance, resolving some replication issues of physical testbeds and attaining a certain degree of realism. They simulate ICS operations and are connected to a small hardware scale, making them cost-effective and efficient. However, hardware sharing poses a challenge for widespread research.

In summary, achieving higher fidelity through more electrical, mechanical, or IT components in physical testbeds is possible but requires high investment cost, resources, and expertise. The costs and complexities associated with accessing physical testbeds can hinder cybersecurity research. Also, it is important to recognize that any testbed type cannot perfectly replicate every component of a full-scale production environment. Thus, exploring the development and application of virtual or hybrid testbeds as more cost-effective options to physical testbeds is necessary.

3 TRIST: Design and Proof of Concept

This section provides the methods for designing, setting up and operating TRIST.

3.1 Design Requirements of Virtual Testbeds

Table 3 presents design requirements for virtual testbeds, aiming to cope with the difficulties in building physical testbeds. These principles serve as the basic requirements of virtual testbeds to ensure effective implementation.

Table 3 Requirements for virtual testbed

Factor	Explanation of virtual testbed requirements
Fidelity	Critical components are accurately replicated
Cost	Low cost of testbed development, maintenance, and changes
Time	Efficient and ease of testbed setup, operations, and dismantling
Hardware skills	Minimal know-how on hardware acquisition and installation
Accessibility	Access to the testbed is unrestricted by location and time
Reproducibility	Enable consistent, sharable, and repeatable testbed setup
Scalability	Allow expansion and customization of testbed

Considering the reproducibility requirement, modularization capabilities and application deployment, TRIST has opted for containerization technology for the development of virtual testbeds. While traditional virtualization methods such as Virtual Machines (VMs) offered by VirtualBox [25] or VMware [26] and container-based solutions such as Docker [27] both enable isolation of environment in a sandboxed manner, the reasons below explain why containers are used in our testbed.

1. **Reproducibility**: Containers enable consistent and repeatable setup of testbeds in different platforms through configuration in Dockerfile. To start threat simulation in container-based testbed, the only prerequisite is to have Docker Desktop installed on the operating systems such as Linux, Windows, or Mac.
2. **Modularization Capabilities**: Containers allow separation of individual CPS sub-processes for simulation of different industrial processes and cyber threats. For a more sustainable use of testbeds, containerization can have a more versatile simulation by isolating or aggregating containers through addition or removal of containers and editing its underlying scripts. This requirement enables efficient adjustment to testbed components for adapting to fast-changing threat landscape.
3. **Application Deployment**: While the current focus of TRIST is threat simulation and detection, the use of Docker containers shows a strategic advantage for future deployment of trained machine learning models in computing environments with resource constraints. Its lightweight architecture enables models developed in testbeds to be portable to platforms where computing resources are a concern.

3.2 Proof of Concept

TRIST is inspired by MiniCPS [4], a toolkit developed by Singapore University of Technology and Design, which uses a Linux-based CPS simulation tool with Mininet [19] to emulate CPS network on VMs. Like MiniCPS, TRIST adopts a software-based approach to simulate CPS. However, TRIST specifically uses Python scripts with Docker containers to modularize PLC control logic and communication, working as software-defined PLCs to eliminate the need for physical hardware.

Unlike MiniCPS, which focuses on network behavior, TRIST is process-oriented and specifically designed to simulate CPS processes for examining the system dynamics and impacts of targeted attacks. The goal is to explore the use of simulated data for the optimization of anomaly-based threat detection model.

To ensure realism or fidelity of testbed, TRIST has adapted the SWaT sample case on MiniCPS toolkit as a basis to simulate the raw water supply and storage process in water treatment plant. This case was selected for its representative nature of typical ICS challenges and complexities. Figure 1 illustrates TRIST's system architecture. The entire raw water supply and storage process is containerized with Docker Engine. Development of TRIST involves configuring Dockerfiles for required tools and dependencies and integrating a data collection script. Docker-compose manages resource consumption and allocation for each container using default settings, with

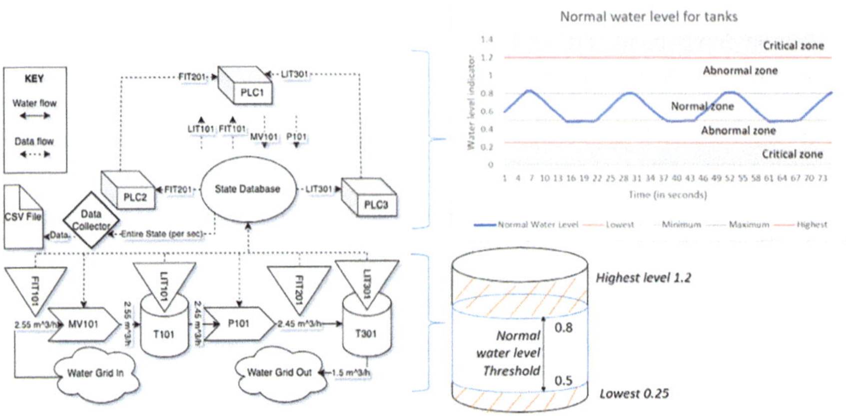

Fig. 1 TRIST consists of six Docker containers simulating water supply and storage process, with process state captured into a database

network communications via Common Industrial Protocol (CIP). A PCAP file showing these network communications is available in our GitHub repository [34].

The objective of the sample case is to maintain the tanks' water levels within predefined thresholds. The ideal threshold is between 0.5 to 0.8 m^3. The lowest and highest levels of the threshold are 0.25 m^3 and 1.2 m^3 respectively. If the reading of water level sensor is greater than 1.2 m^3, the tank overflows, and the simulation will stop. Nonetheless, it is important to note that the thresholds used in the experiment could indeed be adjusted in the Python script (utils.py) to customize the water tank requirement. As illustrated in Fig. 1, this industrial process is controlled by three PLCs (PLC1, PLC2, PLC3) using MiniCPS' API Device module to continuously monitor, set, send, and receive process state to SQLite State Database. Raw water is collected from the source and stored into raw water tanks (T101, T301). To maintain water levels within predefined thresholds, PLCs constantly monitor water flow sensors (FIT101, FIT201) and water levels (LIT101, LIT301) in tanks. Then, PLC decides to open or close the motorized valve (MV101) and pump (P101) to regulate water flow to avoid tank empty or full.

In summary, container-based testbed demonstrates the ease of use over traditional testbed setup. The implementation of the water supply and storage process in our testbed involves several Python scripts, each designed to simulate different components of the aforesaid process. As the Python scripts are collectively containerized, running the simulation is straightforward as users can initiate the entire process with one simple "*docker-compose up*" command in Docker Desktop. After executing this command, six Docker containers are created by Docker as follows:

- Container 'plc1-1': PLC1 monitors water level (LIT101), reads sensor data from PLC2 and PLC3, and then make decision to manage actuators MV101 (open/ close valve) and P101 (start/ stop pump).
- Container 'plc2-1': PLC2 monitors water flow (FIT201), sends data to PLC1.

- Container 'plc3-1': PLC3 monitors water level (LIT301), sends data to PLC1.
- Container 't101-1': tank T101 stores water with water level indicator LIT101.
- Container 't301-1': tank T301 stores water with water level indicator LIT301.
- Container 'data-collector-1': create logs and stores as data files in.csv format.

4 Preliminary Evaluation

This section provides preliminary evaluation on TRIST, including threat simulation (Sect. 4.1), threat detection (Sect. 4.2), and resources consumption (Sect. 4.3).

4.1 Threat Simulation

Evaluation was performed on a Windows 11 laptop with an AMD Ryzen 7 5800H processor and 16 GB of memory. To assess whether TRIST replicates outcomes comparable to those from SWaT physical testbed, we selected two documented attack scenarios conducted in [28]. TRIST-Case#1 and TRIST-Case#2 denote SWaT-Attack#1 and SWaT-Attack#21 respectively—attacker aims to cause tank T101 to overflow. The attacks, shown in Table 4, are akin to Dead Man's PLC technique for which attacker has full control on the control logic, and it is hard to be detected [29].

TRIST-Case#1: plc1 script was modified to keep motorized valve (MV101) permanently open (i.e., value of MV101 should be "0" or "1", but the value is forced to be "1"). As a direct result of PLC code modification, water level sensor indicator LIT101 keeps increasing until it exceeds its highest level of 1.2, i.e., tank T101 overflows at 57th seconds of the simulation. The event is visualized in Fig. 2 as a vertical red line**.** Once the tank overflowed, the simulation stopped as expected.

We analyzed the data generated from the testbed during simulation; PLC1 tried to empty tank T101 by using pump P101 between 39th and 53rd seconds. However, pump P101 is turned off by PLC1 to avoid tank T301 overflow.

TRIST-Case#2: Same as TRIST-Case#1, plc1 script is further modified to keep motorized valve (MV101) open. Also, the water level sensor indicator (LIT101) is set to 0.7 to keep it within the normal zone. This scenario simulates the attacker has

Table 4 Attack cases selected for threat simulation

Scenario	Attacker objective	Attacker method	Expected result
TRIST-Case#1	Make tank T101's water level greater than its highest level of 1.2	Simulate a Denial-of Service (DoS) attack on MV101 to keep MV101 open	Tank T101 overflow
TRIST-Case#2		Keep motorized valve MV101 open, and keep water level sensor indicator LIT101 at a constant level of 0.7	

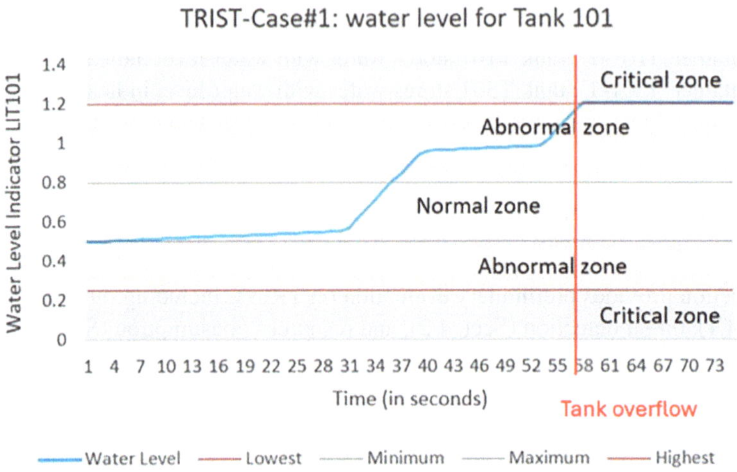

Fig. 2 Water level indicator for tank T101 during simulation of TRIST-Case#1

manipulated the values stored in state database, causing PLC1 to read incorrect sensor value and make inappropriate decision. Since PLC1 was unaware of the actual water level of tank T101, tank T101 overflowed as the script stopped working as expected.

Result: Data generated during two simulated attacks indicated that tank T101 has overflown, achieving the same result as in SWaT-Attack#1 and SWaT-Attack#21 which were conducted in SWaT physical testbed [28].

4.2 Threat Detection

To evaluate whether the proposed testbed could facilitate better threat detection on CPS, we validate whether the time series anomaly detection model trained with the simulation datasets could early detect the two unseen attacks simulated in Sect. 4.1.

While anomaly detection is commonly used to detect unknown threats, the model often suffers from a high false-positive rate (FP). Thus, anomaly detection optimization is needed if the model is to be deployed in production systems [30].

For unsupervised model training, we collected 1-hour of normal water supply and storage process from the Python-based simulation. Given our assumption that the generated data remains unaffected by physical environments and unknown attacks, extending the time for data collection will not affect model training. The dataset, named 'normal_1hr.csv', representing typical system dynamics, was used to train the anomaly detection model with Isolation Forest algorithm in PyCaret library [31].

For anomaly detection optimization, we trained the model with 10 contamination parameters ranging from 1 to 10% to test out which contamination parameter could early detect the two simulated attacks in Sect. 4.1, as shown in Table 5. The

Table 5 Anomaly detection optimization for threat detection

Contamination parameter (%)	Number of records assigned as 'anomaly records' in the 1-hour normal process dataset for model training	Timing of detection by testing the model with unseen attack TRIST-Case#1	Timing of detection by testing the model with unseen attack TRIST-Case#2
10	360 of 3595 records	1st seconds	1st seconds
9	**324 of 3595 records**	**1st seconds**	**1st seconds**
8	288 of 3595 records	5th seconds	1st seconds
7	252 of 3595 records	6th seconds	1st seconds
6	216 of 3595 records	40th seconds	1st seconds
5	180 of 3595 records	40th seconds	40th seconds
4	144 of 3595 records	53rd seconds	40th seconds
3	108 of 3595 records	54th seconds	63rd seconds (tank overflowed)
2	72 of 3595 records	54th seconds	Not able to detect
1	36 of 3595 records	62nd seconds (tank overflowed)	Not able to detect

datasets for two attack cases, specifically TRIST-CASE#1 ('attack1.csv') and TRIST-CASE#2 ('attack21_recorded.csv') are used for assessment. To facilitate validation and reproducibility of our work, all datasets used are available on GitHub [34].

On the one hand, a small contamination parameter assigns fewer normal records as 'anomaly' during model training and is less sensitive to unseen attacks during model testing. On the other hand, a large contamination parameter assigns much more normal records as 'anomaly' for model training, can early detect both simulated attacks during model testing, but could induce many false positive cases.

Based on the experiment, the optimized parameter is 9% if the two attacks carried out in Sect. 4.1 are to be detected at the 1st seconds of simulation. Although setting a lower contamination parameter such as 1 to 5% for anomaly detection model training could significantly reduce false positives, based on the experiment, it is not likely to timely detect the simulated attacks for effective incident response.

Result: With the data captured from TRIST, we can find the optimized contamination parameter for time series anomaly detection model, offering more effective threat detection and the possibility of reducing false positives.

4.3 Resources Consumption

The resources consumption analysis aims to provide an understanding of the potential impact if the testbed is scaled up and to be deployed in computing environment with resource limitations. For simplicity, we extend the testbed by simply reusing the

Table 6 Comparison of resource usage

Item	Water-Small testbed (5 processes)	Water-Medium testbed (10 processes)
CPU	11.84% ± 0.2% variation	23.23% ± 0.5% variation
Memory	0.11 Gigabytes	0.23 Gigabytes

script twice to double the workload of the testbed (i.e., increased number of tanks, motorized valves, pumps etc.).

1-hour simulations are separately conducted on Water-Small testbed (with 5 processes) and Water-Medium (with 10 processes), both simulating normal water supply and storage process. During simulation, we use "*docker stats*" command to collect data about CPU and memory usage, recorded the average values in Table 6.

Result: During the 1-hour simulation, both testbeds exhibit consistent CPU and memory usage, except for a reduction of memory usage during 33rd—38th minutes for Water-Small testbed, and during 15th—36th minutes for Water-Medium testbed. Nonetheless, the drop in memory usage does not adversely affect the simulation process. Overall, it is noticed that the resource consumption exhibits a direct proportion to the number of processes operating in simulation.

5 Discussion

While MiniCPS framework [4] focuses on simulating CPS networks on Unix-based systems, our proposed TRIST testbed adopts a container-based approach to simulate the CPS processes for understanding the system dynamics under targeted attacks. Using the MiniCPS's SWaT testbed as a case study, TRIST shows how CPS-targeted attacks can be executed, with data being collected for anomaly detection and optimization. While SWaT water treatment process serves as our initial test case, it is worth noting that its modular design can be expanded to simulate other CPS settings.

Although not currently within the scope of this research, containers can be easily isolated or aggregated by editing the configurations in Dockerfile to simulate individual CPS subprocesses. Also, the data collection Python script can be modified to capture data for other Machine Learning security analytics experiments. Therefore, the flexibility of container-based testbed enables a more versatile and efficient approach to simulating different threats for training and optimization of anomaly-based detectors, particularly focusing on detecting threats targeting CPS processes.

Without relying on or interacting with physical hardware, our software-based approach to CPS testbeds overcomes the costs associated with physical testbeds. While testbeds such as SWaT [9] allow for more realistic simulation of CPS behavior, the costs associated with it typically prevent them from being used to test out various targeted cyber-attack simulation scenarios. The proposed approach allows for the simulation of targeted attacks in a sandbox manner, using Docker [27] along with

Python [32]. Using containers facilitates reproducible research via consistent and repeatable testbed setups across different platforms by configuring Dockerfile.

In addition, using open-source technologies such as Docker are more cost-effective approach than existing virtual testbeds like GRFICS [20]. While the use of Unity in the latter allows for realistic visualization of the simulated CPS infrastructure, it requires high computing resources. As evidenced by the low computational costs associated with its deployment, TRIST provides a user-friendly, lightweight and low-cost alternative to GRFICS [20], thus lowering the barrier for virtual CPS testbed adoption. TRIST testbed overcomes the costs and computational requirements of both existing physical and virtual CPS testbeds. Given its performance, further capabilities, such as integration with physical devices, will be explored in future.

However, one of the limitations of TRIST testbed is that the physical parameters associated with SWaT subprocesses are hardcoded. While this still allows for the simulation of targeted attacks, making them more dynamic would have provided a more realistic view of the CPS system. Future iterations of TRIST will address this, by including variable flow rate and capacity.

6 Conclusion and Future Work

In this paper, a container-based ICS testbed simulating a water treatment plant is proposed for the enhancement of threat detection. Inspired by the SWaT ICS testbed, the proposed testbed allows for the safe simulation of ICS-targeted attacks in a sandboxed and resource-constrained environment. In addition, it also allows for the generation and collection of cyber threat data essential for enhancing unsupervised time series anomaly detection. Anomaly detection, while effective for identifying novel threats, typically suffers from high false-positive rates. Calibrating the sensitivity of anomaly detection models is required for timely and accurate threat detection: excessive sensitivity can lead to an overwhelming number of false positives, while insufficient sensitivity leads to increased risk of undetected threats. This fine-tuning is crucial for the development of a reliable threat detection system. Our proposed testbed offers a risk-free testing ground to validate and optimize this essential detection control before migrating to production systems for deployment.

The future development of this research will involve adding more container modules which simulate other aspects of the SWaT ICS testbed. In addition, it will involve further exploration of simulating other ICS systems, such as power grids and smart manufacturing. Along with this, further attempts will be undertaken to run more sophisticated ICS-targeted attacks on the testbed and integrating with technologies, such as federated learning and machine learning for automated cyber threat detection.

We encourage further experimentation and repeatability of our approach from the research community, by making all TRIST source code and related data sets available via the GitHub repository: https://github.com/RedClouud/ICS-test-bed-PoC [34].

References

1. Cybersecurity & Infrastructure Security Agency (CISA).: https://www.cisa.gov/sites/default/files/recommended_practices/NCCIC_ICS-CERT_Defense_in_Depth_2016_S508C.pdf. Last accessed 20 Feb 2024
2. MITRE ATT&CK.: https://attack.mitre.org/. Last accessed 20 Feb 2024
3. Dehlaghi-Ghadim, A., Balador, A., Moghadam, M.H., Hansson, H., Conti, M.: ICSSIM—A framework for building industrial control systems security testbeds. Comput. Ind.. Ind. **148**, 103906 (2023)
4. MiniCPS.: A framework for cyber-physical systems real-time simulation. https://github.com/scy-phy/minicps. Last accessed 20 Feb 2024
5. MacKinnon, I.: The biggest offshore wind 'living lab' in the world to be developed in the Humber. https://ore.catapult.org.uk/press-releases/the-biggest-offshore-wind-living-lab-in-the-world-to-be-developed-in-the-humber. Last accessed 20 Feb 2024
6. Florida International University.: Smart grid test-bed laboratory. https://energy.fiu.edu/smart-grid-test-bed-laboratory. Last accessed 20 Feb 2024
7. Alves, T., Das, R., Morris, T.: Virtualization of industrial control system testbeds for cybersecurity. In: ACM International Conference Proceeding Series, ICSS '16, pp. 10–14. ACM, New York (2016)
8. Gov.UK: 5G Testbeds and Trials Programme. https://www.gov.uk/guidance/5g-testbeds-and-trials-programme#about-%20the-programme. Last accessed 20 Feb 2024
9. Mathur, A.P., Tippenhauer, N.O.: SWAT: A water treatment testbed for research and training on ICS Security. In: 2016 International Workshop on Cyber-physical Systems for Smart Water Networks (CySWater). IEEE, pp. 31–36 (2016)
10. Ahmed, C., Palleti, V., Mathur, A.: WADI: A water distribution testbed for research in the design of secure cyber physical systems. In: Proceedings—2017 3rd International Workshop on Cyber-Physical Systems for Smart Water Networks, CySWATER 2017, series CySWATER '17, pp. 25–28. ACM (2017)
11. Neema, H., Koutsoukos, X., Potteger, B., Tang, C., Stouffer, K.: Simulation testbed for railway infrastructure security & resilience evaluation. In: Proceedings of 7th Symposium on Hot Topics in Science of Security, HotSoS 2020, pp. 1–8. ACM, New York (2020)
12. RACE UK.: Connected autonomous vehicle testbed, https://race.ukaea.uk/test-facilities/cav. Last accessed 20 Feb 2024
13. Rolls-Royce Testbed facilities.: https://www.rolls-royce.com/innovation/testbed-facilities.aspx. Last accessed 20 Feb 2024
14. MacKinnon, I.: Energy independence and security act (public law 110–140). https://www.epa.gov/laws-regulations/summary-energy-independence-and-security-act. Last accessed 20 Feb 2024
15. Candell, R., Stouffer, K.A., Anand, D.: A cybersecurity testbed for industrial control systems. In: Proceedings of the 2014 Process Control and Safety Symposium, Process Control and Safety Symposium 2014, National Institute of Standards and Technology, Houston, TX, pp. 6–9 (2014)
16. Stouffer, K., Pillitteri, V., Abrams, M., Hahn, A.: NIST SP800–82 Guide to Industrial Control Systems Security. https://nvlpubs.nist.gov/nistpubs/SpecialPublications/NIST.SP.800-82r2.pdf
17. Green, B., Le, A., Antrobus, R., Roedig, U., Hutchison, D., Rashid, A.: Pains, gains and PLCs: Ten lessons from building an industrial control systems testbed for security research. In: 10th USENIX Workshop on Cyber Security Experimentation and Test, CSET 2017, Vancouver, BC, Canada, 14 Aug (2017)
18. MathWorks: MATLAB.: https://www.mathworks.com. Last accessed 20 Feb 2024
19. Mininet.: https://github.com/mininet/mininet. Last accessed 20 Feb 2024
20. Formby, D., Rad, M., Beyah, R.: Lowering the barriers to industrial control system security with GRFICS. In: ASE 2018—2018 USENIX Workshop on Advances in Security Education, co-located with USENIX Security 2018. USENIX (2018)

21. SCADA simulator.: https://github.com/cmu-sei/SCADASim. Last accessed 20 Feb 2024
22. Koganti, V.S., Ashrafuzzaman, M., Jillepalli, A.A., Sheldon, F.T.: A virtual testbed for security management of industrial control systems. In: 2017 12th International Conference on Malicious and Unwanted Software (MALWARE), pp. 85–90 (2017)
23. Khan, M., Rehman, O., Rahman, I.M.H., Ali, S.: Lightweight testbed for cybersecurity experiments in SCADA-based Systems. In: 2020 International Conference on Computing and Information Technology (ICCIT-1441), pp. 1–5 (2020)
24. Marino, D.L., Wickramasinghe, C.S., Singh, V.K., Gentle, J., Rieger, C., Manic, M.: The virtualized cyber-physical testbed for machine learning anomaly detection: a wind powered grid case study. IEEE Access **9**, 159475–159494 (2021)
25. VirtualBox.: https://www.virtualbox.org. Last accessed 20 Feb 2024
26. VMware.: https://www.vmware.com. Last accessed 20 Feb 2024
27. Docker.: https://www.docker.com. Last accessed 20 Feb 2024
28. Goh, J., Adepu, S., Junejo, K., Mathur, A.: A dataset to support research in the design of secure water treatment systems. In: Proceedings of 2016 Conference, pp. 1–8 (2016)
29. Derbyshire, R., Green, B., van der Walt, C., Hutchison, D.: Dead man's PLC: Towards viable cyber extortion for operational technology (2023). arXiv preprint arXiv:2307.09549
30. Al Jallad, K., Aljnidi, M., Desouki, M.S.: Anomaly detection optimization using big data and deep learning to reduce false-positive. J Big Data **7**, 68 (2020)
31. PyCaret.: https://pycaret.org. Last accessed 20 Feb 2024
32. Python.: https://www.python.org. Last accessed 20 Feb 2024
33. Fujimoto, M., Itani, Y., Mitsunaga, T.: Lessons learned for practical penetration test against industrial control systems. In: 2023 IEEE International Conference on Computing (ICOCO), Langkawi, Malaysia, pp. 59–64 (2023)
34. ICS-test-bed-POC.: https://github.com/RedClouud/ICS-test-bed-PoC

Radio Frequency (RF) Cyber Threats and a Novel RF Cyber Kill Chain

Carolyn J. Swinney and John C. Woods

Abstract Cyber threats are defined as those that target a system's availability, integrity, or confidentiality. As the threat landscape increases with 5G and IoT connectivity, so does the potential to directly affect or access connected 'system of systems' using Radio Frequency (RF) connections. The open architecture and low-cost nature of today's Software Defined Radio (SDR) technology present opportunities to easily exploit RF access points with cyber threats in a growing connected cyber terrain. SDRs can provide low-cost, simple discrete attacks on a system's confidentiality, integrity, and availability, without a requirement for physical access. The Cyber Kill Chain, which focuses on external complex and persistent threats, is not fit for purpose for an RF-based threat. While other researchers suggest that traditional cyber attacks may be increasing in complexity and time to render, it is suggested in this paper that RF-enabled cyber-attacks are becoming less complex and easier to implement due to the availability of low-cost SDR technology and associated high-power amplifiers and antennas. In this paper, a new RF Cyber Kill Chain is proposed to aid penetration testers and improve the cybersecurity of connected systems. The RF Cyber Kill Chain is shown through a case study to accurately map three stages of attack: Reconnaissance, Preparation, and Execution. SDR technology is also shown in this paper to have enormous potential when combined with machine learning (ML) to provide early warning of cyber-attacks utilising the RF spectrum.

C. J. Swinney (✉) · J. C. Woods
Computer Science and Electronic Engineering, University of Essex, Colchester, UK
e-mail: cjswin@essex.ac.uk

M. G. Jaatun et al. (eds.), *Proceedings of the International Conference on Cybersecurity, Situational Awareness and Social Media*, Springer Proceedings in Complexity,
https://doi.org/10.1007/978-981-96-0401-2_14

1 Introduction

1.1 Overview of Software Defined Radios (SDRs)

Radio systems historically have provided capabilities through hardware components with limited ability to change their purpose without re-design or changing physical parts [1]. In the early 1990s, Dr. Mitola proposed the term Software Defined Radio (SDR), a radio that allowed a change of purpose through software re-configuration [2]. SDRs today can be seen in low-cost variants and are used for the signal processing of radio frequency (RF) signals. Purposes vary from enabling cellular mobile networks to radio astronomy and RF spectrum monitoring. The advantages of SDRs are their ability to change purpose entirely or receive updates which can be achieved through software configuration alone [3].

Conceptually the first programmable radio was considered over 50 years ago by the United States government [4]. The very first use cases were military ground forces (US and UK) who used a microprocessor and an analogue to digital converter as a re-programmable radio [5]. In the early 1980s companies began experimenting with software radios as baseband receivers for adaptive filtering [6]. In the early 1990's the United States Defence Advanced Research Projects Agency publicised a radio which could support multiple protocols within an open architecture and over a wider frequency range. The term 'software radio' was used at conferences to describe a radio that could both transmit and receive [2]. In 1996 the United States Department of Defence (DoD) set up a forum called the Modular Multifunction Information Transfer System Forum. Its purpose was to partner with industry and academia to create an open architecture for software radios [7]. The DoD also had another program called the Joint Tactical Radio System which tried to combine hardware-based radios with software radios to improve overall capability [8].

During this period commercial vendors also started looking at applications for cellular networks [9]. In 2000 a hardware platform was developed which allowed for significantly faster prototyping including simulation and execution [10]. From that point on platforms such as GNU Radio were developed as free tool kits [11]. These frameworks and easy-to-access platforms have allowed enthusiasts all over the world to develop SDR use cases. However, the open architecture and low-cost nature of SDRs also open up opportunities for the development of their use for cyber-attacks. Equally, their usage to detect anomalies in the RF spectrum as an early warning mechanism has also evolved.

1.2 Proliferation of Internet of Things (IoT) and 5G Technology

When traditional cyber threats are considered, access to the target device is often achieved through the internet, and mitigation from the threat by disconnecting a system from that internet connection. However, in many modern systems with IoT devices, the internet is not the only access vector and often these systems can have many RF connections which are not secure access points.

The US market for IoT connectivity in 2021 was $5.91 billion and in a market report looking out to 2030, this number is expected to grow with an expected increased demand for the automation of industrial systems [12]. With the rollout of 5G expected to significantly increase the landscape and attack surface for cyber-attacks, including smart transportation, water, power, and gas systems, security issues remain a concern for the European Commission. Of particular concern, as detailed in the '2022 Special Report 5G roll-out in the EU', is the economic impact of an attack that could disrupt services or critical infrastructure such as power grids [13]. The World Economic Forum Global Risks Report 2024 ranked cyber insecurity as the 4th highest risk in terms of severity in the next 2 years [14].

5G networks and the integration with IoT allows any connected device to transmit and receive data. This dawns a new era where cybersecurity enters the home and presents security, ethical, and social challenges [15]. 5G technology is a key enabler for the increased deployment of IoT devices [16]. The cyber threat landscape increases significantly as users and connectivity increase, accelerated by non-compliance with standards. Many applications are planning to use 5G services including vehicle connectivity (vehicle to vehicle, vehicle to everything, vehicle to building/infrastructure), digital twins, video streaming, and augmented/virtual reality applications [17]. Cyber threats are defined as those that target a system's availability, integrity, or confidentiality. As the threat landscape increases with 5G and IoT connectivity, so does the potential for exploitable RF vulnerabilities which could affect the confidentiality, integrity, and availability of those systems.

This paper will consider the use of SDRs for RF-enabled cyber threats. The authors will then propose a simple and novel RF Cyber Kill Chain methodology to map the stages of an RF-based attack, amplified with a case study. Lastly, the paper will look at how SDRs can also be used for anomaly detection and early warning before presenting conclusions.

2 RF-Enabled Cyber Attacks and Early Warning Detection

2.1 RF-Enabled Cyber Attacks

The paper introduces a novel RF Cyber Kill Chain model which addresses the limitations of traditional cyber kill chain models like the one from Lockheed Martin. This new model is tailored for RF-based threats, which are becoming increasingly significant with the proliferation of IoT devices and 5G technology. To aid the reader we have added a table below with the main threats and associated citations (Table 1).

Cyber-attacks against wireless communication networks can be facilitated with low-cost SDRs. Bybyk et al. [18] in their paper 'SDR Receivers as a New Challenge

Table 1 RF-enabled cyber threats

Threat type	Description	Example	Citation
Sniffing	Using an SDR to listen to RF signals to compromise the confidentiality of the information and data on a network	Sniffer programs are designed to monitor and analyse traffic on a network to help identify problems. An attacker can use this to capture data such as credentials. Other literature discusses the passive analysis of blockchain data to identify users via a sniffing attack	Bybvk et al. [18] Science Direct [19] Burra and Maity [20]
Side-channel attack	Attaining confidential information by analyzing differences in parameters such as the noise of processing operations (information which is physically measurable)	Dubrova et al. demonstrate a side channel attack on an implementation of CRYSTALS-Kyber on an ARM Cortex M4 CPU (typically used in embedded systems for industrial control or IOT devices). Power traces were measured during a side channel attacks and deep learning was trained to recover message bits	Bybvk et al. [18] Dubrova et al. [21]
Jamming	A denial of service (DoS) where a legitimate signal is overpowered, denying the availability of the legitimate signal	Overpowering a signal to prevent access and disrupt the reception of wireless communication signals. Pirayesh and Zeng describe the use of a $20 USB dongle easily programmed as a jammer with a 20 MHz bandwidth under 6 GHz with 100mW transmit power	Bybvk et al. [18] Pirayesh and Zeng [22]
Spoofing	Spoofing involves mimicking a legitimate signal to make the receiver believe the incoming data is valid, potentially altering data or control signals	GPS spoofing used to change the date/time of a device and the location status with a BladeRF SDR and the GPS-SDR-SIM project which is an open source signal generator for GPS baseband signals. The work is shown to spoof the positioning data on an iPhone and an apple watch. They further show that apps such as Uber and Wechat trust that data implicitly Ferreira et al. produce a SDR based mobile system which can jam and spoof unmanned aerial vehicles	Wang et al. [23] Ferreira et al. [24]

(continued)

Table 1 (continued)

Threat type	Description	Example	Citation
Replay	Collecting a legitimate signal and replaying it to fool the receiver or to cause a denial of service attack against the network	Replay attacks can be a result of sniffing whereby network traffic data is captured while a user is authenticating to a service. The data can then be replayed for the attacker to gain access to the service. This can also be possible even when the traffic is encrypted as shown by Prowell et al	RTL-SDR [25] Prowell et al. [26]
Reverse engineering	Capturing signals to study, demodulate, decrypt, and decode, allowing an attacker to create new attack methodologies or to attempt to replay or spoof new signals	Pozniak et al. describe the process of reverse engineering using an SDR by recording the signal and processing it, with the processing being more challenging Larouche et al. use an SDR and MATLAB to perform reverse engineering on a Guardian 400 wireless modem with no need for firmware access or technical information. The Guardian 400 associated interoperability with several discontinued radios are widely used today in SCADA systems	Pozniak et al. [27] Larouche et al. [28]
Data injection	Injecting false data into systems, such as inserting steps into a fitness tracker or modifying the data transmitted by a system	Although not specifically using an SDR, Trippel et al. prove they can increase steps on a Fitbit through acoustic interference Kim et al. use a WARP SDR to inject malicious frames on hypertext transfer protocol communication using transmission control protocol. Applications for the attack include services such as domain name service and file transfer protocol	Trippel et al. [29] Kim et al. [30]

(continued)

Table 1 (continued)

Threat type	Description	Example	Citation
Meaconing	Repeating real signals to confuse systems, typically used in navigation or communication systems to mislead about the actual position or status	Zhu et al. produce a real GNSS signal using a GNSS simulator and use an SDR to transmit/meacon the signal with more power than the legitimate signal and with a delay	Zhu et al. [31]
GPS Spoofing	Using an SDR to perform a GPS spoofing attack, altering the perceived position of the target	Zheng and Sun show three different methods for hijacking a drone using a HackRF SDR and spoofing GPS data. The authors show that the drone can be landed, flown to a specified area and thirdly landed in that specified area. They use an open software project called GPS-SDR-SIM to generate the GPS signal data using the information for the spoofing and then use the SDR to transmit this information to the platform	Zheng and Sun [32]

to Cybersecurity Wireless Technology' detail several different attack vectors caused by SDR threats. These include sniffing, described as using an SDR to listen to RF signals, compromising the confidentiality of the information, along with details of the signal such as sender identity, recipient, and modulation type. The next attack described is a side-channel attack which is used to attain confidential information by analysing differences in parameters such as the noise of processing operations. Jamming is a denial of service (DoS) whereby a legitimate signal is overpowered and this can be easily achieved using an SDR that denies the availability of the legitimate signal. Spoofing is where a legitimate signal is spoofed making the receiver think the information being received is the real signal, when the data may have been altered or the signal used to control the receiver. SDRs can be used to collect a legitimate signal and then replay it which can be used to either fool the receiver or to cause a DoS attack against the availability of the network.

An example of a replay attack occurred in 2017 whereby 156 tornado sirens were compromised causing a city-wide false alarm in Dallas, Texas. It is believed that the person responsible recorded the signal during one of the monthly tests and then replayed it later [25], setting off all alarms across the city for 90 min. Another prominent example of replay attacks that have been highlighted in the media is vulnerabilities with remote keys for cars. Research has shown cars can be successfully hi-jacked by jamming at a slight deviation of the frequency of the unlock button being triggered. The signal from the key is then recorded using a filter so as not to capture any of the jamming. The user will likely press the button twice but the car will not unlock due to the jamming and the attacker will therefore have captured two codes. If the first code is replayed by the attacker, they will see the car being unlocked, then the second code is still usable to hijack the car at a later time. Ibrahim et al. design attack strategies that are effective against robust remote keyless systems for buildings and vehicles [33].

Pozniak et al. [27] show how an SDR can be used to reverse engineer signals. Once the signal has been captured, it can be studied, de-modulated, and potentially decrypted and decoded. This can allow an attacker to not only devise brand new attack methodologies but also attempt to replay or spoof new signals. Trippel et al. [29] show data injection where logged steps are increased on a Fitbit through intentional acoustic interference via the insertion of digital values into an embedded system. Automatic Dependent Surveillance-Broadcast (ADS-B) is where aircraft broadcast information such as GPS, altitude, etc. to allow air traffic control and other aircraft to be aware of position. Moser et al. [34] show ADS-B to be vulnerable to spoofing attacks even with mitigation techniques in place that try to verify the signal integrity via multi-lateration.

The operation of Satellites and Satellite communication form another concern for cybersecurity professionals. Wang et al. [23] use GPS spoofing and an SDR to change the date/time of the device they were targeting. Zheng and Sun [32] show that a small unmanned aerial system (UAS) could be hijacked using a low-cost SDR to perform a GPS spoofing attack. They show the attacks easy to control and inexpensive to conduct. Wang et al. [35] suggest that cyber security in satellite systems should be categorised using the terminology of confidentiality, availability, and integrity. An example of a cyber threat which denies availability to a link or system is a Denial of Service attack which disrupts a user's ability to operate the system. Cybersecurity professionals commonly ensure confidentiality through link encryption, preventing an adversary from gaining access to the information. Wang et al. suggest that the main concern lies around integrity and understanding whether data manipulation has occurred. Unintentional interference in satellite communication links are common, for example, ionosphere and Doppler effects. Intentional interference includes Spoofing attacks whereby the satellite receiver accepts the spoofed signal as legitimate. Meaconing is where an attacker repeats real signals, used to confuse; and jamming which normally involves overpowering the genuine signal.

Wang et al. use a tool called Metasploit [36] to demonstrate false data injection onto a satellite link, fooling the system into thinking there were more users and therefore changing the available bandwidth. Wang et al. conclude that mitigation techniques such as cross-layer design with multi-path routing and channel coding were required to ensure the transmission of reliable information. However, there are many legacy systems employed today that were not designed with these mitigation techniques in mind. Anomaly detection in the RF spectrum can inform decision-makers early of potential cyber-attacks. The use of SDR for this purpose makes it a cost-effective venture.

Another interesting development in the cybersecurity realm has been the use of small unmanned aerial systems (UAS) to gain closer proximity to target devices.

Staniforth in his article 'Countering Drone—an evolving cybersecurity requirement' describes UAS equipped with Wi-Fi Pineapples (devices that have been developed to penetration test W-Fi networks). An attacker can use the device to perform man-in-the-middle attacks, set up rogue access points, credential harvest, and gather information about a network. Researchers have also shown more complex attacks whereby the proximity of a UAS hovering outside an office block allowed malware to be installed onto a system via a Bluetooth mouse connection [37].

2.2 RF Anomaly Detection and Early Warning

An anomaly is defined as a variation in the data outside of what is considered normal [38]. Edgeworth in 1887 was the first person to write about anomalies in data and statistical approaches to the identification of those anomalies. He defined discordant observations as "those which present the appearance of differing in respect of their law of frequency from other observations with which they are combined" [39]. AD techniques are important for many different applications from Intrusion Detection Systems on networks, to fraud detection, radio astronomy, and video surveillance.

A common statistical approach to AD is the Cumulative Sum Control Chart (CUSUM). CUSUM is a form of sequential analysis which looks at the cumulative sum of variations from the dataset. Detection occurs through the comparison of an adaptive threshold and the cumulative sum but there are limitations concerning its ability to detect small changes [40]. The stock market uses another technique to predict change called Exponential Weighted Moving Average (EWMA). It is popular in the stock market as it puts more weight on the most recent data, believed to be the best prediction of future trends [41]. EWMA is not just used in the stock market, Zhou and Tang [42] consider the use of a modified version of EWMA for AD in satellite imagery. The issue comes with finding a statistical approach that can detect small changes as well as large ones. The effectiveness of EWMA has been shown for the prediction of cyber-attacks such as Denial of Service and Brute Force attacks within a network [43]. These are known as burst attacks and each attack has different characteristics. Rather than trying to find one statistical method to fit all types of attacks, future research could look to vary statistical methods based on attack characteristics.

Another common approach is for statistical properties of modelled data to confirm whether samples are from the same distribution or exceed a threshold. Change detection techniques used within imagery and video often use histograms to represent the distribution as an intensity value for every pixel. The range of values is divided into intervals called bins and this forms a Histogram. The height of each bin is proportional to the number of values that have fallen within that interval. Histogram Based Outlier Scores is an unsupervised statistical algorithm that creates a histogram for each feature of a dataset. The algorithm assumes all features are independent of each other which is often deemed a weakness of the methodology. However, research shows that HBOS AD methods outperform machine learning approaches in network intrusion experiments. Goldstein and Uchida [44] tested HBOS as one of 19 algorithms on 10 datasets from network application domains. HBOS outperformed the other 18 algorithms on 4 out of 10 datasets. HBOS proved very fast computationally and therefore good for large datasets. Paulauskas and Baskys [45] took this work further in 2019 and showed HBOS superior in the detection for 3 out of 4 of the following attacks compared with machine learning techniques; denial of service, probing a network, remote access from an unauthorised source, and unauthorised access as a super user. The work further showed that dynamic bin sizes could gain better performances for rare events such as unauthorised super user access.

ML is defined as "a tool used to automate decision-making based on patterns found in available data" [46]. Traditional computing takes data with a program and produces an output. ML takes the data and the output and produces a program. There are 4 types of machine learning. Within Supervised learning, the training data contains the output. Unsupervised is where the training data does not include the output for example clustering. Semi-supervised lies somewhere in between the two with some of the training data including the output which is desired. Lastly reinforcement learning works by actions before reinforced when they are correct. Supervised learning is the most advanced and mature form of ML [47].

To detect threats in the RF spectrum a very large amount of data needs to be monitored. ML can provide a way of automating responses to AD with large datasets. The ML uses training data as learning material to establish a normal. The network can then give a decision or prediction as to whether an anomaly is present based on its training for 'normal' [48]. The 2017 CRFS White Paper [49] considers machine learning techniques for RF spectrum intelligence gathering. Specifically, it considers the use of a neural network to identify signals based on modulation and protocol. It further suggests that treating the signal as an image allows for the use of image recognition techniques to perform pattern recognition. Tandiay et al. [50] use an existing ML-based solution PredNet [51] for video prediction on a series of generated spectrograms and spectral correlation functions (SCF) stored as 2D images. This technique successfully detects anomalies such as jamming, and spectrum hi-jacking by comparing real-time data with a model trained on 'normal'. The use of spectrograms was not instantaneous due to high correlation levels. SCF proved fast enough for real-time operations but needed much more computational power. Researchers could extend this work through processing multiple transforms of raw data or using pre-processed data.

Sreeraj et al. [52] stressed the current research difficulty of monitoring the spectrum with an automated approach over frequency, space, and time. Sreeraj et al. use changes in Power Spectral Density to signify anomalies through captured datasets using RTL-SDR and Raspberry Pi (part of a project called Electrosense [53]), along with Hack RF SDRs. The ML model uses a Convolutional Neural Network (CNN) and is set apart from previous work as the model trains unsupervised over multiple bands, rather than training one model per frequency band of interest. However, the research does not look at using raw I&Q samples from the SDR as an input to the model, only pre-processed data. O'Shea et al. [54] do look at raw spectral data again using a CNN to characterise the error in the prediction. However, the spectrum is only captured for between 4-10 ms and therefore it struggles to fully realise protocols that communicate in bursts such as GSM uplink and Wi-Fi. Longer capture times could potentially solve this issue. In this paper, classification is also explored but the model is unable to separate signals which are near each other in frequency. The work highlights the fact that there is no standardised dataset available for signals which are manually labeled but the production of this would be a very large task.

Li et al. [55] consider the use of Deep Neural Networks (DNN) to baseline spectrum use and detect anomalies in cellular networks. Li specifically considers the issue of scalability by creating a model that applies transfer learning across the network. Future work includes the consideration of other DNNs and the identification of corrupted or modified signals. Feng et al. [56] show that the deeper the architecture of a model, the better its performance with AD in the RF spectrum. Swinney and Woods considered the application of transfer learning and CNN feature extraction for RF signals to aid detection and classification [57, 58]. They take this a stage further in [59] where a low-cost Raspberry Pi-based detection and classification system is produced using machine learning for early warning. Although this research is primarily conducted for the detection and classification of small UAS it has use cases for GPS jamming detection and the detection and classification of other RF signals.

3 Kill Chains and Case Study

3.1 Alternative Cyber Kill Chains

The Lockheed Martin Cyber Kill Chain details a set of steps that an adversary must complete to meet their objectives for a cyber-attack. The model, taken from the Lockheed Martin website [60], has seven steps detailed below:

- Reconnaissance—gaining background information from activities such as harvesting email addresses, conference information, etc.
- Weaponisation—coupling the Exploit with a backdoor into deliverable payload.
- Delivery—delivering weaponised bundle to the victim via email, web, USB, etc.
- Exploitation—exploiting a vulnerability to execute code on a victim's system.
- Installation—installing malware on the asset.
- Command and control—command channel for remote manipulation of the victim.
- Actions on objectives—with 'hands on keyboard' access, intruders accomplish their original goals.

NCSC simplifies the model into four stages; Survey; Delivery; Breach; and Affect [61]. 'Survey' is where the attacker looks to find the information required to understand vulnerabilities (technical, procedural, or physical) that could be exploited. 'Delivery' is where the attacker gets into a position where they could carry out an attack such as giving away USB sticks to employees or sending a phishing link. The

'Breach' stage is where the attacker has gained access to the system or changed its operation, depending on the type of exploit. Lastly, the 'Affect' stage is where the attacker carries out the activities to enact their goals such as causing a disruption or gaining persistence on the network.

Malone [62] challenged the Lockheed Martin model at the Black Hat conference as it only caters to external threats that require a breach of the enterprise network perimeter. Malone proposed a new model that incorporated the insider threat and focused on developing the 'internal' part of the kill chain [62]. Kim et al. [63] take this a stage further and look specifically at the insider threat for the multi-media environment by securing visibility to endpoints. Fadzil et al. [64] suggest that cyber-attacks are becoming more complex and time-consuming and further detail an improved cyber kill chain framework. While traditional cyber-attacks could well be increasing in complexity and time consumption, Swinney and Woods suggest that RF-enabled cyber-attacks on a system's confidentiality, integrity, and availability are becoming less complex and easier to implement due to the availability of low-cost SDR technology and high power amplifiers.

3.2 *RF Cyber Kill Chain*

The Lockheed Martin and various improved versions have been considered. The following reasons are identified as to why the Lockheed Martin Kill Chain methodology is weaker for RF-enabled cyber attacks:

- Advances in SDR, power amplifiers and directional antenna technology and ease of accessibility simplify RF cyber threats at a low cost to the attacker. Pirayesh and Zeng [22] describe the use of a $20 USB dongle easily programmed as a jammer with a 20 MHz bandwidth under 6 GHz with 100mW transmit power.
- We observe that the majority of RF cyber threats detailed in Sect. 2 are not trying to obtain persistence but rather are discrete and finite [21, 24, 26, 28, 31, 32].
- Physical access to a system is not necessarily required for RF cyber threats which can be carried out at a distance depending on the equipment parameters or the use of novel techniques such as small UAS fitted with SDR technology. Jayaraj et al. [65] describe RF attacks as those that "do not require physical contact with the target".
- Persistence is not required for more complex RF-enabled tasks such as reverse engineering which can be carried out as soon as a signal has been collected. Pozniak et al. [27] describe two parts of reverse engineering, recording the signal and then processing it.

Swinney and Woods propose a 3-stage RF Cyber Kill Chain presented in Fig. 1. The methodology incorporates a 'Reconnaissance' stage that focuses on researching the target device properties and could involve activities such as spectrum scanning with an SDR. The second stage is 'Preparation' where signals are either collected/ listened to or generated/created. In the final 'Execution' stage signals are either

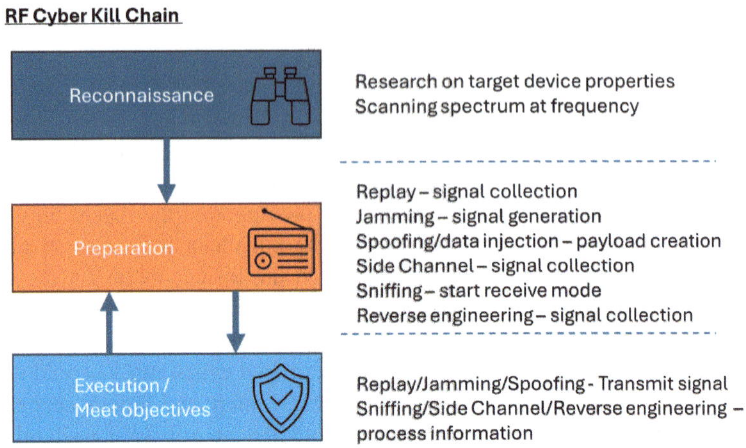

Fig. 1 RF cyber kill chain

transmitted or information is processed to meet the objectives of the attack. This simple model allows for reconnaissance and preparation information to be used again, such as the repeated use of collected signals, signified by the loop between stages two and three.

3.3 *Case Study*

The case study aimed to locate the frequency of a wireless doorbell, to record the signal and replay it, hence activating the doorbell with the attacker signal. The case study will be considered under the model of the 3-stage RF Cyber Kill Chain. For this case study the following equipment was used:

- HackRF SDR
- Skyscan Desktop Scanner Antenna (25–2000 MHz)
- Tecknet Wireless Doorbell
- Laptop running Ubuntu 20.02 and GNURadio.

RF Cyber Kill Chain Stage 1—Reconnaissance

The back of the Tecknet doorbell details the operating frequency. The frequency can also be located via the Federal Communications Commission (FCC) who certify equipment for use in the US. Figure 2 presents the information gained from the FCC website during the reconnaissance stage.

Gaining the exact frequency information allows the pinpointing of the spectrum analyser software on the SDR. Without knowing the frequency it is still possible to

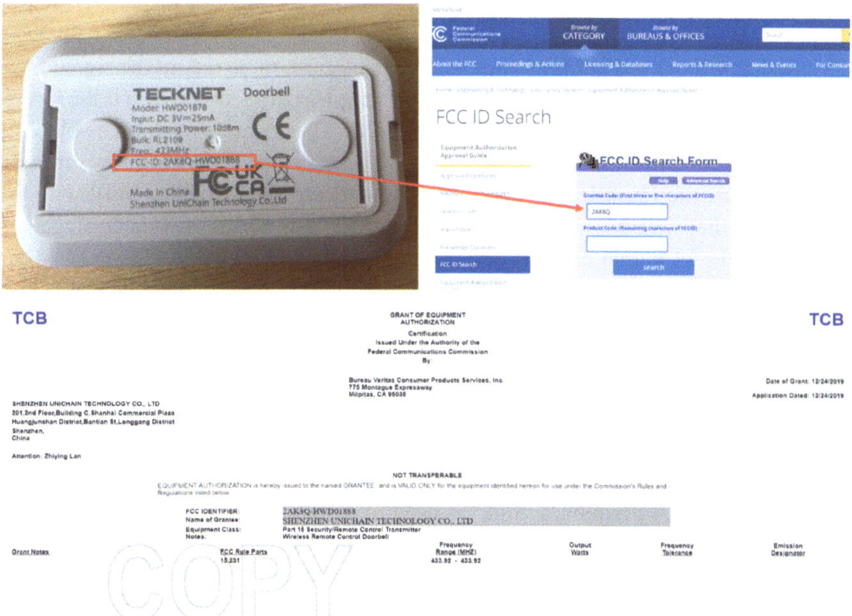

Fig. 2 Information gathering from FCC [66]

locate the signal of interest but it would take longer to find by manually searching the spectrum.

Different applications can be used to view the spectrum using the HackRF SDR. Figure 3 shows the HackRF on the left connected to an Android mobile phone running the application 'RF Analyser'. On the right of Fig. 3 is a Linux application called GQRX whereby you can do the same functions as the mobile application which includes numerous demodulation options for different types of signals.

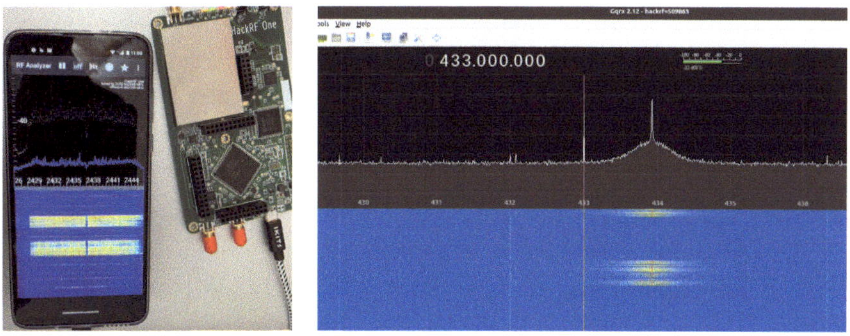

Fig. 3 Spectrum analyser applications

RF Cyber Kill Chain Stage 2—Preparation

GNURadio is displayed in Fig. 4 showing the configuration to record the signal from the HackRF. The sample rate is set to 2 million samples per second which is the lowest recommended sample rate for the HackRF SDR. The ‘osmocom Source’ block is used to take the samples off the SDR via an API library that supports many different SDRs, not just the HackRF. The device arguments are set to ‘hackrf = 0’ with the 0 indicating that we only have one HackRF in use. In the reconnaissance stage, we found from the FCC that the doorbell frequency was 433.92 MHz. However, in the experiment, the HackRF is set to 433.5 MHz to allow avoidance of the DC spike which is an artifact present due to the imbalance of complex samples taken in by the SDR. It is possible to filter the DC spike but then there is a risk that a legitimate part of the signal will also be filtered out. There are options to reduce the spike via the DC offset mode function. The RF gain is set to 0 as we are not using an amplifier, and the IF and BB gain are both set to 16 dB which is the gain value recommended on the HackRF Great Gadget site [67].

Next a ‘QT GUI Frequency Sink’ is used to display the signal in the frequency domain and a ‘QT GUI Time Sink’ is used to display the signal in the time domain. This is so we can observe the signal. The blocks are connected from the output of the source to the input of the frequency and time sink, the blue connectors represent the fact it is a complex signal. The ‘File Sink’ block is used to save the complex samples into a file.

Figure 5 shows the time and frequency domain displays. On the bottom in the frequency domain, we can see the DC spike at the centre frequency and when we press the doorbell we see the signal peak at 433.92 MHz highlighted inside the red box. In the time domain, the signal is displayed with the increase in amplitude.

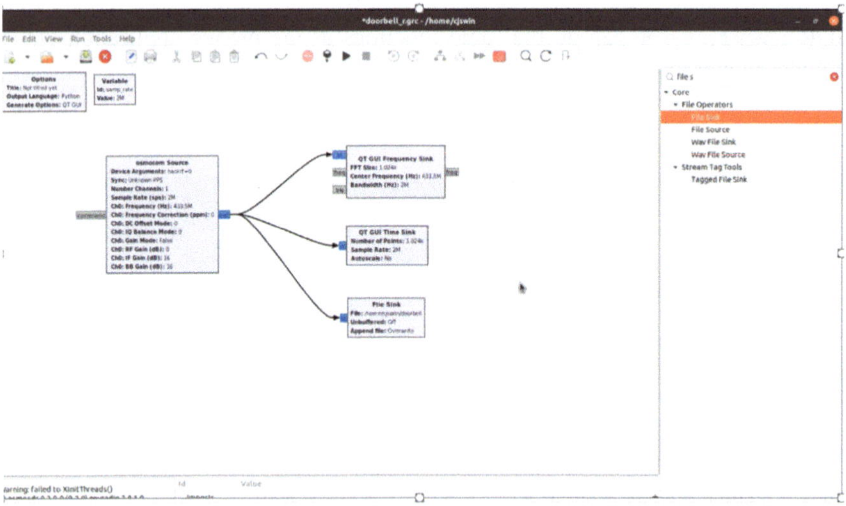

Fig. 4 GNURadio blocks for receive

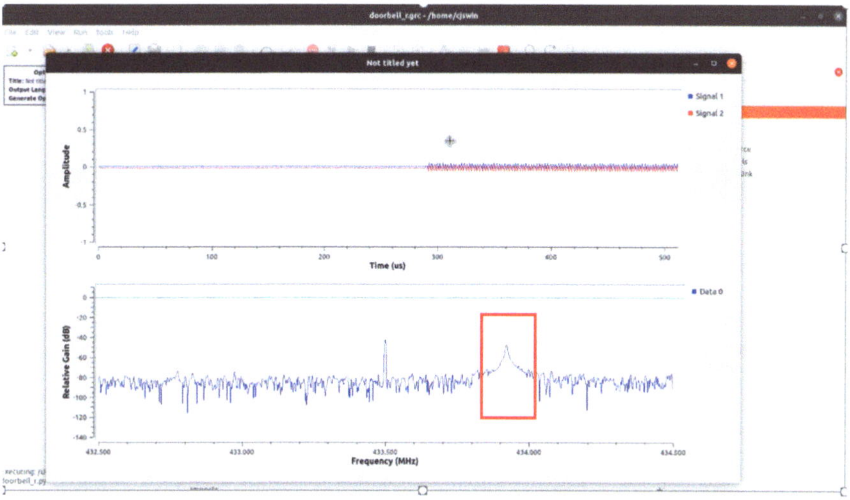

Fig. 5 GNURadio time and frequency domain transmit capture

RF Cyber Kill Chain Stage 3–Execution

Now that the signal has been captured into the file using the file sink, a new GNURadio setup is created to replay the signal.

Figure 6 shows the GNURadio blocks for the first attempt to replay the signal using a 'File Source' to open the recorded samples and an 'osmocom Sink' to transmit the signal. The frequency is set to 433.5 MHz and the IF gain to 20 dB as there is no BB gain on the transmitter side. Again a frequency and time sink are used to visualise the signal.

Figure 7 shows the signal replaying in the time and frequency domain but it does not trigger the doorbell. When we look at the amplitude in the time domain it is only using a very small part of the dynamic range which could extend from -1 to 1. This means isn't transmitting much higher in amplitude than the baseline noise level. To compensate for this we need to increase the dynamic range in the digital domain using a multiply constant block.

Figure 8 shows the signal again and this time we can see that the dynamic range of the amplitude in the time domain has increased significantly. This has the effect we are looking for and the doorbell triggers when we replay the signal. The RF Cyber Kill Chain captured the stages of the attack accurately and the replay signal file could be saved and used in execution again at an alternative time.

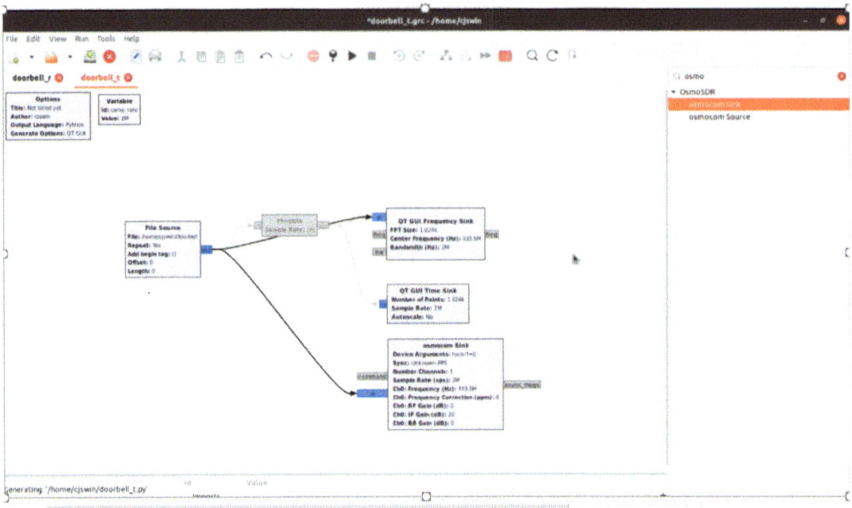

Fig. 6 GNURadio blocks for replay

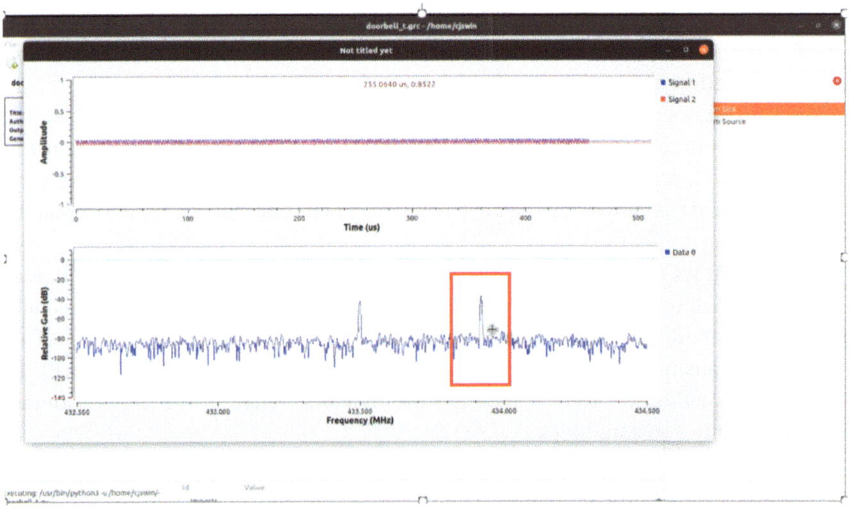

Fig. 7 GNURadio time and frequency domain replay capture

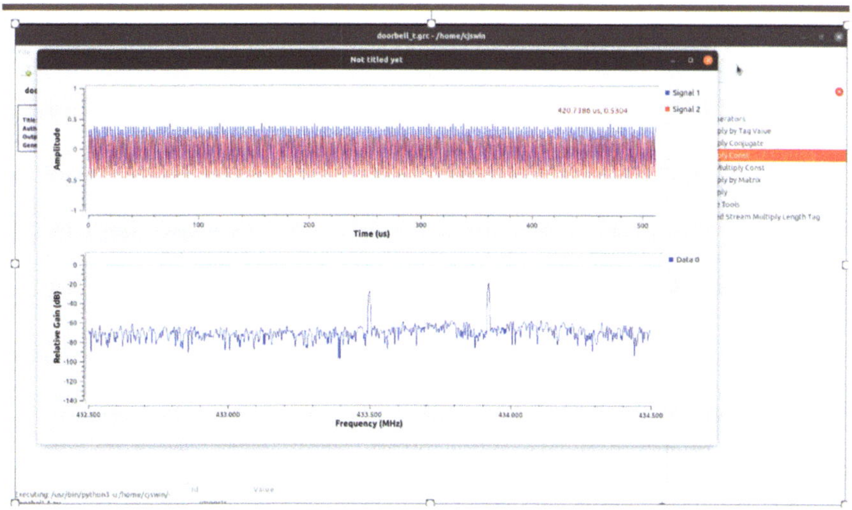

Fig. 8 GNURadio time and frequency domain replay capture amplified

4 Conclusion

Overall, this paper has highlighted that RF cyber threats will likely become more prominent as 5G and IoT connectivity, and the associated cyber threat landscape, increase. SDRs, coupled with high power amplifiers and antennas, present a low-cost, easily accessible technology threat that can produce fast and effective RF cyber-attacks that affect the confidentiality, integrity, and availability of systems. This paper has shown that the Lockheed Martin Cyber Kill Chain is not well suited for this type of threat due to its focus on complex, persistent cyber threats. This paper has proposed a novel RF Cyber Kill Chain. Using a case study the paper has shown the model can accurately map the stages of an RF-based cyber-attack. SDRs were also considered as an effective early warning tool for anomaly detection in the RF spectrum. Future work could look to address the detection and mitigation of attacks utilising the stages within the RF Cyber Kill Chain model.

References

1. Tribble, A.C.: The software defined radio: fact and fiction. In: 2008 IEEE Radio and Wireless Symposium, RWS. 2008, pp. 5–8 (2008). https://doi.org/10.1109/RWS.2008.4463414
2. Mitola J.: Software radios-survey, critical evaluation and future directions. In: Proceedings NTC-92: National Telesystems Conference, pp. 13/15–13/23. https://doi.org/10.1109/NTC.1992.267870
3. Machado, R.G., Wyglinski, A.M.: Software-defined radio: bridging the analog-digital divide. Proc. IEEE **103**(3), 409–423 (2015). https://doi.org/10.1109/JPROC.2015.2399173

4. Hussain, W., Isoaho, J., Nurmi, J.: The evolution of software defined radio. Computing platforms for software-defined radio, pp. 1–4 (2016). https://doi.org/10.1007/978-3-319-49679-5_1
5. Sdr, Software defined radio implementation (2010)
6. Arya, N., Sharma, K., Singh, G.: Analysis of channels for software defined radio using LabVIEW. Int. J. Recent. Res. Asp. **1**, 1–6 (2014)
7. Fritz, A.K., Froehlich, E.: The Froehlich/Kent Encyclopedia of Telecommunications. The Froehlich/Kent Encyclopedia of Telecommunications: vol 18–Wireless, p. 143, Accessed 06 Nov 2019. https://books.google.co.uk/books?id=mj1o_v17jQAC&pg=PA143&lpg=PA143&dq=The+Modular+Multifunction+Information+Transfer+System+(MMITS)+Forum&source=bl&ots=D107sVp6Fs&sig=ACfU3U2K5d7fY9KNwXGETuw7tpI41xHYaQ&hl=en&sa=X&ved=2ahUKEwjyqbPXrNXlAhVgThUIHYAsAG0Q6AEw
8. Francis P.L.: Challenges and Risks Associated with the Joint Tactical Radio System Program. (2003), Accessed 06 Nov 2019. https://books.google.co.uk/books?id=UX6ovAEACAAJ&dq=Joint+Tactical+Radio+System&hl=en&sa=X&ved=0ahUKEwjfuZfPsNXlAhVxVBUIHfhhCRkQ6AEILjAB
9. Wyglinski, T.H.A.M., Nekovee, M.: Cognitive radio communications and networks: principles and practice (2009)
10. Moy, C., Raulet, M.: High-level design methodology for ultra-fast software defined radio prototyping on heterogeneous platforms. Adv. Electron. Telecommun. **1**(1) (2010)
11. GNU Radio, About GNU Radio, GNU Radio, Accessed 12 Nov 2019. https://www.gnuradio.org/about/
12. Emergen Research, Internet of things connectivity market, by component, by application & home automation, by deployment type, by end-use, and by region to 2030, 2022. Accessed 23 Mar 2024. https://www.emergenresearch.com/industry-report/iot-connectivity-market
13. European Commission, Special Report 5G roll-out in the EU: delays in deployment of networks with security issues remaining unresolved (2022)
14. World Economic Forum, The Global Risks Report 2024, vol. 19th Edition. (2024). www.weforum.org
15. Kizza, J.M.: Evolving Cyberspace: The Marriage of 5G and the Internet of Things (IoT) Technologies, pp. 259–280 (2019). https://doi.org/10.1007/978-3-030-03937-0_12
16. Shafique, K., Khawaja, B.A., Sabir, F., Qazi, S., Mustaqim, M.: Internet of things (IOT) for next-generation smart systems: a review of current challenges, future trends and prospects for emerging 5G-IoT Scenarios. IEEE Access **8**, 23022–23040 (2020). https://doi.org/10.1109/ACCESS.2020.2970118
17. Mohan, J.P., Sugunaraj, N., Ranganathan, P.: Cyber security threats for 5G networks. In 2022 IEEE International Conference on Electro Information Technology (eIT), IEEE, pp. 446–454 (2022). https://doi.org/10.1109/eIT53891.2022.9813965
18. Bybyk, R., Opirskyy, I., Mcintosh, M.: SDR receivers as a new challenge to cybersecurity wireless technology. Cybersecur. Provid. Inf. Telecommun. Syst (2022)
19. Science Direct, Web-based Mail Issues, E-Mail Virus Protection Handbook. pp. 119–145 (2000). https://doi.org/10.1016/B978-192899423-7/50008-2
20. Burra M.S., Maity, S.: Characteristics, advances, and challenges in blockchain-enabled cyber-physical systems. Blockchain technology for emerging applications: a comprehensive approach, pp. 67–89 (2022). https://doi.org/10.1016/B978-0-323-90193-2.00008-9
21. Dubrova, E., Ngo, K., Gärtner, J., Wang, R.: Breaking a fifth-order masked implementation of CRYSTALS-kyber by copy-paste. In: Proceedings of the 10th ACM Asia Public-Key Cryptography Workshop, Association for Computing Machinery, Inc, pp. 10–20 (2023). https://doi.org/10.1145/3591866.3593072
22. Pirayesh, H., Zeng, H.: Jamming attacks and anti-jamming strategies in wireless networks: a comprehensive survey. In: IEEE Communications Surveys and Tutorials, vol. 24, no. 2. Institute of Electrical and Electronics Engineers Inc, pp. 767–809 (2022). https://doi.org/10.1109/COMST.2022.3159185

23. Wang, K., Chen, S., Pan, A.: Time and position spoofing with open source projects. https://blackhat.com/docs/eu-15/materials/eu-15-Kang-Is-Your-Timespace-Safe-Time-And-Position-Spoofing-Opensourcely-wp.pdf
24. Ferreira, R., Gaspar, J., Sebastião, P., Souto, N.: A software defined radio based anti-uav mobile system with jamming and spoofing capabilities. Sensors. **22**(4) (2022). https://doi.org/10.3390/s22041487
25. RTL-SDR, Explaining the Dallas Siren Hack, RTL-SDR. Accessed 23 Mar 2024. https://www.rtl-sdr.com/explaining-dallas-siren-hack/
26. Prowell, S., Kraus, R., Borkin, M.: Man-in-the-Middle. Seven deadliest network attacks, pp. 101–120 (2010). https://doi.org/10.1016/B978-1-59749-549-3.00006-7
27. Pozniak, M., Sadhukhan, D., Ranganathan, P.: R.F exploitation and detection techniques using software defined radio: a survey. Inst. Electr. Electron. Eng. (IEEE) 345–350 (2019). https://doi.org/10.1109/eit.2019.8834203
28. Larouche, J.B., Roy, S., Mailhot, F., Tardif, P.M., Frappier, M.: SCADA radio blackbox reverse engineering. In: Lecture Notes in Computer Science (including subseries Lecture Notes in Artificial Intelligence and Lecture Notes in Bioinformatics), Springer Science and Business Media Deutschland GmbH, pp. 287–302 (2023). https://doi.org/10.1007/978-3-031-30122-3_18
29. Trippel, T., Weisse, O., Xu, W., Honeyman, P., Fu, K.: WALNUT Waging Doubt on the Integrity of MEMS Accelerometers with Acoustic Injection Attacks. Accessed 15 Nov 2019. https://spqr.eecs.umich.edu/walnut/
30. Kim, W., Kim, S., Lim, H.: Malicious data frame injection attack without seizing association in IEEE 802.11 wireless LANs. IEEE Access **9**, 16649–16660 (2021). https://doi.org/10.1109/ACCESS.2021.3054130
31. Zhu, X., Lu, Z., Hua, T., Yang, F., Tu, G., Chen, X.: A novel GPS meaconing spoofing detection technique based on improved ratio combined with carrier-to-noise moving variance. Electronics (Switzerland) **11**(5) (2022). https://doi.org/10.3390/electronics11050738
32. Zheng, X.C., Sun, H.M.: Hijacking unmanned aerial vehicle by exploiting civil GPS vulnerabilities using software-defined radio. Sens Mater**32**(8), 2729–2743 (2020). https://doi.org/10.18494/SAM.2020.2783
33. Ibrahim, O.A., Hussain, A.M., Oligeri, G., Di Pietro, R.: Key is in the air: hacking remote keyless entry systems. In: Lecture Notes in Computer Science (including subseries Lecture Notes in Artificial Intelligence and Lecture Notes in Bioinformatics), Springer Verlag, pp. 125–132 (2019). https://doi.org/10.1007/978-3-030-16874-2_9
34. Moser, D., Leu, P., Ranganathan, A., Ricciato, F., Capkun, S.: Investigation of multi-device location spoofing attacks on air traffic control and possible countermeasures (2016). https://doi.org/10.1145/2973750.2973763
35. Wang G., et al.: Cyber security with radio frequency interferences mitigation study for satellite systems. Sensors and Systems for Space Applications IX, vol. 9838, p. 98380K (2016). https://doi.org/10.1117/12.2224632
36. Metasploit | Penetration Testing Software, Pen Testing Security | Metasploit. Accessed 03 Apr 2020. https://www.metasploit.com/
37. Staniforth, A.: Countering drones an evolving cybersecurity requirement. Defence IQ. Accessed 30 May 2022. https://www.defenceiq.com/cyber-defence-and-security/articles/countering-drones-an-evolving-cybersecurity-requirement
38. Callegari, C., Giordano, S., Pagano, M.: Anomaly detection: an overview of selected methods. In: Proceedings 2017 International Multi-Conference on Engineering, Computer and Information Sciences, SIBIRCON 2017, Institute of Electrical and Electronics Engineers Inc, pp. 52–57 (2017) https://doi.org/10.1109/SIBIRCON.2017.8109836
39. Edgeworth, F.Y.: XLI On discordant observations. Lond. Edinb. Dublin Philos. Mag. J. Sci. **23**(143), 364–375 (1887). https://doi.org/10.1080/14786448708628471
40. Christodoulou V., Bi, Y.: A combination of CUSUM EWMA for anomaly detection in time series data. In: Proceedings of the 2015 IEEE International Conference on Data Science and Advanced Analytics, DSAA 2015, Institute of Electrical and Electronics Engineers (2015). https://doi.org/10.1109/DSAA.2015.7344899

41. Adewuyi, A.W.: Modelling stock prices with exponential weighted moving average. J. Math. Financ. **06**(01), 99–104 (2016). https://doi.org/10.4236/jmf.2016.61011
42. Zhou, Z.G., Tang, P.: Improving time series anomaly detection based on exponentially weighted moving average of season-trend model residuals. In: International Geoscience and Remote Sensing Symposium (IGARSS), Institute of Electrical and Electronics Engineers Inc, pp. 3414–3417 (2016). https://doi.org/10.1109/IGARSS.2016.7729882
43. Silva, A. Pontes, E., Zhou, F., Guelf, A., Kofuji, S.: PRBS EWMA based model for predicting burst attacks in computer networks. In: 2014 9th International Conference on Digital Information Management, ICDIM 2014, Institute of Electrical and Electronics Engineers Inc, pp. 194–200 (2014). https://doi.org/10.1109/ICDIM.2014.6991410
44. Goldstein, M., Uchida, S.: A comparative evaluation of unsupervised anomaly detection algorithms for multivariate data. PLus One (2016). https://doi.org/10.1371/journal.pone.0152173
45. Paulauskas, N., Baskys, A.: Application of histogram based outlier scores to detect computer network anomalies. Electronics Switzerland **8**(11), 1–8 (2019). https://doi.org/10.3390/electronics8111251
46. Murphree, J.: Machine learning anomaly detection in large systems. In: AUTOTESTCON (Proceedings), Institute of Electrical and Electronics Engineers Inc. (2016). https://doi.org/10.1109/AUTEST.2016.7589589
47. Brownlee, J.: Basic concepts in machine learning. Machine learning mastery. Accessed 30 Mar 2020. https://machinelearningmastery.com/basic-concepts-in-machine-learning/
48. Chandola, V.: Anomaly detection : a Survey (2009)
49. CRFS, Machine learning and RF spectrum intelligence gathering (2017)
50. Tandiya, N., Jauhar, A., Marojevic, V., Reed, J.H.: Deep predictive coding neural network for RF anomaly detection in wireless networks (2018)
51. PredNet by coxlab. Accessed 02 Apr 2020. https://coxlab.github.io/prednet/
52. Rajendran, S., Meert, W., Lenders, V., Pollin, S.: SAIFE: unsupervised wireless spectrum anomaly detection with interpretable features
53. ElectroSense, Collective Spectrum Monitoring. Accessed 02 Apr. 2020. https://electrosense.org
54. Shea, T.J.O., Roy, T., Erpek, T.: Spectral detection and localization of radio events with learned convolutional neural features. In: 25th European Signal Processing Conference, EUSIPCO 2017, vol. 2017-Janua, pp. 331–335 (2017). https://doi.org/10.23919/EUSIPCO.2017.8081223
55. Li, Z., Xiao, Z., Wang, B., Zhao, B.Y., Zheng, H.: Scaling deep learning models for spectrum anomaly detection. p. 10 (2019). https://doi.org/10.1145/3323679.3326527
56. Feng, Q., Zhang, Y., Li, C., Dou, Z., Wang, J.: Anomaly detection of spectrum in wireless communication via deep auto-encoders. J. Supercomput. **73**(7), 3161–3178 (2017). https://doi.org/10.1007/s11227-017-2017-7
57. Swinney, C.J., Woods, J.C.: GNSS jamming classification via CNN, transfer learning and the novel concatenation of signal representations. In: 2021 International Conference on Cyber Situational Awareness, Data Analytics and Assessment (CyberSA), pp. 1–9 (2019). https://doi.org/10.1109/CYBERSA52016.2021.9478250
58. Swinney, C.J., Woods, J.C.: Unmanned aerial vehicle operating mode classification using deep residual learning feature extraction. Aerospace **8**(3), 79 (2021). https://doi.org/10.3390/aerospace8030079
59. Swinney, C.J., Woods, J.C.: Low-cost raspberry-pi-based UAS detection and classification system using machine learning. Aerospace **9**(12), 738 (2022). https://doi.org/10.3390/AEROSPACE9120738
60. Martin, L.: Cyber Kill Chain®. Lockheed Martin. Accessed 23 Mar 2024. https://www.lockheedmartin.com/en-us/capabilities/cyber/cyber-kill-chain.html
61. NCSC, How cyber attacks work. NCSC. Accessed 23 Mar 2024. https://www.ncsc.gov.uk/information/how-cyber-attacks-work
62. Malone, S.: Using an expanded cyber kill chain model to increase attack resiliency. In Black Hat (2016) Accessed 23 Mar 2024. https://www.blackhat.com/docs/us-16/materials/us-16-Malone-Using-An-Expanded-Cyber-Kill-Chain-Model-To-Increase-Attack-Resiliency.pdf

63. Kim, H., Kwon, H.J., Kim, K.K.: Modified cyber kill chain model for multimedia service environments. Multimed Tools Appl **78**(3), 3153–3170 (2019). https://doi.org/10.1007/s11042-018-5897-5
64. Fadzil, L.M., Manickam, S., Al-Shareeda, M.A.: A review of an emerging cyber kill chain threat model. In: 2023 Second International Conference on Advanced Computer Applications (ACA), IEEE, pp. 157–161 (2023) https://doi.org/10.1109/ACA57612.2023.10346959
65. Jayaraj, I.A., Shanmugam, B., Azam, S., Samy, G.N.: A systematic review of radio frequency threats in IoMT. J. Sens. Actuator Netw. **11**(4) MDPI (2022). https://doi.org/10.3390/jsan11040062
66. Federal Communications Commission, FCC ID Search, Federal Communications Commission. Accessed 23 Mar 2024. https://www.fcc.gov/oet/ea/fccid
67. Great Scott Gadgets, Software Defined Radio with HackRF, Great Scott Gadgets. Accessed 23 Mar 2024. https://greatscottgadgets.com/sdr/

Cybersecurity Governance, Risk Management and Assurance

Perception of Cyber Risk for Assisted Living Technologies in the Norwegian Healthcare Sector—Key Stakeholders' Perceptions

Alvhild Skjelvik and Bian Yang

Abstract Facing a demographic shift and healthcare personnel shortage, Norway and other European nations anticipate increased healthcare demands. Assisted living technologies are seen as potential solutions to manage demands, yet they carry inherent cybersecurity risks. While the majority of previous research within cybersecurity for assisted living technology has focused on technical risks, this study examines cybersecurity risk perception among key stakeholders in the Norwegian healthcare sector. Utilizing the BCISQ-questionnaire and incorporating a behavioral simulation, we analyzed risk perception patterns among 185 survey respondents and 33 interviewees. Despite observable differences in risk perception among stakeholders, attributed to factors like human error, organizational affiliation, and technology familiarity, the sample size limits generalizability. Key findings highlight the risk perception related to assisted living technology amongst five stakeholder groups, where their perception of risk is formed by likelihood and consequence considering confidentiality, integrity and availability. This study not only capture stakeholders' cybersecurity perception but also identify the differences between the groups, and factors influencing their perceptions.

1 Introduction

Cybersecurity has in recent years received more and more attention. One explanation of the increased focus is the number and impact of cybersecurity attacks, followed by an industry growing within the field of cybercrime [1]. Cybersecurity risks poses a threat to companies worldwide who make use of information- and communication technology (ICT), a risk that increase when technology is connected to the internet. A

A. Skjelvik (✉) · B. Yang
Norwegian University of Science and Technology, Gjøvik, Norway
e-mail: alvhild.skjelvik@ntnu.no

B. Yang
e-mail: Bian.yang@ntnu.no

M. G. Jaatun et al. (eds.), *Proceedings of the International Conference on Cybersecurity, Situational Awareness and Social Media*, Springer Proceedings in Complexity,
https://doi.org/10.1007/978-981-96-0401-2_15

growing trend within healthcare is the adoption of different ICT solutions for assisted living and remote care, such as Internet of Medical things (IoMT) and wearable devices [2]. The healthcare sector is already benefiting greatly from making use of technology, such as electronic medical records (EMR), advanced medical technology such as MRI/CT systems, and data sharing through the use of internet. This means that the healthcare sector, similar to other sectors, is frequently exposed to cybersecurity risks, materialized through attacks or data breaches [3]. The WannaCry-attack in 2016 severely impacted the National Health Services (NHS) in the UK, an example of how healthcare services can be affected by non-targeted attacks in the cyber-domain [4]. Previous research emphasize a need for strong built-in security mechanisms in devices and make strong suggestions for the implementation of technical security measures [5–10].

Cybersecurity risks does not materialize solely through exploitation of technical vulnerabilities, however previous research tends to focus more on technical security mechanisms in the context of medical devices and healthcare technology. A less examined area is the human and organizational aspects. In this study, we explore how stakeholders perceive cybersecurity risk for assisted living technologies. In the Nordics, assisted living technology is referred to as '*welfare technology*', a term coined to describe technologies that provides assistance, which in turn contributes to increased safety and security for individuals, and enabling the individual to be independent, regardless of illness, social-, or physical disability. In Norway, welfare technology is divided in four categories (1) safety and security technology, (2) compensation and wellbeing technology, (3) technology for social contact (4) technology for care and treatment [11]. Within these four categories, several different technologies are included, such as IoMT's, personal medical devices and smart devices. Remote care solutions refer to technological solutions that enables the patient to interact with healthcare staff remotely, outside traditional healthcare facilities such as clinics and hospitals. In these solutions, patients can share data about their health and clinicians can provide medical guidance and support [12]. Remote care solutions can include devices that are classified as medical devices, such as continuous glucose monitoring or cardiac defibrillators. However, it can also be technologies that are not classified as medical devices, such as digital self-reporting forms and video consultations, which are viewed as assisted living technologies. In the following, we will refer to welfare technology and remote care solutions as *assisted living technology* as this is a more broadly adopted term. Previous research related to assisted living technologies in the Nordics have focused on adoption barriers, stakeholder satisfaction and effect on care provision [13–15]. A variety of different stakeholders are included, but the most critical stakeholders are rarely aligned in their understandings, for example of the experienced benefit of using assisted living technologies [13–15]. Hence, the acceptance criteria for technology varies greatly between stakeholders. The previously identified differences in adoption and acceptance criteria amongst stakeholder made us curios to explore if these differences persist when exploring cybersecurity risk perception. The question we attempt to answer in this paper is.

How does key stakeholders in the Norwegian healthcare sector perceive cybersecurity risks for assisted living technologies?

With cybersecurity risk perception, we refer to the perceived risk related to the use of information and communication technology (ICT) and apply the case of assisted living technologies to explore different stakeholder perceptions. Within this context, cybersecurity risk perception encompasses assisted living technologies ability to function as intended, manage the privacy of individuals, and uphold the confidentiality, integrity and availability of devices and data being collected, transmitted, stored and managed. Building on previous research where key stakeholders in the remote care system has been identified, we have chosen to investigate the perspective of mainly care providers, namely healthcare professionals from both primary and specialist healthcare. In addition, we have included vendors of technology, health authorities/administration and experts relevant to the domain, resulting in five stakeholder groups.

2 Background

Cybersecurity has the recent years received more and more attention, and is a topic of importance on the global, national, and individual level. Risk is also a widely used term and much research has been conducted on different types of risk, such as economical risk, environmental risk, and safety risk to mention a few examples [16]. Risk, however, has different representations depending on the context it is expressed. In phycology, political science, and sociology, risk and risk perception has been thoroughly explored to create understanding of different aspects of risk. In SCADA-systems, the technological risk has different implications, depending on the involved stakeholders [17]. Cybersecurity risk and cybersecurity risk perception have become a topic of interest in many scientific disciplines, from the technical aspects in computer science to the behavioral aspects in phycology. Cybersecurity risks can have various negative consequences if not managed sufficiently, which increases the need to understand these potentialities – especially in critical sectors such as energy, military, and health care.

There are various definitions of the term "cybersecurity risk", we build our understanding on exiting attempts to define cybersecurity [18]. Cybersecurity risks are often assessed to the degree to which confidentiality, integrity, and data availability is preserved, and can be defined as the following "the protection of computer systems and networks from attack by malicious actors that may result in unauthorized information disclosure, theft of, or damage to hardware, software, or data, as well as from the disruption or misdirection of the services they provide" [19]. Alike cybersecurity, the term "risk" has numerous definitions. The definition by [20] still stands as one of the most used in literature, where risk is "the probability of an adverse future event multiplied by its magnitude" (p.59) [20]. This definition of risk postulates a cohesion between probability and consequence, which can be expressed as the following equation $R = P\ x\ C*$ ($r = $ risk, $P = $ probability, $C = $ consequence, $* = $ uncertainty).

In recent years, the uncertainty-dimension related to risk has received more attention and has been explored in terms of the effect uncertainty has on risk and risk perception [21]. The dimension of trust is a factor included in this work, as trust and distrust are seen as implicating factors. Risk can be understood as a future event yet to materialize, where some argue that risk is a social and culturally constructed phenomenon [22, 23]. [24,25] argues that risk is an invented concept to help humans understand and cope with danger and uncertainty, which brings us to the concept of risk perception [24].

Slovic [25–28] is known for his extensive work on risk perception, as a part of the work in social phycology. Risk perception, according to the work of Slovic, view risk in the psychometric paradigm – risk means different things to different people. In the psychometric paradigm, it is argued that both dread and familiarity have effect on how one understands risks. Knowledge, also referred to as familiarity and competence, is argued to have effect on risk perception, which can help explain why experts and laymen hold different opinions on risk [29]. In the context of knowledge and familiarity, dread can also affect how risk is perceived [30]. The conceptual framework "the social amplification of risk" (SARF), argue that risk is amplified and affected by psychological, social, political, and cultural factors [27]. In addition, research has found that feelings and affectivity are important for risk perception [31]. Supported by previous research, risk perception can depend on feelings and subjective understanding of consequences, which makes it interesting to explore whether different stakeholder perceive cybersecurity risk differently in the context of healthcare [32].There are several theories and models that address elements of risk perception, such as the protection motivation theory [33]. The protection motivation theory (PMT) views the individual's behavior to protect themselves against perceived threat through the threat appraisal, the assessment of the severity of the threat *and* the vulnerability affiliated with the threat, and the coping appraisal, the belief they have in their ability to mitigate (response efficacy) or perform protective behaviors (self-efficacy). Together, the threat appraisal and coping appraisal shapes the perception and behavior of the individuals to either reduce or avoid the threat [33].

The inclusion of specific stakeholder groups in this study is based on previous research related to assisted living technology and remote care. Previous studies published in the PubMed-database focus on many different stakeholders (28 unique terms) within the assisted living/remote care research domain. These can be summarized in the following seven groups 1) Care receivers [34–50], 2) Caregivers (including informal caregivers and family) [36, 37, 39, 43, 46, 48–51], 3) Healthcare professionals [34–37, 42–44, 46–49, 52–54], 4) Government and policymakers [34, 44, 46, 47, 49, 55–57], 5) Researchers [34, 44, 48, 55, 56], 6) Technology companies/developers [34, 44, 46, 48, 54, 55], and 7) Others [44, 52, 54, 57]. This study will focus on five key stakeholders which coincides with the stakeholder groups, however in the context of their role in the Norwegian healthcare sector. Norway has a publicly funded healthcare sector, where the population have a legal right to healthcare services. Subsequently, the Norwegian healthcare sector is fragmented, divided in two main parts: primary healthcare and specialist healthcare. These are organized differently and have different responsibilities. The primary healthcare service is managed by the local municipalities, which results in 356 different instances. Here, services such as nursing homes, care facilities, general practitioners and special needs institutions are delivered. Specialist healthcare is run more centralized, through four regional health authorities/enterprises who manages different health organizations. A hospital is an example of a health organization managed by the regional health authorities. Within the specialist healthcare service, all hospitals, acute care facilities and specialized services are included. This results in a healthcare sector consisting of a range of different stakeholders who each have their own roles, responsibilities, needs, and motivations.

3 Methodology

In this study, we use a dual approach to gather a comprehensive data material on stakeholders' perception of cybersecurity risk related to assisted living technologies, building on the identified stakeholder groups from existing literature. Due to constrains in retrieving patient information, the stakeholder group that includes care receivers and patients have not been included in this study. Our methodology belongs to the qualitative approach where our desire is to create understanding of different stakeholders' perception of risk. Further, the topic being explored is complex and resides within a complex ecosystem. We believe our dual approach to gather data has benefited us greatly in generating data that help examine stakeholders' perception in depth. Two main methods were used to collect data – survey and interviews. The interviews and survey were developed independently, but with the goal of collecting data related to same topic so that findings could be analyzed in parallel. Hence, the two data collection methods inform each other, where the survey provide data from a larger population, whilst interviews provide data from a smaller population but more describing, in-depth data. The research project has received ethical approval by the Norwegian Agency for Shared Services in Education and Research (SIKT, previously NSD). In the following we will describe how we applied these two approaches for data collection.

3.1 Survey

Survey is a powerful tool often used to collect data from a larger population [58]. A desire in this study is to generate knowledge on the stakeholder's perception. This can be accomplished by comparing what respondents say they do with what they actually do, through a behavioral self-assessment. Then, to analyze results quantitatively. The Behavioral-Cognitive Internet Security Questionnaire (BCISQ) by Velkie et al. (2019) is an instrument that tries to measure and capture both the behavior and cognitive awareness of individuals [59]. The BCISQ-questionnaire was used to gather knowledge on the risk perception of employees working with operational technology (OT) in the Norwegian energy-sector, demonstrating its effectiveness in the Norwegian context [60].

Our survey is divided in two main parts. The first part of the survey follows the BCISQ-questionnaire consisting of two behavioral scales that measure behavior (how often do you lend your password, and password-simulation) and two cognitive scales that capture information security risk perception (term used in survey is probability and importance) [59]. All responses are measured using a five-point Likert-scale. To fit the target group of the survey, we translated the original BCISQ-questionnaire to Norwegian. The BCISQ-questionnaire is a validated questionnaire to capture risk perception and behavior, which is why we evaluated it to be suitable for our purposes. This enabled us to first capture insights on the respondent's general behavior and perception of risk when using the internet, and then the second part of the survey allowed us to narrow the focus to our main topic. Therefore, for the second part of the survey, we developed questions to capture cybersecurity risk perception, benefiting from the same pattern as the BCISQ-questionnaire, using a five-point Likert scale. Further, the focus was related to specific technologies, namely (1a) medicine dispensers, (1b) digital self-reporting forms and (2) smart watches, where we captured the respondents understanding of "likelihood" and "consequence" related to confidentiality, integrity, and availability, for example *"How would you rate data in the digital self-reporting form was altered without your knowledge?"*. Based on the respondents' background and knowledge, we provided the option to choose between (1a) medicine dispensers and (1b) digital self-reporting forms, while all respondents received questions related to (2) smart watches as it is a broadly adopted technology without a dedicated healthcare function. We chose to include smart watches as it is a wellbeing technology that does not serve a specific function within the healthcare system, even though such devices can provide information related to the users health.

The survey was distributed in several channels, mainly by emails to municipalities, hospitals and vendors who have agreed to be contacted for research purposes and collaboration. The Norwegian solution "Nettskjema" hosted the survey (ID: 360,090/ https://nettskjema.no/a/360090), which made it possible to collect survey answers anonymously and non-traceable as it does not collect IP or identifying information. The inclusion criteria for the survey was simple, the respondents had to have experience with assisted living technology and work in/or with the healthcare sector. A total of N = 185 responded to the survey in the time-period between September 2023–February 2024. Of the respondents, 74,6% identify as women, 24,3% as men and 1,1% did not want to provide information about their gender identification. The distribution of women and men are representative for the Norwegian healthcare sector, as 82% of the healthcare sectors workforce are women [61].

3.2 *Interview*

Interviews were conducted in the period from August 2021 to February 2024. The respondents eligible for interview had to belong to one of the five stakeholder groups previously identified, and have experience with the use, management, operation, or development of assisted living technology. This resulted in a total of 33 respondents and the completion of 31 interviews, as one of the interviews was with two respondents due to convenience. For our purpose, investigating the perception of risk, we evaluated that individual interviews would be more suitable than focus-group interviews, as this would allow for more in-depth follow up of each respondent [62]. The interviews followed a semi-structured approach, lasting between 45–75 min. The distribution between the genders is near equal and evaluated to be representative (Men: N = 16/Women: N = 17). A set of predefined categories and questions formed the baseline for interviews, although some leeway was provided allowing follow-up questions of interesting topics raised throughout the interview. Such an approach help generate a more natural and open conversation related to a complex subject.

In preparation for interviews, four interview-guides were developed and tested. The development of questions builds on theoretical aspects of cybersecurity, risk-, and risk perception and protection motivation theory, resulting in the following topics: familiarity of assisted living technologies, cybersecurity risk understanding, cybersecurity related to welfare technology, risk assessment- and risk management process, management- and organization of cybersecurity, and protection motivation specific questions such as perceived vulnerability, perceived severeness (threat appraisal), response efficacy, self-efficacy (coping appraisal) and adoption of preventative behavior [33]. To increase the validity, the interview guide was tested in advance. Although we developed questions that are rooted in theory, we ensured that all questions were formulated simple and did not require respondents to be familiar with theory in advance. When necessary, we explained words and context to the respondents. After completing interviews, recordings were transcribed and played-back twice for verification, resulting in transcriptions ready for analysis. The tool NVivo was used to analyze interviews, following the inductive-deductive approach [62]. This led to the identification of overarching themes which embodied 72 codes.

4 Analysis

This section will provide the results of our study, we will first present some descriptive statistics from the survey, followed by analysis of the BCISQ-scales and risk perception related to assisted living technologies included in this study. Lastly, we enrich the survey findings with results from the analysis of interviews. In the data material specialist healthcare is overrepresented (N = 121) versus primary healthcare (N = 50). To make up for this, we have adjusted the further analysis to enable comparison (Table 1).

Table 1 Respondent distribution

Stakeholder group	Respondents (%)
Primary healthcare	27
Specialist healthcare	65,4
Others	7,6

80,8% of all respondents (N = 147) answered "yes" when asked if they used information technology as a part of their daily work, meaning that most of our respondents are familiar with information technology. Specifically for assisted living technologies, a total of 84,3% answered yes to being familiar with these, where the majority had this knowledge by identifying as healthcare personnel (58,9%). Interestingly, 10,7% (N = 19) of our respondents answered that they were familiar with the technology as a patient, meaning that some of the respondents have a dual background for using the technology, both as healthcare personnel (initial identification) and patient.

4.1 BCISQ-Subscales

The BCISQ-subscales measure respondents' behavior in two ways, first through self-assessment of situations, secondly through simulation by asking them for their most frequently used password. Information security awareness is captured on two scales using two sets of questions, one referring to risk-awareness by using the word likelihood [20], while the second set relates to the evaluation of importance [59]. Before filtering the analysis between the stakeholder groups, the Cronbach's alpha is calculated to $\alpha = 0.765$, which is an acceptable level of internal consistency (Table 2).

In the simulation, a total of 46 (25% of respondents), provided an answer to the voluntary question "To check the quality of your password, write your most used password". The majority provided answers that was evaluated to be false passwords (e.g. "IWillNotGiveYouMyPassword", "Nei", "What?", "Never"). However, 17 of the passwords was evaluated to be real as they suit standard password requirements – resulting in 9.4% of the respondents revealing their real password.

Table 2 BCISQ-subscale for all respondents

BCISQ-subscale	Min	Med	SD
Risky behavior self-assessment scale	1.0	1.75	0.557
Cognitive likelihood (risk) scale	1.0	4.00	0.907
Cognitive importance scale	1.0	4.0	1.089

Table 3 BCISQ-subscale differentiated

Primary healthcare			
BCISQ-subscale	Min	Med	SD
Risky behavior self-assessment scale	1	1	0.52
Cognitive likelihood (risk) scale	1	3	1.04
Cognitive importance scale	2	4	1.13
Specialist healthcare			
BCISQ-subscale	Min	Med	SD
Risky behavior self-assessment scale	1	1	0.50
Cognitive likelihood (risk) scale	1	3	0.84
Cognitive importance scale	1	4	1.02

4.2 BCISQ-Subscales Differentiated by Stakeholder Group

The dataset was overrepresented by healthcare workers, namely from the primary healthcare services and specialist healthcare services, therefore we chose to include only these two in our analysis. The Cronbach's alpha for the responses based on who answered that they worked in the "primary healthcare services" is calculated to $\alpha = 0.810$, while for "specialist healthcare services" the Cronbach's alpha is $\alpha = 0.801$, indicating a strong internal consistency for both groups with a minimal difference. Moreover, the BCISQ subscales demonstrate that primary healthcare workers and specialist healthcare workers have a nearly identical self-assessment of behavior and awareness. The only difference is the measurement of the importance, where respondents belonging to specialist healthcare services evaluated importance lower than respondents belonging to primary healthcare services (Table 3).

4.3 Risk Perception for Assisted Living Technologies Devices Amongst Healthcare Workers

One of the key elements of this study is to capture the cybersecurity risk perception related to assisted living technologies and investigate if there are differences between stakeholders' perception of risk. First, we performed an analysis of the whole dataset without differentiating between stakeholders, to detect if there were significant differences between technologies. There are only minor differences between the perceived likelihood of confidentiality, integrity, and availability (CIA) being compromised and the consequences related to them. The median for likelihood is between 3–4 for the majority of results, which indicates that most respondents find it neutral to unlikely that either of the devices are compromised. Perceived consequence is evaluated slightly higher, where the median of each technology is 4 "major consequences". For medicine dispensers, the consequence of unavailability and integrity is perceived

Table 4 Cyber risk perception for assisted living technology

Medicine dispenser	Mean	Med	SD	Digital self-reporting	Mean	Med	SD
P (unavailability)	3.12	3.0	1.01	P (unavailability)	3.28	3.0	0.82
P (integrity)	2.66	3.0	0.91	P (integrity)	2.63	3.0	0.91
P (confidentiality)	2.48	3.0	0.90	P (confidentiality)	2.46	3.0	0.87
C (unavailability)	3.48	4.0	0.77	C (unavailability)	2.99	3.0	0.78
C (confidentiality)	3.40	3.5	0.95	C (confidentiality)	3.70	4.0	0.81
C (integrity)	3.86	4.0	0.86	C (integrity)	3.85	4.0	0.77

slightly higher than for confidentiality, while for digital self-reporting forms consequences related to confidentiality and integrity are perceived more serious. When calculating the Cronbach's alpha a for internal consistency, medicine dispensers (Cronbach's alpha $\alpha = 0.735$)and digital self-reporting (Cronbach's alpha $\alpha = 0.647$). Values above 0.6 are acceptable, although values above 0.7 would indicate a stronger internal consistency. Like the work of Velkie et al. (2022) the Cronbach's alpha is slightly lower for digital self-reporting forms, meaning that there is some room for improvement in the question clarity or the scale we use to measure responses [63] (Table 4).

The next step in this analysis was to filter the dataset based on stakeholder group. For medicine dispensers, the two stakeholder groups have a relatively similar perception. Specialist healthcare evaluated consequences slightly higher than primary healthcare, indicating that specialist healthcare workers perceive the consequences of a cyber incident towards medicine dispensers to be more serious than primary healthcare workers. Both groups perceive integrity compromise as a major consequence. Hence, both perceive the likelihood of incidents to be more unlikely than likely. The Cronbach alpha is above 0.7 for each of the groups (at $\alpha = 0.715$; $\alpha = 0.721$ for primary-, and specialist healthcare). When analyzing the responses for digital self-reporting forms, the same pattern repeat, where specialist healthcare workers have a slightly higher perception of likelihood and consequence. The two largest differences are identified between consequence of either confidentiality or integrity compromise, where specialist healthcare evaluate the consequences higher, meaning they perceive it to be more serious. The Cronbach alpha for each of the groups is calculated $\alpha =$ 0.730 for primary healthcare and $\alpha = 0.612$ for specialist healthcare, indicating a stronger internal consistency amongst primary healthcare workers (Table 5).

4.4 *Risk Perception for Assisted Living Technologies Amongst Stakeholders'*

Building on the findings from the survey, the healthcare workers (both primary and specialist healthcare services) have some concerns related to the use of assisted

Table 5 Cyber risk perception for assisted living technologies differentiated

Medicine dispensers				Digital self-reporting			
Primary healthcare	Mean	Med	SD	Primary healthcare	Mean	Med	SD
P (unavailability)	3.15	3	1.10	P (unavailability)	3.50	3.5	0.53
P (integrity)	2.55	3	0.90	P (integrity)	2.71	2.0	0.95
P (confidentiality)	2.43	2.5	0.84	P (confidentiality)	2.25	2	0.89
C (unavailability)	3.30	3	0.69	C (unavailability)	2.75	3	0.71
C (confidentiality)	3.45	3.5	0.88	C (confidentiality)	3.75	4	0.46
C (integrity)	3.70	4	0.91	C (integrity)	3.63	4	0.52
Specialist healthcare	Mean	Med	SD	Specialist healthcare	Mean	Med	SD
P (unavailability)	3.10	3	0.89	P (unavailability)	3.32	3	0.84
P (integrity)	2.86	3	0.91	P (integrity)	2.63	3	0.93
P (confidentiality)	2.57	3	0.98	P (confidentiality)	2.46	3	0.89
C (unavailability)	3.90	4	0.77	C (unavailability)	3.01	3	0.74
C (confidentiality)	3.48	4	1.03	C (confidentiality)	3.76	4	0.80
C (integrity)	4.19	4	0.60	C (integrity)	3.97	4	0.69

living technologies, especially regarding the potential consequences related to confidentiality, integrity, and availability. These are perceived low in likelihood, but major in consequence. The interviews provide more in-depth insight into the perception of healthcare workers from both stakeholder groups. In addition, the perceptions of vendors, authorities and experts are included, as these were not sufficiently captured in the survey. The following table demonstrates the key findings from the analysis (Table 6).

The results from our analysis demonstrate that our respondents (mainly healthcare workers) have a low risky-behavior as the majority of answers aligned with "never" and a high sense of information security awareness. When looking at the BCISQ-scale for primary healthcare workers and specialist healthcare workers, there are only minor differences in their self-assessments.

The minor differences continue throughout the analysis, which indicates that primary healthcare workers and specialist healthcare workers have coinciding perceptions of cybersecurity risks for medicine dispensers and digital self-reporting forms. The same is confirmed during the interviews, especially related to concerns of consequences for patient safety. A difference identified in interviews is the underestimation/disbelief of cybersecurity affecting assisted living technologies from the primary healthcare workers. Here, specialist healthcare workers have a stronger focus on cybersecurity incidents, especially in terms of data breach and privacy violations. This may indicate a higher perception of cybersecurity risks. Conversely, when analyzing the simulated response, a higher number of specialist healthcare workers provided their passwords than primary healthcare workers. This indicates that there may be some discrepancies that the survey is not able to detect between the risk perceptions (6% primary vs 10% in specialist). The number that provided an "real" password

Table 6 Stakeholder interview analysis

Stakeholder	Perception of Threats	Perception of Consequences	Percepived likelihood	Perception of Risk
Primary healthcare	Technical faults, power outages, DoS, firmware malfunction, less emphasis on cyber incidents	Compromised patient safety, service disruption, downtime data breach	Low	Underestimated focus on cybersecurity, patient safety oriented (human error, lack of competence, lack of regulation and guidelines)
Specialists healthcare	Data breaches, firmware malfunction, cyber incidents	Impact on patient care (patient safety), service disruption, downtime	Low	Patient safety oriented, accountability, outdated technology (legacy systems, updates), complex system management
Vendors	Advanced persistent threats (APT), cyberattacks (DdoS, ransomware, malware)	Data breaches, privacy breach, reputational damage, financial loss/ contractural breach, patient safety	Neutral/Low	Concerned with business challenges, supply chain dependencies, operational dependency/infrastrucutre, human error
Authorities	Cyber attacks, inadequate immediate system security, unauthorized access, insufficient security measures	Compromised patient confidentiality, data breach, service disruption, increased costs	Likely	Vulnerability of sector to attacks, new exposures from new technology and operating models (such as cloud migration, network risk), concentration risk., data transfer/sharing
Experts	Complex threat actor landscape, malware, cryptolocker, ransomware, targeted attacks on personal health devices	Spread of malware, privacy violations, potential for direct patient harm through device manipulation	Very likely	High inevitability of cybersecuity threats, privacy concerns, Standardisation, complexity of ecosystem, systemic weakeness

in the simulation is only 9.4% of the whole population which indicates that there is an understanding of password hygiene. The simulation demonstrate that specialist healthcare workers were marginally more likely to reveal their password, which contradicts the vendors understanding of customer knowledge/competence considering cybersecurity. It is stated several times throughout the vendor interviews that there are significant differences amongst primary healthcare and specialist healthcare, where primary healthcare/municipalities are perceived less knowledgeable in cybersecurity than specialist healthcare/hospitals. One respondent stated, "*if the primary healthcare workers were asked about their passwords to check or verify a salary question, everyone would share it"*. Additionally, one expert mentioned that "healthcare workers share their credentials", where one account can be logged into the computer/ patient journal system for an entire day where all at work have access to one account. These understandings are not consistent with the survey findings, which may indicate the survey is not sufficient to gather the real perception of stakeholders, or that the interviewees have a bias towards primary healthcare service workers.

From the interviews, we found that there are some distinct differences in how the stakeholders perceive cybersecurity risk. The experts demonstrate the highest perception of cybersecurity risks, whereas one of the experts stated, "it is not a question of if (*a cyberattack occurs*) but when" (E1). It is not surprising that experts demonstrate the highest perception of cybersecurity risk, as these are trained professionals within the field of cybersecurity. They also acknowledge that the likelihood of incidents occurring for assisted living technologies is lower than for the sector in general. Hence, experts and authorities highlight the healthcare sector as a critical sector and an attractive target for advanced threat actors. Additionally, vendors acknowledge that healthcare is an attractive target, but mainly in terms of healthcare data, not the functionality of healthcare systems/devices or the sector in general. In terms of perceived consequences, the difference in perceptions continues. Healthcare workers (primary and specialist) tend to focus more on the consequences related to patient safety and service disruption/downtime (consequences related to availability), while vendors focus more on data breaches and reputational damage—emphasizing business risks. By vendors security is seen as a "license to operate". Authorities largely focus on privacy breaches and confidentiality breaches but does not view these in terms of business risk. Experts focus on all these aspects, additionally experts mentions device manipulation that could result in loss of functionality as a risk. Evaluating the statements in the context of the CIA-triad, healthcare workers are most concerned about the availability aspect due to potential consequences, vendors and authorities are most concerned about the confidentiality, especially related to sensitive/health data, whilst experts focus on all aspects, highlighting integrity-issues through the potential of data manipulation. Only one of the informants belonging to the stakeholder "vendor" highlighted issues related to availability of devices.

There is agreement amongst all interviewees that human error (through lack of competence, understanding, awareness) represent a noteworthy risk in the cybersecurity domain. When asked about the "biggest security risk", 31 answered that a human-related issue was the largest. One vendor stated, *"Humans are without doubt the biggest risk, and lack of competence*" (1AD_V). Another interesting statement from primary healthcare was *«we don't have the capacity, competence or understanding to do a lot of this (cybersecurity) stuff*". Human error can also be connected to the low capacity that exists in both primary-, and specialist healthcare, where one of the primary healthcare workers stated, *"there is a lot, so it's complicated, we have to do this on top of everything else (…) You're not able to stay updated on absolutely everything"*. One of the experts connected human error and competence to culture, where it was emphasized that the healthcare workers are not formally trained or educated in cybersecurity–thereby it cannot be expected of them to have a strong cybersecurity culture or awareness. One of the experts also highlighted the lack of cybersecurity training in the healthcare sector. In addition to identifying the perception of risk amongst stakeholders, we identified factors that have effect on how cybersecurity risks were perceieved, namely trust, competence (education/background), organization (management, responsibility), healthcare context (device function, device criticality, care provision) and holistic approach (sectorial, standardization/regulatory).

5 Discussion

In the following, we will apply theory of risk perception to discuss the results presented in the previous section. First, looking at similarities in risk perception, then we apply theoretical perspectives to explain key differences in risk perception. Lastly, we will analyze results using the protection motivation theory.

5.1 Similarities in Risk Perception

The results demonstrate that there are variations in risk perception amongst stakeholders. In some areas, they have coinciding perceptions, while on others there are larger differences. The corresponding risk perceptions amongst stakeholders can be summarized as follows:

- There are shared concerns for patient safety and patient care, albeit the severity/ impact is perceived differently.
- Stakeholders perceive technical malfunctions and loss of functionality as a risk.
- For risk perception, a high degree of uncertainty is expressed related to likelihood and consequence as several other factors have the potential to implicate the security of assisted living technologies.
- There is consensus amongst stakeholders that human error is a crucial aspect of risk.
- The organizational structure and the environment where technology is deployed has influence on the perception of a risk materializing. Again, the uncertainty dimension is relevant as there is unclear lines of responsibility and management, which is evaluated a risk increasing factor.
- Risk is seen in relation to dependencies towards vendors, third-parties, or other dependencies such as national components, national infrastructure, governmental bodies, and regulators.
- Risk perception impacted by competence is mentioned by all stakeholders, either as an aid in how they understand risk, or because the lack of sufficient competence represents a risk by itself.

Even though there are coinciding perceptions, there are variances in how cybersecurity risks are perceived, and stakeholders express different concerns on the matter. The stakeholders included in our study have various roles within healthcare, moreover, they have different competencies and background. In terms of familiarity referring to knowledge of the risk, the capability to understand the risk is based on knowledge, and ability to effect or control potential consequences [29, 30]. Competence is a highly relevant aspect of the stakeholders' risk perception, as competence can enable or disable the ability to understand information provided on cybersecurity risks and threats [30]. Several of the healthcare workers acknowledge that lack of competence and training in the field of cybersecurity resulting in difficulties to comprehend potential cybersecurity incidents, how incidents can occur and why anyone would want to target technology that provides care for innocent individuals. Malicious intent is hard to comprehend, especially among healthcare workers.

Another interesting finding is that most of the stakeholders view it as more unlikely than likely that incidents are caused with malicious intent, which result in a low-risk perception. There is limited reflection over the potential that technologies can be affected by run-by-attacks, which was how the NHS in the UK became a victim of the WannaCry-attack in 2016, which lead to detrimental consequences [4]. Previous research in cybersecurity risk perception have focused on specific hazards that can occur, rather than the potential negative affect on confidentiality, integrity or availability [64]. In our research, hazards do not appear to be in the mind of most stakeholder, except for data breach. The majority of respondents referred to risks related to data loss and data breach, not including incidents related to manipulation of device functionality, information/vitals, modification, or unavailability.

5.2 *Differences in Risk Perception*

The results for the survey found only minor differences between the healthcare workers risk perception related to assisted living technologies. However, during the interviews the perception of these two groups became more evident. As demonstrated through the survey utilizing the BCISQ-questionnaire, the healthcare workers score high on the scale measuring behavior and awareness–which slightly deviates from the interview findings where perceptions are communicated more diversely. As with identified similarities, we have summarized the differences in risk perceptions:

- Stakeholders perceive likelihood of incidents differently.
- The risk considering the CIA-triad varies among stakeholders, indicating different perceptions of risk.
- The results identified different perceptions of the consequence of cybersecurity risks for assisted living technologies amongst stakeholders.
- There are substantial deviations in technical competence and awareness amongst stakeholders.

- Stakeholders have conflicting views on the responsibility for cybersecurity, resulting in deviations of perceived risk.

These bullet points illustrate key differences in how the five stakeholder groups perceive cybersecurity risks, with notable deviations particularly around the perceived likelihood of incidents, focus on specific aspects of the CIA triad, and the responsibility for managing cybersecurity risks.

Healthcare workers and vendors perceive the likelihood as low, which differs from experts and authorities who express that healthcare services are likely to be affected by cybersecurity attacks, albeit they do not attribute this to assisted living technologies alone. Further, healthcare workers perceive the availability of assisted living technology to be most important as loss of availability may have direct impact on patient care which is also supported when viewing the survey results. Vendors highlight the confidentiality dimension with emphasis on data breaches and privacy concerns, while experts on the other hand, focus on both confidentiality and availability, and additionally on integrity through data manipulation potential. Healthcare workers in general are more concerned for patient safety and care as a result of cybersecurity incidents and service disruptions, while vendors acknowledge this as a risk their primary concern is business risk. This may be explained by different understandings of severity and benefit which is argued in the psychometric paradigm [25, 26]. Vendors are concerned for data and privacy breaches, as these can lead to violation of regulations and contractual requirements, negatively impacting their reputation and financial prospects. Similarly, authorities focus on data breaches and privacy while experts are more concerned for the societal and patient related implication of cybersecurity risks targeting healthcare. From the perspective of SARF, it is argued that risk perception is amplified or attenuated through different social processes [27]. The concern healthcare workers have related to patient safety may be amplified by direct interaction with patients and adverse patient events, while vendors are more affected by the business objectives they should meet.

While vendors state that primary healthcare workers lack sufficient cybersecurity knowledge in comparison to specialist healthcare workers, the survey does not capture these differences. On the other hand, interviews highlight a gap in knowledge, indicating that primary healthcare might underestimate cybersecurity risks compared to specialist healthcare. Such differences can be attributed both to perceived control and familiarity with cybersecurity [30]. Experts express clearly that the responsibility of cybersecurity must belong to vendors of assisted living technologies, while vendors themselves express a shared responsibility, where the healthcare sector hold a large part of the responsibility. Primary healthcare workers view responsibility as "someone else's" but are uncertain where the responsibility should be placed. Both primary and specialist healthcare workers have an expectation that the technologies they use in care provision should be secure by default. The different views of responsibility may indicate gaps in addressing cybersecurity risks effectively.

5.3 Protection Motivation Theory

The protection motivation theory can be applied to analyze the interview findings. Perceived vulnerability refers to the susceptibility of being harmed by a specific threat, where especially primary healthcare workers display a lower awareness of vulnerability than the other groups as they do not understand why threat actors would be motivated to attack or compromise assisted living technologies. Further, one of the healthcare respondents acknowledged that limited knowledge in cybersecurity and technology made it difficult to comprehend the potential vulnerabilities, and the high trust in vendors resulted in a low perceived vulnerability as security was a topic "someone else" was responsible for. Vendors also perceive the vulnerability to be low, due to the implemented security mechanisms. Experts and authorities, on the other hand, demonstrate a higher awareness in terms of perceived vulnerabilities and perceived threat through acknowledging the weakness in the system and management of technology, the value of the data the technology (and healthcare sector) have access to, and the inevitability of attacks occurring. There is mutual alignment in the perceived severeness, as all stakeholders express their concern related to the potential consequence of a cybersecurity incident. Experts, primary healthcare workers and specialist healthcare workers express the severity with emphasis on patient security consequences, vendors and authorities mention these, but vendors focus mainly on business consequences related to business reputation and privacy breach.

There are identified differences in coping appraisal. Coping appraisal refers to response efficacy (belief that protective behavior will reduce threats) and self-efficacy (belief in one's ability to perform the protective behavior) [33]. Vendors and experts express a higher coping appraisal than primary- and specialist healthcare workers. Interestingly, the survey demonstrates that primary healthcare workers have a stronger protective behavior (response-efficacy) in terms of password hygiene than specialist healthcare workers, however, when asked about their own protective behavior the gap in ability such as lack of competence and capacity to cope with cybersecurity risks is highlighted, resulting in low self-efficacy. Moreover, specialist healthcare workers acknowledge challenges with updating and securing systems, which impacts the self-efficacy as they are unable to perform these activities themselves. Vendors overall have high belief in their own behavior in terms of implementing security mechanisms, completing certifications of their companies, established information security management systems and good practice for cybersecurity, resulting in a high coping appraisal. Another element that effects coping appraisal is the dependency on vendors and third-parties, which is prevalent amongst primary- and specialist healthcare workers and vendors. This may lower the response efficacy slightly, as these three groups are highly dependent on the protective behavior of others. Experts and authorities demonstrate a stronger coping appraisal, supported by the belief in protective behaviors and ability to enact these, especially compared to the other three groups expressing dependency on vendors and third parties. Another aspect that seems to affect the coping appraisal amongst especially primary healthcare workers is their emphasis on lack of competence to manage cybersecurity risks effectively. Nearly all of respondents are aware of their lack of competence, which creates uncertainty and reduces the belief in their own capacities.

6 Conclusion

Cybersecurity risk perception is complex and multifaceted, the same can be stated for the healthcare sector in Norway. This study has shed light on the cybersecurity risk perceptions of different stakeholders for assisted living technology. Notably, the distinctions between healthcare workers, vendors, experts, and authorities underline the complexity of comprehending *and* of managing cybersecurity risk for emerging technology such as assisted living technologies in the healthcare setting. The protection motivation theory highlights the importance of understanding behavior and motivation in light of cybersecurity and provides guidance on how we can enhance protective behaviors amongst the different stakeholder groups. Moreover, we have identified some factors which seem to have implication on how risk is perceived, whom we frame as contextual factors such as competence, educational background, organizational structure, device function/criticality and trust.

The novelty of this research lies in combining the BCISQ-questionnaire with the interviews to gather more data on the risk perceptions of specific stakeholder groups. Additionally, previous research in the domain of healthcare technology does not sufficiently cover the socio-technical aspects of cybersecurity risk, as it often tends to focus only technical security or implementation/adoption challenges. We believe we contribute to address this gap in our research. This work demonstrates the applicability of risk perception theory to domains that are fairly "new" in comparison to when theories were established and in which domains they have previously been applied. In the domain of healthcare where cybersecurity risk has been the key research topic, this study has found distinctions between risk perceptions which can be explained by the stakeholders' roles and engagement in the healthcare sector. Moreover, findings demonstrate a clear need to strengthen the capacity and competence of especially healthcare workers, if they are expected to make use of more technology in care provision the coming years. The goal of this study to identify if stakeholders held different perceptions of cybersecurity risks in the context of assisted living technologies in Norway is thereby met.

However, more research is needed on the topic. Further research should look more in depth into contextual factors, and view how prominent theories within the risk perception field highlight these. Another interesting aspect is to view how the specific context of healthcare devices affects risk perception in contrast to other type of smart devices deployed in a home-setting. This can allow the exploration of device function and criticality for patient care, whether this effects how risks are perceived, and potentially accepted. Subsequently, future work should explore in-depth why these stakeholders hold different perceptions, and if the identified contextual factors have significant effect to explain these. Another topic of interest is in the incentives that can be applied to harmonize risk perceptions. We believe this is important as cybersecurity incidents often materialize through human error, therefor generating a harmonized perception in this environment can help identify how to manage and prevent risks proactively. Mere technological security measures are not enough in a sector heavily reliant on human behavior and practice.

References

1. Europol: Cybercrime. https://www.europol.europa.eu/crime-areas/cybercrime. Accessed 21 Mar 2024
2. Tarikere, S., Donner, I., Woods, D.: Diagnosing a healthcare cybersecurity crisis: the impact of IoMT advancements and 5G. Bus. Horiz. **64**, 799–807 (2021). https://doi.org/10.1016/j.bushor.2021.07.015
3. Mxoli, A., Gerber, M., Mostert-Phipps, N.: Information security risk measures for Cloud-based personal health records. In: International Conference on Information Society, i-Society 2014. pp. 187–193. Institute of Electrical and Electronics Engineers Inc. (2015)
4. Ghafur, S., Kristensen, S., Honeyford, K., Martin, G., Darzi, A., Aylin, P.: A retrospective impact analysis of the WannaCry cyberattack on the NHS. NPJ Digit Med. **2** (2019). https://doi.org/10.1038/s41746-019-0161-6
5. Van Deursen, N., Buchanan, W.J., Duff, A.: Monitoring information security risks within health care. Comput. Secur. **37**, 31–45 (2013). https://doi.org/10.1016/j.cose.2013.04.005
6. Aceto, G., Persico, V., Pescapé, A.: The role of information and communication technologies in healthcare: taxonomies, perspectives and challenges (2018)
7. Buchanan, W.J., Van Deursen, N.: HI-risk: A method to analyse health information risk intelligence. In: 2016 IEEE 18th International Conference on e-Health Networking, Applications and Services, Healthcom 2016. Institute of Electrical and Electronics Engineers Inc (2016)
8. Ahmad, R., Samy, G.N., Ibrahim, N.K., Bath, P.A., Ismail, Z.: Threats identification in healthcare information systems using Genetic Algorithm and Cox regression. In: 5th International Conference on Information Assurance and Security, IAS 2009, pp. 757–760 (2009)
9. Samy, G.N., Ahmad, R., Ismail, Z.: Threats to health information security. In: 5th International Conference on Information Assurance and Security, IAS 2009, pp. 540–543 (2009)
10. Ayatollahi, H., Shagerdi, G.: Information security risk assessment in hospitals. Open Med Inform J **11**, 37–43 (2017). https://doi.org/10.2174/1874431101711010037
11. Woll, A.: Use of Welfare Technology in Elderly Care (2017)
12. omsorgsdepartementet, H.: NOU 2011:11. Innovasjon i Omsorg. regjeringen.no (2011)
13. Rostad, H.M., Stokke, R.: Integrating welfare technology in long-term care services: nationwide cross-sectional survey study. J. Med. Internet Res. **23**, e22316 (2021). https://doi.org/10.2196/22316
14. Stokke, R., Hellesø, R., Sogstad, M.: Hvorfor er det så vanskelig å integrere velferdsteknologii omsorgstjenesten? Tidsskrift for omsorgsforskning. **5**, 7–20 (2019). https://doi.org/10.18261/issn.2387-5984-2019-03-02
15. Isaksen, J., Paulsen, K.B., Skarli, J., Stokke, R., Melby, L.: Hvilken nytte har hjemmeboende med hjelpebehov av velferdsteknologi? Tidsskrift for omsorgsforskning. **3**, 117–127 (2017). https://doi.org/10.18261/issn.2387-5984-2017-02-09
16. Engen, O.A., Pettersen, K.A.G., Kruke, B.I., Lindøe, P.H., Olsen, K.H., Olsen, O.E.: Perspektiver på samfunnssikkerhet. Cappelen Damm Akademisk (2021)
17. Cherdantseva, Y., Burnap, P., Blyth, A., Eden, P., Jones, K., Soulsby, H., Stoddart, K.: A review of cyber security risk assessment methods for SCADA systems (2016)
18. Hubbard, D.W., Seiersen, R.: How to Measure Anything in Cybersecurity Risk. Wiley (2016)
19. Andress, J. ed: The basics of information security: understanding the fundamentals of InfoSec in theory and practice. Syngress

20. Adams, J.: Risk. Press, London, UCL (1995)
21. Fjaeran, L., Aven, T.: NC-ND license creating conditions for critical trust-How an uncertainty-based risk perspective relates to dimensions and types of trust. Saf. Sci. **133**, 105008 (2021). https://doi.org/10.1016/j.ssci.2020.105008
22. Weinstein, N.D.: Optimistic biases about personal risks. Science **1979**(246), 1232–1233 (1989). https://doi.org/10.1126/science.2686031
23. Spencer, T.: Risk Perception: Theories and Approaches. Nova Science Publishers, inc (2016)
24. Slovic, P.: The Perception of Risk. Routledge (2016)
25. Slovic, P.: The perception of risk. In: Scientists Making a Difference: One Hundred Eminent Behavioral and Brain Scientists Talk about their Most Important Contributions. pp. 179–182. Cambridge University Press (2016)
26. Fischhoff, B., Slovic, P., Lichtenstein, S., Read, S., Combs, B.: How safe is safe enough? A psychometric study of attitudes towards technological risks and benefits*. Policy **9**, 127–152 (1978)
27. Kasperson, R.E., Renn, O., Slovic, P., Brown Halina, S., Emel, J., Goble, R., Kasperson, J.X., Ratick, S.: The Social amplification of risk: a conceptual framework. Risk Analysis. **8**, 177–187 (1988)
28. Slovic, P., Fischhoff, B., Lichtenstein, S.: Why Study Risk Perception? Risk Analysis. **2** (1982)
29. Slovic, P.: Perception of risk. Science **1979**(236), 280–285 (1987)
30. Fischhoff, B., Bostrom, A., Quadrel, M.J.: Risk perception and communication. Annu. Rev. Public Health **14**, 183–203 (1993). https://doi.org/10.1146/annurev.pu.14.050193.001151
31. Moen, B.E.: Risk perception, priority of safety, and demand for risk mitigation in transport (2008)
32. García del Castillo, J.A., García del Castillo-López, A., López-Sánchez, C., Dias, P.C.: Risk perception and psychosocial vulnerability in health behaviours. In: Spencer, T. (ed.) Risk Perception (2016)
33. Rogers, R.: Cognitive and physiological processes in fear appeals and attitude change: a revised theory of protection motivation. In: Cacioppo, J., Petty, R. (eds.) Social psychophysiology, pp. 153–177. Guilford Press, New York (1983)
34. Abdullah, N.N.B., Clancey, W.J., Raj, A., Zain, A.Z.M., Khalid, K.F., Ooi, A.: Application of a double-loop learning approach for healthcare systems design in an emerging market. In: Proceedings International Conference on Software Engineering, pp. 10–13. IEEE Comput. Soc (2018)
35. Haydon, H.M., Caffery, L.J., Snoswell, C.L., Thomas, E.E., Taylor, M., Budge, M., Probert, J., Smith, A.C.: Optimising specialist geriatric medicine services by telehealth. J. Telemed. Telecare **27**, 674–679 (2021). https://doi.org/10.1177/1357633X211041859
36. Alwan, M.: Passive in-home health and wellness monitoring: Overview, value and examples. In: Proceedings of the 31st Annual International Conference of the IEEE Engineering in Medicine and Biology Society: Engineering the Future of Biomedicine, EMBC 2009. pp. 4307–4310. IEEE Comput. Soc (2009)
37. Gallegos-Rejas, V.M., Thomas, E.E., Kelly, J.T., Smith, A.C.: A multi-stakeholder approach is needed to reduce the digital divide and encourage equitable access to telehealth. J. Telemed. Telecare **29**, 73–78 (2023). https://doi.org/10.1177/1357633X221107995
38. Alwan, M., Turner, B., Kell, S., Penberthy, K.J., Cohn, W., Felder, R.: Development of survey instruments to guide the design of health status monitoring systems for the elderly: Content validity evaluation. Presented at the October 24 (2006)
39. Nabhani-Gebara, S., Kayyali, R., Philip, N., Pierscionek, B., Sobnath, D., Kaimakamis, E., Perantoni, E., Chang, J., Davies, N., Siva, R., D'Arcy, S., Vaes, A., Spruit, M., Consortium, W.: WELCOME project: What do Stakeholders want? In Depth Analysis of COPD Patients, Carers and Healthcare Professional Views. Presented at the December 5 (2014)
40. Nilsen, E.R., Stendal, K., Gullslett, M.K.: Implementation of eHealth technology in community health care: the complexity of stakeholder involvement. BMC Health Serv Res **20** (2020). https://doi.org/10.1186/s12913-020-05287-2

41. Wrede, C., Braakman-Jansen, A., van Gemert-Pijnen, L.: How to create value with unobtrusive monitoring technology in home-based dementia care: a multimethod study among key stakeholders. BMC Geriatr. **22**, 1–19 (2022). https://doi.org/10.1186/S12877-022-03550-1/FIGURES/6
42. Shin, M.H., McLaren, J., Ramsey, A., Sullivan, J.L., Moo, L.: Improving a mobile telepresence robot for people with Alzheimer disease and related dementias: semistructured Interviews With Stakeholders. JMIR Aging. **5**, e32322 (2022). https://doi.org/10.2196/32322
43. Gilson, A., Gassman, M., Dodds, D., Lombardo, R., Ford Ii, J.H., Potteiger, M.: Refining a digital therapeutic platform for home care agencies in dementia care to elicit stakeholder feedback: focus group study with stakeholders. JMIR Aging. **5**, e32516 (2022). https://doi.org/10.2196/32516
44. Freiesleben, S.D., Megges, H., Herrmann, C., Wessel, L., Peters, O.: Overcoming barriers to the adoption of locating technologies in dementia care: a multi-stakeholder focus group study. BMC Geriatr. **21**, 1–17 (2021). https://doi.org/10.1186/S12877-021-02323-6/TABLES/2
45. Landau, R., Auslander, G.K., Werner, S., Shoval, N., Heinik, J.: Who should make the decision on the use of GPS for people with dementia? Aging Ment. Health **15**, 78–84 (2011). https://doi.org/10.1080/13607861003713166
46. Sultana Nipa, N., Alam, M., Sanaul Haque, M.: Identifying Relevant Stakeholders in Digital Healthcare View project (2017)
47. van Boekel, L.C., Wouters, E.J.M., Grimberg, B.M., van der Meer, N.J.M., Luijkx, K.G.: Perspectives of stakeholders on technology use in the care of community-living older adults with dementia: a systematic literature review. Healthcare (Switzerland). **7** (2019). https://doi.org/10.3390/healthcare7020073
48. Chien-I Chen, S., Liu, C., Wang, Z., McAdam, R., Brennan, M., Davey, S., Cheng, T.Y.: How geographical isolation and aging in place can be accommodated through connected health stakeholder management: Qualitative Study with Focus Groups. J Med Internet Res. **22** (2020). https://doi.org/10.2196/15976
49. Read, E.A., Gagnon, D.A., Donelle, L., Ledoux, K., Warner, G., Hiebert, B., Sharma, R.: Stakeholder perspectives on in-home passive remote monitoring to support aging in place in the province of new Brunswick, Canada: rapid qualitative investigation. JMIR Aging. **5**, e31486 (2022). https://doi.org/10.2196/31486
50. Guise, V., Anderson, J., Wiig, S.: Patient safety risks associated with telecare: a systematic review and narrative synthesis of the literature. BMC Health Serv Res. **14** (2014). https://doi.org/10.1186/s12913-014-0588-z
51. Cooper, J., Burrow, S., Pusey, H.: What are the perceptions of people living with dementia, family carers, professionals and other potential stakeholders to the use of global positioning systems to promote safer outdoor walking?: a qualitative literature review (2021)
52. Mantzana, V., Themistocleous, M., Irani, Z., Morabito, V.: Identifying healthcare actors involved in the adoption of information systems. European J Infor Syst. 91–102. Palgrave Macmillan Ltd (2007)
53. Serrano, A., Garcia-Guzman, J., Xydopoulos, G., Tarhini, A.: Analysis of barriers to the deployment of health information systems: a stakeholder perspective. Inf. Syst. Front. **22**, 455–474 (2020). https://doi.org/10.1007/s10796-018-9869-0
54. Barlow, J., Bayer, S., Curry, R.: Implementing complex innovations in fluid multi-stakeholder environments: experiences of telecare. Technovation **26**, 396–406 (2006). https://doi.org/10.1016/j.technovation.2005.06.010
55. Greenhalgh, T., Procter, R., Wherton, J., Sugarhood, P., Shaw, S.: The organising vision for telehealth and telecare: Discourse analysis. BMJ Open. **2** (2012). https://doi.org/10.1136/bmjopen-2012-001574
56. Nilsen, E.R., Dugstad, J., Eide, H., Gullslett, M.K., Eide, T.: Exploring resistance to implementation of welfare technology in municipal healthcare services - a longitudinal case study. BMC Health Serv. Res. **16**, 1–14 (2016). https://doi.org/10.1186/s12913-016-1913-5
57. Sponselee, A.M., Schouten, B., Bouwhuis, D., Willems, C.: Smart home technology for the elderly: perceptions of multidisciplinary stakeholders (2008)

58. Braun, V., Clarke, V., Boulton, E., Davey, L., McEvoy, C.: The online survey as a qualitative research tool. Int. J. Soc. Res. Methodol. **24**, 641–654 (2021). https://doi.org/10.1080/13645579.2020.1805550
59. Velkie, T., Solic, K.: Development and validation of a new measurement instrument: the behavioural cognitive internet security questionnaire (BCISQ). Int. J. Electr. Comput. Eng. Syst. **10** (2019)
60. Kannelønning, K., Katsikas, S.: Cybersecurity-Related Behavior of Personnel in the Norwegian Industry. Presented at the (2023)
61. sentralbyrå, S., Jia, Z., Kornstad, T., Martin Stølen og Geir Hjemås, N.: Arbeidsmarkedet for helsepersonell fram mot (2040)
62. Tjora, A.: Kvalitative forskningsmetoder i praksis. Gyldendal (2021)
63. Velkie, T.: Psychologists as information-communication system users: Is this bridge between information-communication and behavioral science enough to prevent risky online behaviors? In: International Convention on Information, Communication and Electronic Technology (MIPRO) (2022)
64. Huang, D.L., Rau, P.L.P., Salvendy, G.: A survey of factors influencing people's perception of information security. Lecture Notes in Computer Science (including subseries Lecture Notes in Artificial Intelligence and Lecture Notes in Bioinformatics). 4553 LNCS, 906–915 (2007). https://doi.org/10.1007/978-3-540-73111-5_100/COVER

Workshop Insights: Navigating Cybersecurity Regulations for Device Manufacturers and Healthcare Operators

Andrea Skytterholm, Lars Halvdan Flå, and Martin Gilje Jaatun

Abstract Both the manufacture and use of medical devices are heavily regulated, but stakeholders have a varying level of maturity, and often struggle to comply with rules and regulations. This paper reports on an empirical elicitation activity that sought to enumerate the challenges faced by (particularly smaller) device manufacturers and device operators (typically hospitals), with a goal to informing the creation of tools that these stakeholders can use to address the challenges. The stakeholders completed a brief survey, and then participated in two focus group sessions delving into perceived challenges. The results confirm a need for increased collaboration between manufacturers and operators, and improved guidance material to ease complying with relevant regulations.

1 Introduction

The healthcare system faces significant challenges in the coming years, with an increasing number of elderly and ill people and the difficulty of recruiting healthcare workers [18]. Connected medical devices play a crucial role in addressing these challenges. These devices enable continuous monitoring of patient health, allowing patients who previously needed hospitalisation to receive treatment at home. This shift towards more home-based treatment and care not only enhances patients' quality of life but also presents economic benefits by reducing the burden on the healthcare system and lowering overall healthcare costs.

However, the integration of connected medical devices into healthcare systems introduces a new set of challenges, particularly concerning cybersecurity. As

A. Skytterholm (✉) · L. H. Flå · M. G. Jaatun
SINTEF Digital, Trondheim, Norway
e-mail: andrea.skytterholm@sintef.no

L. H. Flå
e-mail: lars.flaa@sintef.no

M. G. Jaatun
e-mail: martin.g.jaatun@sintef.no

M. G. Jaatun et al. (eds.), *Proceedings of the International Conference on Cybersecurity, Situational Awareness and Social Media*, Springer Proceedings in Complexity,
https://doi.org/10.1007/978-981-96-0401-2_16

these devices collect and transmit sensitive medical data, ensuring their security is paramount to protecting patient privacy and preventing potential cyber threats. Medical device manufacturers and healthcare operators face specific challenges in identifying and complying with cybersecurity regulations, standards, and guidelines, in the middle of the rapid evolution of technology and the complex regulatory landscape.

This paper aims to investigate these challenges and potential solutions for mitigating them. To guide our work, we have developed the following research questions:

RQ1: What specific challenges do device manufacturers and healthcare operators encounter in identifying and complying with cyber security regulations, standards, and guidelines?

RQ2: What strategies or solutions can be implemented to effectively mitigate these challenges?

By addressing these research questions, the goal is to contribute to increased cybersecurity for connected medical devices, ultimately enhancing the safety, reliability, and effect of healthcare delivery in an increasingly digitalized world.

This paper is structured as follows, Sect. 2 provides the background of our work. Section 3 describes the methodology used, and in Sect. 4 we present our findings. Finally, in Sect. 5 we discuss the results, and Sect. 6 concludes the paper.

2 Background

The World Health Organization (WHO) has produced a report on the healthcare workforce in Europe and the challenges we are facing in the years to come [18].

In 2019 Levine et al. [8] performed a comparison study of home treatment versus hospital care for patients requiring admission. The results of this study show that administering home treatment for acutely ill adults gives significant benefits compared to hospital care. Home treatment reduces costs and healthcare utilization, besides, it also lowers the rates of readmission. Additionally, patients receiving home treatment demonstrate higher levels of physical activity in comparison with those receiving care in a hospital environment. These findings underscore the effectiveness and advantages of implementing home-based treatment approaches for acutely ill adults, emphasizing the potential for improved outcomes and reduced burdens on the healthcare systems.

The Code of Conduct for information security and data protection in the healthcare and care services (In Norwegian: "Normen") [17] is created and maintained by organisations and companies in the health sector in Norway, and is used by many of the Norwegian connected medical device (CMD) operators. Normen provides requirements which detail and supplement current regulations, however, it does not cover all the regulatory requirements. The goal of Normen is to contribute to safe and secure information exchange between the different actors in the healthcare sector.

In the US, the Food and Drug Administration (FDA) [3] is responsible for protecting the public health. This responsibility includes ensuring the safety, efficacy and security of medical devices.

3 Methodology

To answer the research questions, we conducted a workshop with 17 participants from the healthcare sector. The participants were recruited from a list of local players in the health technology ecosystem, specifically targeting individuals with competence in regulations, standards, and guidelines for cybersecurity related to CMDs.

Workshop

The participants represented two distinct roles within the healthcare sector, CMD Operators and CMD Manufacturers. Table 1 describes the two roles and provides an overview of the number of participants in each group accordingly.

Workshop agenda:

- Welcome and lunch
- Menti survey
- Breakout session one
- Break
- Breakout session two

The workshop started with a welcoming session and lunch to allow the participants to warm up before beginning the discussions in the breakout sessions. At the end of the lunch, they were given a survey presented in Menti [10]. Menti is a tool designed for creating interactive presentations, quizzes, or surveys. It enables participants to provide real-time feedback or responses. Once the survey is finished, Menti offers the functionality to download the results in a PDF format. The survey was completed in plenary by the participants using smartphones and invited the participants to give input on what standards and guidelines they used, and on challenges they faced. In

Table 1 Participants

Role	Description	Number of participants
CMD operator	Healthcare service providers such as hospitals and healthcare institutions	11
CMD manufacturer	Manufacturers of CMDs, including developers of software applications	6

addition to collecting data, the survey also served to start the process of reflecting on these topics. The questions from the survey are presented in Table 2.

Following the survey, participants were divided into three distinct focus groups for the initial breakout session. These groups were formed according to their roles. One group was composed of CMD Manufacturers, while the other two groups were made up of CMD Operators. In the second breakout session, the groups were restructured. The new groups were formed to ensure an even distribution of CMD Manufacturers and CMD Operators. We also strived to distribute participants from the same organization across different groups as much as possible. The workshop was structured in this manner to facilitate discussions on challenges and solutions among participants in similar situations as well as those from different parts of the health ecosystem.

The two focus group sessions each lasted 75 min. In both sessions, all groups were presented with 3 to 4 questions to discuss. The questions are presented in Table 3. Participants were instructed to silently write down their responses to each question on a post-it note. Following this, all the responses were collectively reviewed before moving on to the next question. Each group had a facilitator responsible for guiding the discussions, taking notes and keeping track of time.

Data analysis

The data set used for analysis in this work consists of the responses to the survey, the facilitators' notes and the notes from the post-its. After the workshop, the notes taken by the facilitators during the two breakout sessions were processed and completed, and the notes from the post-its were added. The results from the survey were downloaded to a PDF file.

Based on the topics touched upon during the workshop, five main categories, or 'codes', were identified to help analyze the collected data. Some of these main codes also had smaller related codes, or 'sub-codes', under them. All codes and sub-codes are listed in Table 4. After the codes were decided, the data was reviewed and parts were marked based on which code they fit into. This process is known as 'coding'. Once the coding was completed, the coded data was examined more closely to understand what it was indicating. The findings from this examination are written down and can be found in the results section.

4 Results

In this section we present the results from the workshop, structured according to the codes shown in Table 4.

An overview of which regulations and guidelines the participants fulfil/use is shown in Fig. 1, as reported by the participants in Menti. For readability, we have excluded the regulations/guidelines which none of the participants reported as relevant. The full overview of the regulations and guidelines included in the Menti questionnaire can be found in Table 2.

Table 2 Questions from the workshop survey presented in Menti

Questions	Options
Which actor do you represent?	Device manufacturer, Integrator, Healthcare provider, Others
Which of the following do you use/fulfil?	GDPR, Medical Device Regulation, In Vitro MDR, Cybersecurity Act, NIS2, Cybersecurity Enhancement Act, Consolidated Appropriations Act, MDCG—Guidance on Cybersecurity for Medical Devices, IMDRF—Principles and Practices for Medical Device Cybersecurity, ENISA—Procurement Guidelines for Cybersecurity in Hospitals, ANSM—Cybersecurity of Medical Devices Integrating Software During Their Lifecycle, eHealth Suisse—Guide for App Developers, Manufacturers and Distributors, BSI—Cyber Security Requirements for Network Connected Medical Devices, Health Canada—Pre-market Requirements for Medical Device Cybersecurity, FDA—Cybersecurity in Medical Devices: Quality System Considerations and Content of Premarket Submissions, FDA—Postmarket Management of Cybersecurity in Medical Devices, TGA—Medical Device Cyber Security Guidance for Industry, TGA—Medical Device Cyber Security Information for Users, Others
To what extent do standards and regulations …	Facilitate awareness of cyber security among owners/users of medical equipment, establish a common understanding of legislation and requirements among all parties involved, ensure that the cost of security is in line with the benefit, have purpose and goals that correspond to their needs, are flexible enough to take account of new technology and changes in the threat landscape and risk
Have you experienced contradictions/conflicts in legislation and guidelines between cyber security, ethics and treatment? If so, which ones?	
What challenges have you encountered when using/fulfilling regulations, standards and guidelines regarding …	Asset management, differences across countries and markets, risk assessments, intrusion detection and logging, incident management and training, complex procedures for certification, different processes in different countries, conduct state-of-the-art analysis, outdated documents and regulations, privacy regulations, lacing standards for OT equipment, regulations lacing focus on security, regulations lacing focus on privacy, how cybersecurity is considered in classification, lack of assessment of the cybersecurity of equipment, lack of protection mechanisms for cybersecurity, others
What are the biggest challenges you experience? when do they occur, why do they occur, and how can they be resolved? Give one or two examples	

Table 3 Questions from breakout sessions

Questions from section 1
1. What standards, guidelines and regulations do you use, and what are they missing?
2. How do you work with cybersecurity?
3. What challenges do you face when following regulations, standards, and guidelines?
4. How do you handle these challenges? What is needed for them to be solved?
Questions from section 2
1. What tensions/conflicts do you observe between cybersecurity, ethics, and clinical care when using standards, regulations and guidelines?
2. How do you handle these conflicts? What is needed for them to be solved?
3. Which area should be focused on in standards, regulations, and guidelines in order to support/realize future scenarios in healthcare?

Table 4 Codes and subcodes used for analysis

Codes	Sub-codes
Regulations	MDR FDA privacy and GDPR
Guidelines	MDCG Normen filing and certification
Cybersecurity	NIS2 cybersecurity versus patient safety
Operators	–
Manufacturers	–
Supply chain	–

As this workshop was carried out in the context of the Horizon Europe project NEMECYS, a project which also plans to organize similar workshops in other European countries, the questionnaire in Menti focused on international regulations and guidelines. However, during the focus groups, Normen [17], The Act on Medical Devices [13], and NO-18 [15] were also discussed. These are all specific to Norway. The Act on Medical Devices serves as the legal foundation for national regulations and for the execution of directives from the European Commission [13]. NO-18 is a requirements specification document and should be used for the evaluation of the Supplier's offered solution within the areas of ICT and information security [15].

4.1 Regulations

Workshop participants describe the healthcare sector as an extremely regulated sector. Regulatory requirements are one of the major barriers to entry into the market for CMD manufacturers. Workshop participants reported that there is an extensive set of regulations, from many different parties. They report that it is difficult to navigate the landscape and difficult to get an overview of what they need to consider.

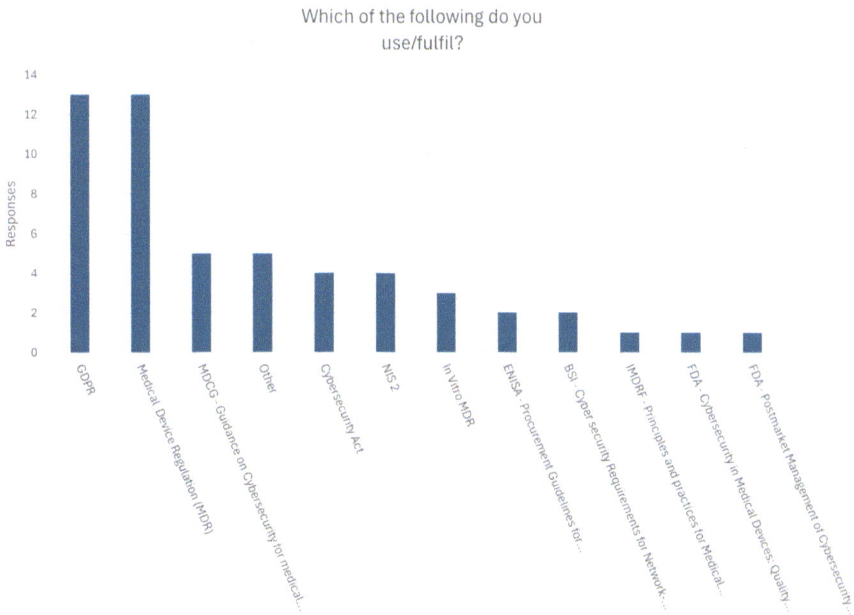

Fig. 1 Standards, regulations and guidelines workshop participants follow

Consequently, many new companies struggle to enter the market, and they rarely manage to do so. Not only is it difficult to get an overview, but the lack of resources makes it difficult to keep up with the changes that are made to the regulations and who are affected by such changes. However, although there is an extensive set of regulations, it was reported that there is not much legislation on cyber security. The question raised was if this will become better with NIS2 [6].

> How can we know which regulations apply to our product? Very segregated laws and requirements, especially when thinking beyond national borders. —Workshop participant

Developments of medical equipment are happening faster than the regulations can keep up with. Hence, CMD operators have to say no to new equipment because the regulations do not allow it yet. An example of this challenge is regulations for precision medication and medical genetics algorithms. Machine learning models cannot be changed after they have been approved, because then they are no longer approved.

Another challenge raised by the participants is the challenge of assessing *must* against *should*. There may be different interpretations in situations where the regulations are a bit unclear, and in such situations, decisions from court proceedings become leading. However, it was also said that nobody follows the rules completely, and with medical equipment, you always break some rule or another.

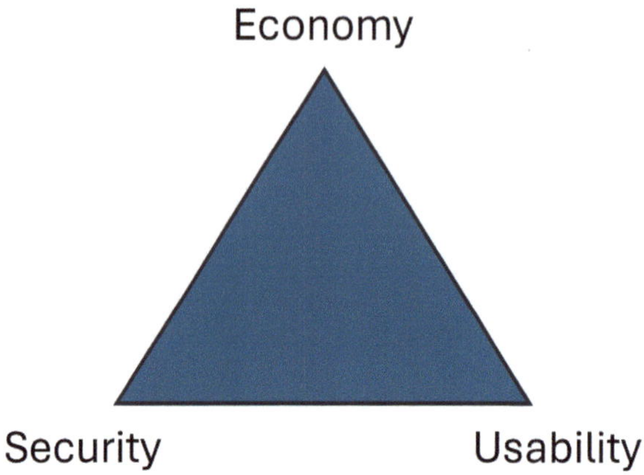

Fig. 2 The security, economy and usability triangle

> Health legislation, privacy legislation, and cybersecurity requirements require a balanced approach, which requires expertise, and often needs to be assessed on a case-by-case basis.—Workshop participant

4.1.1 MDR

The starting point of the the Medical Device Regulation (MDR) [5] is "to not harm the patient". However, the MDR is highly relevant to CMD manufacturers but has limited content relevant to the users of CMDs.

One of the challenges concerning the MDR is that they require medical equipment to be developed with regard to the state-of-the-art. Several of the CMD manufacturers participating in the workshop said that it is difficult to understand what is meant by the state-of-the-art, and also what the bar is for something being good enough.

> [Challenge:] Identify state-of-the-art in cybersecurity for a specific medical device.—Workshop participant

Workshop participants stated that some security often leads to reduced usability, which conflicts with the MDR that has a high focus on usability and clinical benefit. The participants used a triangle of security, economy and usability to illustrate this example, and said that you can only get two things. The triangle is illustrated in Fig. 2. The participants found it challenging to know where the bar is for when one has to prioritize, and security can be viewed as a barrier or cost.

Regulations apply differently for welfare technology and medical technology. Another finding from the workshop was that it can be challenging for product owners to identify whether their product falls under the MDR regulation or not.

4.1.2 FDA

The FDA [3] guidelines were described as more concrete. However, it was uncertain how they could use standards in and towards the FDA process.

4.1.3 Privacy and GDPR

Workshop participants described privacy as very strict, which could be both positive and negative. Further, they said that compliance with GDPR [4] trumps everything and that it is not always easy to follow the regulations. However, it was mentioned that it is not always necessary to take the worst path and that there are differences in how pragmatic different regions are. Decisions in court proceedings are leading when the regulations are unclear. Yet, other participants mentioned that privacy legislation is always trumped by health legislation and that there is always a trade-off between privacy and confidentiality, and ease of use.

Another comment made during the workshop was that "GDPR is not a big deal until the Data Protection Authorities come to visit, but this is also the reason why it needs to be kept up to date".

Secondary use of data is another challenge that was mentioned by several of the workshop participants, and especially by the CMD operators. Using data for research and to develop better treatment for patients is of course important, but more importantly, they need to have control over what the data is used for. Another challenge related to this topic is that CMD operators have the impression that many of the suppliers do not have a good enough understanding of what anonymization of data entails. In some situations, suppliers might want to use data to develop better products which can offer patients better treatment, but it is however important that the CMD operators know what they say yes to. An example from Norway shows that it is not always easy to get an overview of the data collected by the supplier and what it is used for. The Norwegian Ministry of Health procured continuous glucose monitors (CGMs) from an American vendor, without being aware that the associated application collected data about the users which was sent back to the vendor and used for various and unclear purposes. It was mentioned that it is important to address who is the owner of the data.

It is not only suppliers who struggle to understand the concept of anonymization, also researchers have made the mistake of using personal data beyond the original purposes. The local hospital participating in the workshop is a university hospital, and many of the employees have positions at both the hospital and the university. In research projects, data have been collected from patient journals and used for purposes other than first intended.

Access control is another challenge. It was mentioned during the workshop that there is an inherent conflict in the GDPR, and one of the workshop participants feared that this might result in people not having the access they should. Access control was also said to be challenging for small companies.

As with the MDR, the GDPR also has requirements for state-of-the-art. However, it is unnecessary to fill out a comprehensive state-of-the-art documentation.

4.2 Guidelines

4.2.1 MDCG

The workshop participants opined that the MDCG 2019-16 guidance [9] is good, but that it should be updated.

4.2.2 Normen

Every operator using services from Norsk Helsenett (a governmental entity providing digital services and infrastructure to the health sector) is required to follow Normen [17]. Many of the operator representatives participating in the workshop follow Normen. Normen provides a set of requirements for suppliers, and one of the main challenges here is that foreign suppliers have no relation to this norm, and also it lacks an English translation of the full text. The technical specification requirements can also be challenging to understand, and some suppliers report that they do not understand them and that they do not know what the operators need from them.

Before Normen, operators had a fact sheet that was formed as a questionnaire, and workshop participants claimed that this was easier to understand. Now, Normen comes with an Excel sheet that replaces the old fact sheet, however, this is not as easy to understand. It is unclear what advice in the sheet applies to them, and it does not use the questionnaire format as in the former document.

It was said during the workshop that nobody fulfils Normen, but that they use a risk-based approach [11]. A risk-based approach is commonly mandated by Norwegian regulatory authorities, and means that the scope of implemented measures according to a given regulation increases where the risk is higher than usual. Conversely, in situations where the risk is lower than usual, the scope of measures can be reduced [11]. This basically the opposite of a rule-based approach. There is a need to balance and consider what is the weightiest consideration. Normen fosters more dialogue between CMD operators and CMD manufacturers.

4.3 Filing and Certification

Participants in all groups mentioned that there is an extensive set of regulations from many different parties. Notified bodies are not allowed to advise or consult, but they are the ones responsible for approval. It is difficult to get an answer from the authorities, they prefer not to deal with practical issues or problems.

There might also be a need for a road map to certification. The best thing the EU could have done would be to publish previous filings. We don't know if different notifying bodies have the same practice, and consider the same things when approving a product.

> Often there is a lack of practical examples of how the requirements can be met.—Workshop participant

Very few provide advice on how to do certification. There are major barriers in place for new actors, where is the bar for certification? Not a single example exists of what an approved filing/process looks like. The process is also complicated by the fact that sometimes one only has to comply with parts of a standard.

4.4 Cybersecurity

4.4.1 NIS2

One challenge mentioned is that cybersecurity costs money, but there is no return of investment here. NIS2 [6] makes it possible to fine cybersecurity violations, which can result in cybersecurity no longer being seen only as an expense. No one would like to buy equipment which can cause your organisation fines in the order of millions.

4.4.2 Cybersecurity Versus Patient Safety

Balancing cybersecurity and patient safety is a real challenge, and there are several reasons for that. Clinicians and patients most often care about clinical benefit, and their primary focus is not on regulations and security. In clinical settings, the primary focus is on patient care and security requirements will be viewed as something cumbersome and can be experienced as a barrier for the clinicians. One participant claimed that it is ethically problematic to stop good projects due to legal requirements. Cybersecurity practices may conflict with patient safety as cybersecurity can affect usability and make it difficult to use the device. One example of this is two-factor authentication. One of the workshop participants mentioned that they had implemented two-factor authentication for their app, however, as their patients are people struggling with anxiety and depression, two-factor authentication increased the threshold for using the app. This example also shows how important it is to follow technological development, and the solution to this problem was solved by replacing BankID (a personal identification method [1]) with biometry, the threshold was thereby lowered.

> [Identified conflicts between cybersecurity, ethics and patient treatment:] There is a challenge in terms of usability—Workshop participant

During the initial discussion, cybersecurity and patient safety were viewed as two different things, however, it was also mentioned that cybersecurity aligns with patient safety. One of the CMD manufacturer participants stated that they regard cybersecurity as part of patient safety but that some hospital people regard cybersecurity as something separate.

> In tenders, ICT security is often deprioritized by clinicians. But information security also provides patient safety. This needs to be made more aware—Workshop participant

A third view on this is the balance of risk and benefit. When developing a product, you need to consider the clinical benefit, and the effect of the product should weigh more than the harm it causes. A risk management plan can be used to report cybersecurity risks, which in turn can translate to patient risk. Cybersecurity can again lead to new risks to patient safety, so this balance is a challenge. It was said that it is usually easier to be better safe than sorry, which in turn affects user-friendliness in some situations.

Another challenge the CMD manufacturers raised was that you do not know who will manage the device. Is it a doctor at the hospital, a home nurse or the patient himself? Depending on who operates the equipment, different challenges will arise. The technological maturity of the user is also something that needs to be considered in this situation.

4.5 Operators

The focus of the CMD operators is on operation and maintenance, and here The Act on Medical Devices [13], MDR [5] and GDPR [4] have a central role. CMD operators focus on making the manufacturers responsible for compliance with regulations and standards.

In a procurement process, CMD operators communicate with the sellers of the product, these are not necessarily the ones with the technical specification knowledge. This extra link in the communication process is perceived as a barrier to obtaining the necessary information.

The total purchases made by CMD operators in Norway are small compared to the purchases made by large hospitals in the U.S., thus, CMD operators find it challenging to put requirements on CMD manufacturers, when they respond that the device is FDA-approved.

Another challenge mentioned is that new systems brought in are not new but legacy systems that do not follow the relevant standards and guidance, nor do they comply with current regulations. However, they must be used because there are no available alternatives on the market. It takes time to get the certifications in place. The solution in such scenarios is to have them operate in a closed network that does not communicate with the hospital network. This allows operators to use it as a standalone solution.

> Cybersecurity should not be seen as an expense, but as an investment—Workshop participant

CMD operators claimed during the workshop that they are good at complying with regulations, however, not enough resources have been allocated to deal with this. As a result, it is easier to say no to new CMDs rather than familiarizing themselves with the regulation and taking the time to keep up with the developments. From a political point of view, the healthcare sector will be digitalized, however, CMD operators find this challenging when they do not have enough resources to build competence within the organisation.

4.6 Manufacturers

CMD manufacturers reported that they find it challenging when the requirements put forward by the hospitals are different from those in standards and regulations. CMD manufacturers need more information about what is required to market and sell products, and it is important to get this information early in the development process to prevent them from ending up on the wrong track.

The findings from the workshop show that it is unclear whether it is the manufacturer or the operator who has the responsibility for the update of the CMDs. Some workshop participants claim that the manufacturer is not required to perform the update, whereas others claim that CMD operators demand over-the-air updates and require equipment to be kept updated by the manufacturer.

4.7 Supply Chain

When CMDs are certified, they are certified by themselves, but one of the challenges mentioned is that this certification does not consider the situation when the CMD is integrated with the hospital systems. One question raised regarding this topic was about the demarcations for responsibility in this scenario. Every integration into a hospital is different, and a risk assessment is needed in each case.

One of the concerns raised by some of the workshop participants was related to the integration of CMDs and the division of responsibility. What is the CMD operator's responsibility and what is the CMD manufacturer's responsibility? In the requirements specification NO-18 there is a self-declaration that covers parts of this responsibility challenge, but NO-18 has more focus on the operational side [15]. It is unclear who is responsible for the configuration of a product so that it is integrated securely, and it is unclear how this is done in practice. Also, the main point of this is who is left with the risk?

5 Discussion and Conclusion

What specific challenges do device manufacturers and healthcare operators encounter in identifying and complying with cyber security regulations, standards, and guidelines? And what strategies or solutions can be implemented to effectively mitigate these challenges?

The healthcare sector is heavily regulated, which causes significant barriers to entry for CMD manufacturers. Navigating the complex landscape of regulations, standards and guidelines from various authorities is challenging, often resulting in new companies struggling to enter the market. Additionally, the rapid technological development exceeds regulatory updates, leaving CMD operators struggling with outdated regulations that prevent them from accepting innovative medical equipment.

One of the main challenges for CMD manufacturers is to get an overview of all the relevant regulations, standards and guidelines they need to consider when developing new medical devices. It is not only resource-demanding to identify the right documents but also to identify which parts of the documents are relevant.

Normen [17] may also be considered a challenge for the CMD operators. It provides a framework for compliance, but with no English translation, it is difficult to convince foreign suppliers to follow the requirements. Also, Norway is such a small country, and some procurements even happen at the hospital level, meaning that the procurements are quite small compared to the ones in the U.S. Furthermore, the lack of clarity on responsibility and integration issues within the supply chain adds another layer of complexity.

The MDR [5] emphasizes patient safety but lacks clarity on state-of-the-art requirements, creating uncertainty for manufacturers. Balancing security, usability, and economic considerations further complicates compliance efforts, especially concerning cybersecurity and privacy regulations such as GDPR [4].

Access control was identified as a challenge, particularly for small companies. One participant expressed concern that many individuals do not have the necessary access rights, and that access control is one of the primary challenges with Norway's newest journal system. However, the issues with access control are two-fold. On the one hand, some healthcare personnel do not have the access rights they need [12]. On the other hand, healthcare personnel have access to data they should not be able to access [14]. This highlights the complexity and importance of effective access control in healthcare systems, especially when implementing new journal systems. As mentioned during the workshop, there is an inherent in the GDPR. The inherent conflict can be viewed as the challenge of balancing the need for access and at the same time limiting the level of access to what is strictly necessary.

Cybersecurity emerges as a critical concern, with the introduction of NIS2 [6] potentially shifting perspectives by imposing fines for violations. However, balancing cybersecurity measures with patient safety remains a challenge, highlighting the need for adaptable solutions that prioritize usability and clinical benefit.

To tackle these challenges effectively, stakeholders need to work together. This means simplifying regulatory procedures, clarifying who's responsible for what, and encouraging innovation—all while prioritizing patient safety and data privacy.

Working together is essential for the healthcare sector to make the most of CMDs and improve patient outcomes. These insights emphasize the need for ongoing communication, updating regulations as needed, and investing in the resources and expertise required to overcome regulatory challenges and fully utilize the potential of CMDs in transforming healthcare delivery.

5.1 Threats to Validity

This paper has reported on just a single workshop with a small number of participants from a limited geographical area in a small European country, and thus cannot lay claims to generalizability in a global or even European context. However, Norwegian manufacturers generally market their devices to a European or global audience, and are thus likely to face similar challenges as experienced by other European manufacturers. Furthermore, Norwegian health care providers (or operators) are subject to MDR and GDPR just like their European counterparts. The results from our workshop are also aligned with our previous work [2].

5.2 Further Work

The results reported in this paper will inform the ongoing work of the Horizon Europe NEMECYS project [7], where we are building tools that will help device manufacturers to better comply with MDR through building security in to connected medical devices, as well as tools that will assist operators in better ensuring security in the use of their connected medical devices, and improving the privacy of patients. The empirical work reported in this paper is being complemented by similar workshops conducted in Spain, Italy and Greece, and results from these will appear in subsequent publications. We have also formed a collaboration with four other Horizon Europe projects working on enhancing cybersecurity of connected medical devices, and this work has already resulted in a initial feedback on how the MDCG 2019-16 guideline can be improved [16].

References

1. BankID: What is BankID? https://bankid.no/en/what-is-bankid
2. Bernsmed, K., Jaatun, M.G.: Security-by-design challenges for medical device manufacturers. In: EICC 2024. ACM (2024). https://jaatun.no/papers/2024/HealthSec_Barriers.pdf, to appear

3. Center for Devices and Radiological Health: Cybersecurity in medical devices: quality system considerations and content of premarket submissions (2024). https://www.fda.gov/regulatory-information/search-fda-guidance-documents/cybersecurity-medical-devices-quality-system-considerations-and-content-premarket-submissions
4. European Commission: Regulation (EU) 2016/679 of the European Parliament and of the Council of 27 April 2016 on the protection of natural persons with regard to the processing of personal data and on the free movement of such data, and repealing Directive 95/46/EC (General Data Protection Regulation) (2016). https://eur-lex.europa.eu/eli/reg/2016/679/oj
5. European Commission: Regulation (EU) 2017/745 of the European Parliament and of the Council of 5 April 2017 on medical devices, amending Directive 2001/83/EC, Regulation (EC) No 178/2002 and Regulation (EC) No 1223/2009 and repealing Council Directives 90/385/EEC and 93/42/EEC (2017). http://data.europa.eu/eli/reg/2017/745/oj
6. European Commission: DIRECTIVE (EU) 2022/2555 of the European Parliament and of the Council of 14 December 2022 on measures for a high common level of cybersecurity across the Union, amending Regulation (EU) No 910/2014 and Directive (EU) 2018/1972, and repealing Directive (EU) 2016/1148 (NIS 2 Directive) (2022). https://eur-lex.europa.eu/legal-content/EN/TXT/?uri=CELEX:32022L2555
7. Jaatun, M.G., Taylor, S., Upstill, C., Farkash, A., Garcia, S., Andoutsos, C.: Nemecys: addressing challenges to building security into connected medical devices. Procedia Comput. Sci. (2024). https://jaatun.no/papers/2023/nemecys_longer_paper.pdf, to appear
8. Levine, D.M., Ouchi, K., Blanchfield, B., Saenz, A., Burke, K., Paz, M., Diamond, K., Pu, C.T., Schnipper, J.L.: Hospital-level care at home for acutely ill adults: a randomized controlled trial. Ann. Intern. Med. **172**(2), 77–85 (2020)
9. Medical Device Coordination Group: MDCG 2019-16—Guidance on Cybersecurity for medical devices (2020). https://ec.europa.eu/docsroom/documents/41863
10. Mentimeter: Mentimeter. https://www.mentimeter.com/. Accessed 03 May 2024
11. Ministry of Finance: Prop. 40 L (2017–2018) (2018). https://www.regjeringen.no/no/dokumenter/prop.-40-l-20172018/id2589604/, publisher: regjeringen.no
12. Patterson, R., Standing, H., Lee, M., Dalkin, S., Lhussier, M., Exley, C., Brittain, K.: Paramedic information needs in end-of-life care: a qualitative interview study exploring access to a shared electronic record as a potential solution. BMC Palliat. Care **18**, 1–8 (2019)
13. Lov om medisinsk utstyr (The Act on Medical Devices) (2020). https://lovdata.no/dokument/NL/lov/2020-05-07-37
14. Støbakk, T., Skjesol, H.: Datatilsynet varsler tilsyn med helseplattformen etter flere brudd på personvernet. https://www.adressa.no/nyheter/trondelag/i/O8Gm4l/datatilsynet-varsler-tilsyn-med-helseplattformen-etter-flere-brudd-paa-personvernet. Accessed 08 May 2024
15. Sykehuspartner HF: No-18 - kravspesifikasjon for medisinsk teknisk utstyr (MTU) med tilhørende IKT-grensesnitt v1.5 (No-18—Requirements specification for medical technical equipment with associated ict interface) (2024). https://www.sykehuspartner.no/490b0f/siteassets/documents/sikkerhet---regionale-bruksvilkar/no-18---kravspesifikasjon-for-mtu-medisinsk-teknisk-utstyr-med-tilhorende-ikt-grensesnitt.pdf
16. Taylor, S., Jaatun, M.G., Bernsmed, K., Androutsos, C., Castillo, A., Frey, D., Favrin, S., Rodrigues, J., Milojevic, D., Karas, D.S., Siachos, I., Gedeon, P., Epiphaniou, G., Moukafih, N., Maple, C., Messinis, S., Rallis, I., Protonotarios, N.E., Matragkas, N., DeLong, R., Arvanitis, T., Katzis, K.: A way forward for the MDCG 2019-16 medical device security guidance. In: Proceedings of PETRA 2024 (2024). https://jaatun.no/papers/2024/PETRA-HLTH-13-Collaboration-Paper-MDCG-final.pdf, to appear
17. The Norwegian Directorate of Health: The code of conduct for information security and data protection in the healthcare and care services (2022). https://www.ehelse.no/normen/documents-in-english
18. WHO Regional Office for Europe: Health and care workforce in Europe: Time to act (2022)

Digital Evidence: A Legal Practitioner's Perspective

Ibeabuchi Egbu, Jacques Ophoff, and Annelize McKay

Abstract Cybersecurity, much like traditional security, requires the collaborative effort of various stakeholders, including government, regulators, tech professionals, and legal practitioners. Each party must play their part effectively to achieve the objective of a cyber environment free from danger or threat. Ironically, legal practitioners who are responsible for prosecuting cybercrimes and administering justice face significant challenges in interpreting and applying the technical aspects of digital investigations. Existing research indicates that this difficulty largely stems from their predominantly non-technical background, which hinders their ability to fully grasp the implications of digital evidence. This research study aimed to investigate how employing analogies to traditional forms of evidence could aid legal practitioners in better understanding and applying the technical terms encountered in digital investigations. Building on existing literature on legal practitioners' digital awareness, this research study adopted a five-step methodological approach. First, a digital forensics report was analysed to identify technical terms. These technical terms were simplified and analogies to traditional evidence were constructed. These analogies were published via a website, and made accessible through a custom-built Microsoft Word add-in. A survey involving 16 participants was carried out to assess the value of the simplification provided by the analogies, and the impact was evaluated against Bloom's taxonomy of cognitive learning. Results show that the analogies improved

I. Egbu (✉) · J. Ophoff · A. McKay
Abertay University, Dundee, UK
e-mail: s.egbu@abertay.ac.uk

J. Ophoff
e-mail: j.ophoff@abertay.ac.uk

A. McKay
e-mail: a.mckay@abertay.ac.uk

J. Ophoff
University of Cape Town, Cape Town, South Africa

A. McKay
University of Pretoria, Pretoria, South Africa

M. G. Jaatun et al. (eds.), *Proceedings of the International Conference on Cybersecurity, Situational Awareness and Social Media*, Springer Proceedings in Complexity,
https://doi.org/10.1007/978-981-96-0401-2_17

understanding of the digital forensics report and the way practitioners interpreted and applied technical terms found in the report.

1 Introduction

Over the past twenty years, the nature of criminal activity has undergone a significant transformation. The conventional crime scene, once characterized by traditional evidence such as fingerprints and footprints, has shifted to a digital landscape where substantial evidence is left in forms of digital footprints of computer interactions [1].

With these advancements in technology and exponential increase in the use of computerized/electronic devices, there is hardly a crime that does not feature a 'computer' or leave some form of digital trail [2]. As the 2020 Digital Forensic Science Strategy developed by the National Police Chief Council attests, *"Over 90% of all crime is recognized as having a digital element"* [3]. When these crimes lead to litigation, the pronouncement of guilt or innocence often relies on the strength of the digital evidence, and the ability of either the procurator fiscal or the defence attorney to effectively communicate the narrative in a clear and comprehensible manner to the judge and jury.

It is expedient that the legal practitioners who are responsible for serving justice in cybercrimes become acquainted with the language and implications of digital evidence. As Daniel and Daniel [4] aptly express,

> *"Attorneys are smart people, but they are not typically experts in all forms of evidence. It is critical to make sure that the attorney understands the evidence being presented and has the proper technical foundation to proceed with the digital evidence portion of the case."*

Investigators and prosecutors have a duty to clearly identify the specific elements or aspects of an offense to correctly frame resulting charges or allegations [5]. Without a foundational grasp of these elements, legal practitioners can draw incorrect conclusions. Understanding digital evidence is as important if not more than the evidence itself. Gottehrer [6] makes an interesting compilation of cases in which digital evidence was crucial in motion practice, was presented in court, and significantly swayed the verdict. Beyond collecting the evidence, the turning point usually comes from effectively leveraging on the understanding during depositions and at trial.

When legal practitioners are unfamiliar with technical concepts, they might struggle to pose challenging questions or misinterpret responses. According to Hamer, (1992, p.30) quoted in [5], many lawyers *"admit that they mentally switch off for the forensic evidence at a trial and that barristers avoid asking detailed questions of forensic witnesses for fear of asking one question too many"*. This deficiency can hinder their ability to undermine the opposing party's digital evidence or expert testimony effectively. It can also make them vulnerable to opposing counsels who might exploit their lack of digital knowledge. Legal practitioners should be at least knowledgeable enough to ask relevant questions [7].

Moreso, evidence that is not easily understood by the court can lead to confusion, potentially impacting the court's decision. The final verdict in a case can greatly depend on how well the court grasps the presented evidence [8]. Recognizing this pivotal role of legal practitioners within the broader context of cybersecurity, this research study seeks to address the following question: *What effect does drawing analogies between digital evidence and traditional evidence have on how legal practitioners interpret and apply technical terms found in digital investigations?* The primary aim of this research study was to simplify technical terms found in digital evidence investigations by comparing them to traditional evidence, such that the facts, responsibilities and intents could easily be established. The secondary aim was then to make these simplified definitions accessible through a convenient and user-friendly medium.

The research study was carried out in two phases. The first phase was to craft these analogies and conduct a survey to gauge the impact of these analogies on how lawyers interacted with a sample digital evidence report. Following the positive sentiments from the survey, these analogies were integrated into Microsoft Word through an add-in. This enhancement aimed to facilitate accessibility, allowing lawyers to readily reference these analogies within their primary workspace while reviewing digital evidence reports in Microsoft Word. The term 'digital forensics' which encompasses the overall process of acquiring, pre-serving, and extracting digital evidence is used interchangeably with digital evidence throughout this paper.

2 Literature Review

The question of whether legal practitioners possess sufficient knowledge and training in digital forensics to competently represent their clients is not a novel concern. Over 10 years ago, Murff et al. [9] conducted a survey of 79 licensed attorneys in the USA. The study found that a mere 6.33% of attorney respondents had undertaken any digital forensics-related courses during their law school tenure, and of these, only 1.27% felt the subject was appropriately addressed. More concerning was that most respondents, a stark 73.5%, reported the expert reports presented by digital forensics professionals as either 'not at all understandable' or merely 'somewhat understandable'.

Parallel themes are echoed in the more recent research by Wilson-Kovacs et al. [10] that examined how criminal defence lawyers in the English adversarial system understood and used digital evidence. One common theme from both research studies is the discernible lack of proficiency among legal practitioners in their comprehension and application of digital evidence. Both studies concur that there is a shortfall in digital forensics education during formal legal training. Murff et al. [9] reveal that only 6.33% of the surveyed legal professionals took digital forensics-related courses during their law school years, while Wilson-Kovacs et al. [10] similarly found that most respondents had not received training in understanding and critiquing digital evidence.

As much as these shortfalls have been identified, there have equally been proposals of strategies to address them. The works of [9, 11, 12] highlight the urgent need for enhanced education in digital forensics tailored to legal professionals, and the subsequent impact this education could have on the effectiveness of the criminal justice system. Murff et al. [9] advocate for an expansion of continuing education offerings in digital forensics to include not only solicitors, but also judges. They suggest that comprehensive courses in digital forensics procedures and analysis should be integral parts of law school curricula.

Similarly, Hasan et al. [11] note that current educational material in security and forensics often require a level of technical knowledge beyond the grasp of most legal professionals. They argue for the development of educational resources that not only present security and forensics concepts in a way that judicial officials can understand, but also use the terminology prevalent in the legal profession. Additionally, they recognize the need for these courses and modules to be concise and flexible enough to fit into the busy schedules of these professionals, while also being adaptable to the rapidly changing field of computing.

A survey by Miller [13] explored the role and influence of digital evidence in criminal proceedings. It specifically focused on prosecutors' comprehension of digital evidence and its subsequent impact on their case-building strategies and interactions with judges, investigators, and other significant legal entities. The findings buttress the essential role of digital evidence across various crime categories. For severe offenses, such as those against children, organized crime, and sex offenses, there is a substantial reliance on digital evidence, underscoring its all-important role in these investigations. The study further informs on the varying levels of digital evidence usage between violent and financial crimes. While digital forensic evidence is used to prosecute violent crimes around half the time or less, there is a more pronounced dependence on digital forensic evidence in financial crimes cases. In the case of organized crime prosecutions, the study found that use ranged from occasional to infrequent. This differential use highlights the versatile nature of digital evidence and its ability to be adapted to the demands of different types of crimes.

Miller's investigation highlights the role of digital forensic evidence in criminal proceedings. It not only draws attention to its varied applications across different crime types, but also emphasizes the need for a thorough understanding and proficient use of digital evidence in the criminal justice system. Away from academia, this need for legal practitioners to engage with digital evidence effectively is globally recognized. This is exemplified by the initiatives undertaken by the European Commission [12]. In response to the increasing impact of digital evidence in the criminal justice system, the Commission implemented a series of training programs for legal practitioners between 2012 and 2017. Specifically, between 2012–2015 in Trier Germany, the Training Centre on Cybercrime for judges and prosecutors was carried out as a project that saw the training of about 500 judges and prosecutors with 'essential skills' for cybercrime prosecution. Also, between 2015–2017 a series of events across five cities including Budapest, Madrid, Lisbon, Cracow, and Trier were held to give legal practitioners a 'basic understanding' of technical concepts like Internet protocol, encryption, cloud computing, etc. These investments, and the overarching objectives of these programs, emphasize the value and need for ongoing, well-structured training programs that accommodate practitioners' needs and expertise levels.

Across these studies and initiatives, there is a common call for more innovative and tailored approaches to training legal practitioners in digital evidence and its attendant technicalities. This research study presents an innovative solution that complements current efforts and opens possibilities for more advanced and usable solutions.

3 Research Methodology

This research study adopted a mixed methods approach, combining qualitative analysis of a case study forensics report with a standard survey involving legal practitioners. Starting with a detailed analysis of the forensics report, technical terms were identified and subsequently translated into simplified analogies. These analogies were then incorporated into a dedicated website for access.

To assess the effect of these simplified analogies, a cross-section of legal practitioners was asked to read the report, referencing the website as needed. Their feedback on the worth and clarity of the analogies in enhancing their understanding of the report was then collected and analysed quantitatively using Likert scale questions. Finally, a Microsoft Word add-in was developed that delivered these simplified definitions directly within the Microsoft Word application.

The digital forensics report used was one that featured in the US 2020 election fraud matter [14]. Efforts to obtain redacted copies of digital investigation reports from Police Scotland through secondary contacts were made between May and June 2023 but proved unsuccessful. Subsequent online searches eventually yielded several reports, including child pornography cases, music copyright infringement, and the US election fraud matter. These reports were scrutinized by the authors, with the report on the US election displaying a relatively denser cluster of technical terms which were extracted for use.

Notably, a significant cluster of these technical terms was located on page 4, Sections 16 and 20 of the report. The following twelve 'technical' terms were extracted: *Security Log; Advanced Persistent Threat; Domain control; Authentication; File Server; Server Log; Network Scanner; Ethernet Port; Packet Capture Log; Application Programming Interface; HTTP; Network Interface Card.*

3.1 Developing Analogies

The standard definitions of the terms were first researched. Primarily, these definitions were derived from the National Institute of Standards and Technology's (NIST) Computer Security Resource Centre glossary [15]. Where a term's definition was not available, a Google search was performed, and results from trustworthy and authoritative sources were chosen.

Arguably, criminal prosecutors often ground their understanding of crime in the foundational concepts of 'mens rea'—the mental intention and 'actus reus'—the physical act [16]. These twin pillars of criminal liability serve as the basis for determining guilt and assigning responsibility. As such, when introducing legal practitioners to these terms and terminologies, it is imperative to bridge their understanding by aligning digital forensic terms with these core legal concepts. By drawing parallels between the digital actions (or inactions) and their associated intent, it is expected that the legal practitioners would better understand the technical terms. This approach not only demystifies these technical terms but positions each term within the framework that lawyers are most familiar with, making it easy to translate digital evidence into legal arguments. The analogies were created over a period of four weeks by one of the authors, aiming to highlight at least one of the following elements:

- Comparison to traditional evidence—what can the term be compared to in traditional evidence?
- Intent—Can any intentions be deciphered from the description of the term?
- Responsibility—Who is responsible for the item described by the term?

These elements were engrafted in a subtle manner rather than in a direct and overt way. This approach was to ensure that the analogies were fluid and not fixed in their interpretations. By doing this, users were given capacity to draw inferences and conclusions when they engaged with and analysed these analogies. The intention was to foster a more thoughtful and interpretive engagement with the content, rather than prescribing a fixed meaning. The following is an example definition and the corresponding analogy:

Security log

Definition—A log is a record of the events occurring within an organization's systems and networks. Logs are composed of log entries; each entry contains information related to a specific event that has occurred within a system or network. When these logs are related to security, they are referred to as security logs [15].

Analogy—Security logs can be compared to the thorough event diaries or incident logs kept by security guards or watchmen in conventional settings.

Imagine a sizable estate or facility with a security officer parading the property. The guard is required to keep a thorough journal as part of their duty. They record every visitor's arrival and departure timings as well as any strange events, disruptions, maintenance tasks, and even their own rounds and checks in this journal. This notebook eventually turns into a timeline of all significant incidents that occurred on the property.

Like physical security logs, digital security logs record every action taken on a device or network. However, unlike physical security logs where entries are manually made by a security guard, digital security logs automatically capture every action taken on a device or network by the system itself. They keep track of a variety of actions, including login attempts, IP addresses, file access, updates, and more. The security logs can assist in tracing unauthorized access, data breaches, or other cyber occurrences, just as the guard's notebook can assist in investigating an incident at the estate, such as a break-in or trespassing.

Just like the incident logs kept by a security guard is a joint responsibility of them and the security department, the responsibility for security logs typically falls on multiple parties within an organization, each playing a specific role in ensuring the proper collection, storage, monitoring, and analysis of these logs. Depending on the size and structure of the organization, this may be the system and network administrators, IT team, security analysts or even third-party service providers.

In this example analogy, the comparison to traditional evidence is achieved by comparing security logs to incident logs kept by security guards. It is reasonable to assume that many legal professionals have met a security guard and at some point, signed their logbook. Also, the parties responsible for the security logs are listed and include system and network administrators, IT team and security analysts. Although this description does not directly indicate intent, it remains consistent with the strategy to address at least one of the three factors: comparison to traditional evidence, intent, or responsibility.

The full glossary of analogies developed in this research study can be found at https://techbenchers.online/glossary/.

3.2 Survey Design

A survey questionnaire was designed using Google Forms and consisted of three sections, as shown in Fig. 1. Following the introductory pages of the questionnaire, a third page contained the digital forensics report excerpt which had a cache of technical terms. Each of these technical terms was hyperlinked to a webpage that contained the analogized definition. Participants were prompted to follow these links to explore the analogies as they read through the document. Following this, participants then moved on to respond to a series of questions before finalizing and submitting their responses.

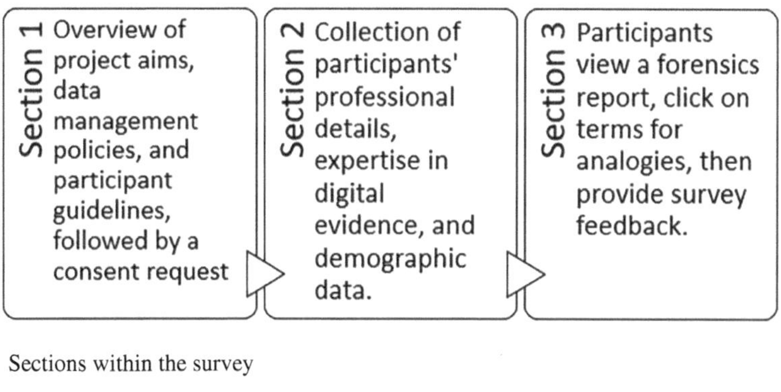

Fig. 1 Sections within the survey

The survey adopted Bloom's Taxonomy to enable a thorough and systematic evaluation of the expected learning outcomes from the analogies. Bloom's Taxonomy offers a hierarchical classification of cognitive abilities and has been significant in influencing educational goals ever since its inception [17]. These were measured using the following questions:

- Knowledge and Recall—*"The analogies on the website make it easier to recall concepts from the report."* This query was designed to determine if the analogies bolstered the participant's ability to recall and remember the technical concepts delineated in the report.
- Comprehension—*"The analogies on the website make the report more understandable."* Here, the intent was to ascertain if, with the help of the analogies, participants not only recalled but also grasped the content better.
- Application—*"The analogies on the website make it easier to use the report in my work."* This was aimed at assessing whether the analogies supported the application of the information or concepts in tangible, real-world contexts.
- Analysis—*"The analogies on the website provide insights into the report that I would not have considered."* This statement was set to determine if the analogies prompted participants to break down the information, identify interrelationships, and pinpoint aspects they might previously have missed.
- Evaluation—*"The analogies on the website effectively clarify the technical terms in the report."* The focus here was to evaluate the quality of the analogies in the context of clarifying complex terms. Participants were asked to assess or make judgments on the effectiveness of the analogies.
- Creation—*"The analogies on the website sparked ideas for further exploration in digital forensics."* This looked to gauge if the analogies stirred participants to consolidate varied pieces of knowledge, suggesting innovative patterns or alternative solutions.

Participants were prompted to respond to the following statements using a 5-point Likert scale, ranging from Strongly Disagree to Strongly Agree. By incorporating questions that span the spectrum of cognitive skills, from rudimentary recall to critical evaluation, the survey aimed to evaluate a holistic understanding of the effectiveness of the analogies.

Prior to data collection the study was approved by the Abertay University Ethics Committee. Participants were recruited through the personal contacts of the authors. Emails soliciting participation were sent in June 2023, to approximately 70 practitioners, including recent law graduates. Since the research study was still in its exploratory phase, it was decided that participation should be open to all classes of legal practitioners. The target sample included legal practitioners who had relevant experience in working with digital forensics reports, final year and recently graduated law students, and academics teaching law-related courses. Email was chosen as the primary method of data collection and links to the survey were sent to potential recipients, with responses collected over a three-week period.

4 Results

A total of 16 people participated in the survey. The participants were first segmented by the level of training in digital forensics. Almost half (46.7%) confirmed they had no prior training or experience. This was followed by 33.3% (five participants) who had taken informal courses, such as online workshops and seminars. Another 13.3% (two participants) were self-taught, relying on online resources or books, while a mere 6.7% (one participant) had pursued formal education, like college or university classes, in this domain.

To better understand their legal experience, the participants were grouped according to professional level. The majority, 57.1%, identified as legal practitioners from outside the UK. Meanwhile, 28.6% were law students from Abertay University, and 14.3% practiced law within the UK. The decision to include student feedback was influenced by the anticipated challenges of accessing the supposedly busier practitioners within the research project's timeframe.

Next, the participants were sorted according to specific areas of legal practice, 40% indicated a specialization in civil or private law, whereas 26.7% were focused on criminal law. The rest 33.3% either had not specialized or felt the provided options were not representative of their specialty. This group could have included student participants still contemplating their future specialization or professionals not aligning strictly with traditional legal bifurcations.

The participants were also categorized by duration in the legal profession. One respondent stood out, having practiced for 6–10 years. On the other hand, 40% were relatively new to the field with less than 2 years of experience. Another significant group, 33.3%, had practiced for a duration of 2 to 5 years. The remaining 20% felt this question was not relevant to their status, possibly indicating the student participants. Finally, the participants were grouped according to familiarity with digital forensics reports. 42.9% admitted to never having engaged with digital evidence. Equally divided, 28.6% each mentioned they rarely or occasionally encountered these reports. It's particularly noteworthy that none claimed regular or frequent interactions with digital forensics reports.

Upon assessing the feedback regarding the use of analogies the following observations were made:

- For the statement concerning the analogies' effectiveness in aiding the recall of concepts, most participants were neutral (43.8%) or agreed (25%). A smaller portion strongly agreed (18.8%), while a negligible number strongly disagreed (12.5%). No participants selected the second-lowest value on the scale. The mean rating here was 3.38.
- When participants were asked whether the analogies made the report more understandable, a significant 43.8% strongly agreed, while 18.8% agreed and 25% remained neutral. Only 12.5% disagreed with the statement. The mean rating here was 3.94.
- In terms of the analogies enhancing the report's usefulness in professional tasks, 43.8% of the participants agreed, and 31.3% strongly agreed. Only 6.3% expressed

low agreement, while 18.8% remained neutral. The mean rating for this statement was 4.

- When asked if the analogies on the website provided insights into the report that they would not have ordinarily considered, 46.7% strongly agreed, and 20% just agreed. No participants disagreed or strongly disagreed, while 33.3% took a neutral stance. The mean rating for this statement was 4.13.
- In response to the question of how well the analogies clarified the technical terms in the report, 33.3% of the participants strongly agreed, while 20% agreed, 13.3% disagreed and 6.7% strongly disagreed. Meanwhile, 26.7% remained neutral on the statement. The mean score for this statement stood at 3.6.
- Lastly, when asked if the analogies spurred further interest in digital forensics, a majority 53.3% of the participants strongly agreed, while 26.7% agreed. Only 20% remained neutral, and no participants disagreed or strongly disagreed. The mean rating for this statement was 4.33.

The box plot of ratings across the cognitive domains are shown in Fig. 2.

Further to assessing the effect of the analogies, the correlations between some demographic features and the feedback was quite enlightening. For instance, practitioners who had had at least two years of experience seemed to find the analogies and simplification provided by the website more valuable than participants who had less experience. This could have been because they had had the opportunity to encounter these technical terms in the course of their duties and are better positioned to make a comparison between how they felt then and now.

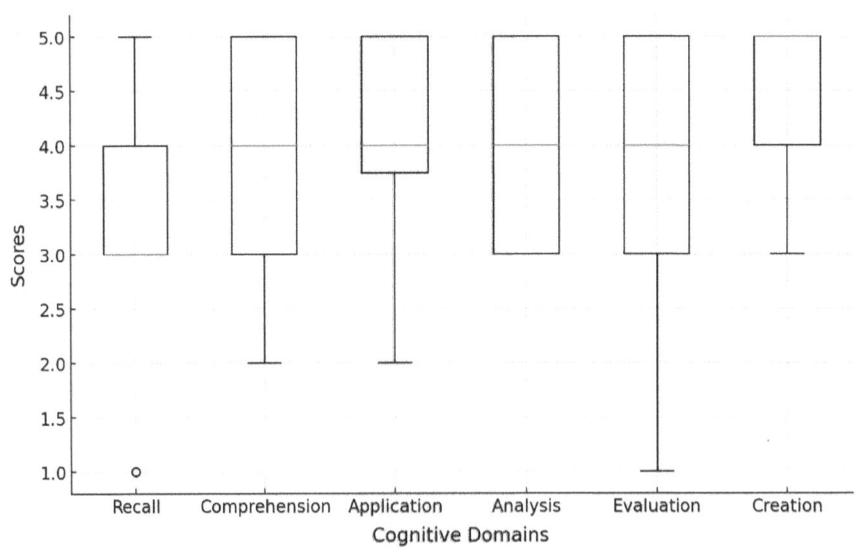

Fig. 2 Box plot of participants' scores across the cognitive domains

On another end, practitioners who specialize in criminal law assigned more value to the analogies as against those who specialize in civil or private law. This was not unexpected, as over 90% of criminal litigation according to [3] contain at least an element of digital evidence. This means that practitioners who specialize in criminal law must have encountered more technical terms than those in civil law, which gives them the proper grounding to assess the value of the analogies.

5 Discussion

This research study set out to investigate how simplified analogies would affect legal practitioners in interacting with digital forensics reports. The existing knowledge gap in IT generally, which trickles down to digital investigations among legal practitioners was the major motivation behind this research study. This knowledge deficit as has been widely discussed by [9, 10, 12], is one that needs to be closed if the global effort towards cybersecurity can be sustained. The cybersecurity chain as we know is not merely about thwarting or detecting cyber breaches, but also ensuring legal recourse against cyber criminals. As established, if the legal professionals responsible for these proceedings grapple with the digital evidence due to its technical jargons, the entirety of global cybersecurity efforts could be jeopardized. This stance is not just one taken by this research study but aligns with prior research on the impact of how legal practitioners' comprehension of digital evidence affected their case building capacities and how digital evidence itself has become a crucial and prevalent aspect of litigation in today's world [6, 13].

5.1 Theoretical Contribution

Comparing the results from this research study to related work, the evident educational gap in the field of digital forensics within the legal profession is emphasized. Murff et al. [9] found that a mere 6.33% of attorney participants in their survey confirmed that they had been exposed to digital forensics-related courses during their legal education. This closely mirrors results from this study, where 6.7% of participants reported that they had undertaken some form of formal education in digital forensics. Such similar outcomes highlight a consistent trend of limited digital forensics education within law curricula. The legal practitioners themselves expressed the desire and willingness to accept this education as part of their training as evidenced by the general comments collected in the survey. The comments advocate for the inclusion of digital forensics education as part of the training for legal practitioners.

On intelligibility of digital forensics reports, prior research shows that a significant 73.5% of participants found reports from digital forensics experts either not understandable or only somewhat understandable because of the level of technicality contained therein [9]. Interestingly, this research study found the possibility that this figure could drastically drop if these terms are explained in analogies that employ familiar language. As seen, after introducing participants to these analogies, 62.6% of participants agree that the analogies made the digital forensics report more understandable, with only 12.5% of participants disagreeing. These figures can only mean that analogies can be adopted as an effective tool in explaining digital forensics terms to legal practitioners.

This research study makes a justified attempt to address the educational needs of legal practitioners in the use of digital evidence which is in resonance with works of other researchers. Hasan et al. [11] suggest that educational resources should present security and forensics concepts in a manner comprehensible to judicial officials, using terminology familiar to the legal profession. They also advocate for the flexibility of such resources to accommodate the busy schedules of legal professionals and the rapidly evolving nature of computing. In response, this study paves the way for more work to be done in creating a framework to interpret security and forensics terms with legal practitioners in mind, and for the development of a usable tool that will make this framework available to them globally.

5.2 *Practical Contribution*

Having achieved the primary objective of the research study, which was to gauge how analogies would impact legal practitioners' interaction with digital evidence, the secondary objective was pursued to extend the accessibility of these analogies. It was deemed fit to incorporate these analogies within Microsoft Word itself, given its widespread use. This integration ensures that the analogies are readily accessible right within Microsoft Word, where these reports are typically created and read.

The Microsoft Word add-in serves as a dynamic, user-friendly utility designed to extend the website's functionality and streamline the user experience. This tool eliminates the need for users to toggle between their digital investigations document and a separate website to decipher complex technical terms. In an open Microsoft Word document, the add-in is activated by clicking on the Add-ins tool in Microsoft Word and selecting the 'Techbenchers' add-in. This intuitive task pane opens adjacent to the main document as seen in Fig. 3.

When the user places the cursor over a word or selects a word, the task pane displays its standard definition and a simplified analogy, provided the word is listed in the glossary. This add-in is directly integrated with the glossary on the website. Consequently, any updates to definitions on the website are automatically reflected in the add-in. This ensures seamless and continuous updates. Key features include:

- Instant Definitions: As users navigate through a document, placing the cursor over a word or highlighting will immediately bring up its standard definition and an analogy. This information is presented in a side panel right beside the document, eliminating the need to leave Word to look up terms.

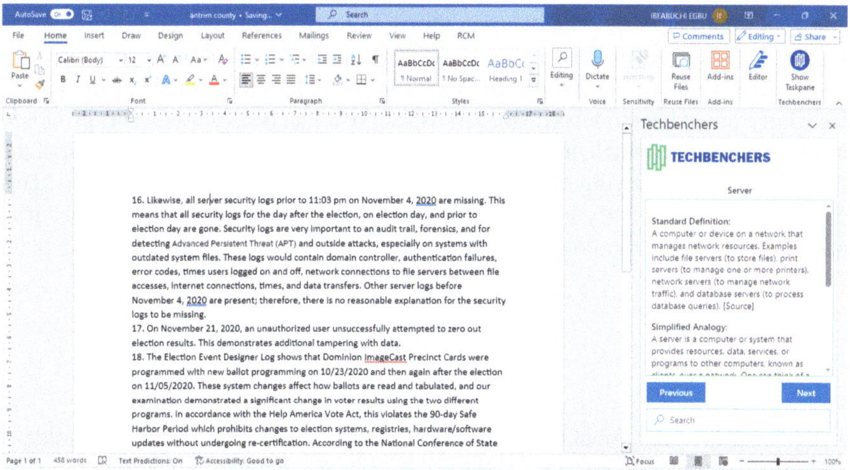

Fig. 3 Microsoft word add-in

- Seamless update: This add-in is directly integrated with the glossary on the website. Consequently, any updates to definitions on the website are reflected on refreshing the add-in. This ensures seamless and continuous updates for the terms that can be interpreted using the task pane.
- Quick Navigation: Users can easily skip to the next or previous technical term in the document using navigation buttons. This makes it easier to find all the technical terms contained in the document and display their meanings.
- Search Functionality: A search box is available for users to quickly find terms in the glossary. The add-in will then navigate the user to the location of that term in the document if it exists.
- User-Friendly: All features are designed to be intuitive and straightforward, requiring no prior training or expertise in using the add-in.

5.3 *Limitations and Future Work*

A limitation of this research study was that the participant pool predominantly consisted of students and practitioners with limited or no prior experience in handling digital evidence. This lack of experience may have skewed their evaluations of the analogies' effectiveness. Future research could greatly benefit from engaging a broader and more diverse sample, particularly including cybercrime prosecutors. Such participants, with their firsthand experience in utilizing digital evidence, would likely offer more insightful feedback on the analogies' relevance and utility. Additionally, the development of these analogies by a team of digital forensics experts could enrich the research by incorporating a wide range of perspectives.

It would also be worthwhile to survey the usability of the Microsoft add-in from the end users' perspective to gain insights for further refinement. Additionally, considering the integration of the Microsoft Word add-in with artificial intelligence, particularly large language models (LLMs) like ChatGPT, can be explored. This proposed enhancement would enable the dynamic generation of analogies as the content of the document changes, moving beyond a static repository of simplified definitions. This adaptability is particularly crucial given the fast-paced evolution of cybersecurity and its associated terminology, offering a tool that evolves in tandem with the field.

6 Conclusion

Recognizing that the global cybersecurity effort must involve several stakeholders, including legal practitioners who play a crucial role in administering justice in cybercrimes, this research study set out to explore how analogies to traditional evidence could help legal practitioners interact with and apply technical terms encountered during digital investigations and litigation. Grounded in extensive literature on the current level of interaction of legal practitioners with technical elements of digital investigations and ongoing efforts to address the shortcomings, this study has successfully developed a valuable tool to improve legal practitioners' interaction with digital evidence.

Survey results, with an average score exceeding 4.0 in five of the six cognitive engagement levels based on Bloom's Taxonomy, suggest that the analogies presented effectively assisted participants in comprehending the section of the case study digital forensics report. This impact extended beyond mere understanding, promoting higher-level cognitive engagement, and even kindling an interest in delving deeper into digital forensics among the participants. These findings are encouraging. They suggest that the tools developed by this study can be adopted or adapted as a vital component in the continuous professional development of legal practitioners. With the aid of the Microsoft Word add-in, legal professionals can engage more effectively with digital forensics experts, witnesses, and their peers in digital investigations. This ensures that digital evidence receives its rightful probative value and is not misinterpreted during litigation. It is hoped that these collective efforts will lead to more informed judgments on matters involving digital evidence and cybercrime, ensuring just outcomes for all parties involved.

References

1. Lee, L.: Cybercrime has evolved: it's time cyber security did too. Comput. Fraud. Secur. **2019**, 8–11 (2019)
2. Muir, R., Walcott, S.: Unleashing the value of digital forensics. Police Found. (2021)
3. National Police Chiefs' Council: Digital forensic science strategy (2020)

4. Daniel, L.E., Daniel, L.E.: Chapter 6 - Digital Forensics at Work in the Legal System. In: Daniel, L.E., Daniel, L.E. (eds.) Digital Forensics for Legal Professionals, pp. 41–52. Syngress, Boston (2012)
5. Brown, C.S.D.: Investigating and prosecuting cyber crime: forensic dependencies and barriers to justice. Int. J. Cyber Criminol. **9**, 55–119 (2015)
6. Gottehrer, G.: "Connected" Discovery: what the ubiquity of digital evidence means for lawyers and litigation. Richmond J. Law Technol. **22**, 8 (2016)
7. Henseler, H., van Loenhout, S.: Educating judges, prosecutors and lawyers in the use of digital forensic experts. Digit. Investig. **24**, S76–S82 (2018)
8. Watt, A.: The challenges of interpreting digital evidence in the courtroom. Precedent (Sydney, N.S.W.). 43–46 (2020)
9. Murff, K., Gardenier, H., Gardenier, M.: Digital forensics and the law. Annual ADFSL Conference on Digital Forensics, Security and Law (2011)
10. Wilson-Kovacs, D., Helm, R., Growns, B., Redfern, L.: Digital evidence in defence practice: prevalence, challenges and expertise. Int. J. Evid. Proof. **27**, 235–253 (2023)
11. Hasan, R., Zheng, Y., Walker, J.T.: Digital Forensics Education Modules for Judicial Officials. In: Choo, K.-K.R., Morris, T., Peterson, G.L., Imsand, E. (eds.) National Cyber Summit (NCS) Research Track 2020, pp. 46–60. Springer International Publishing, Cham (2021)
12. Buono, L.: Updating and diversifying the training offer for EU legal practitioners to meet the challenges posed by the new technologies. ERA Forum. **18**, 291–296 (2017)
13. Miller, C.M.: A survey of prosecutors and investigators using digital evidence: a starting point. Forensic Sci. Int.: Synerg. **6**, 100296 (2023)
14. Allied Security Operations Group: Antrim county forensics report (2020)
15. National Institute of Standards and Technology: Glossary | CSRC, https://csrc.nist.gov/glossary/. Last Accessed 09 May 2024
16. Legal Information Institute: Mens rea, https://www.law.cornell.edu/wex/mens_rea. Last Accessed 09 May 2024
17. Bloom, B.S., Engelhart, M.D., Furst, E., Hill, W.H., Krathwohl, D.R.: Handbook 1: Cognitive Domain. David McKay, New York (1956)

Critical Infrastructures, Systems and Applications

Fuzzing the ARM Cortex-M: A Survey

Silje Marie Sørlien, Åse Marie Solnør, Karin Bernsmed, and Martin Gilje Jaatun

Abstract Security testing low-level connected devices such as Internet of Things (IoT) devices is difficult due to their limited memory, communication bandwidth, and processing power. As a result, traditional testing processes become difficult and time-consuming in such cases. In this paper, we survey recent fuzzers suitable for fuzzing devices with the most common type of microprocessor: the ARM Cortex-M.

1 Introduction

Embedded devices are increasingly connected to critical networks and the internet, emphasizing the need for security testing. This becomes especially challenging with low-level connected devices which have limitations in memory, communication bandwidth, and processing power. As a result, traditional testing processes become difficult and time-consuming in such cases.

We intend to discover, compare and discuss some state-of-the art firmware fuzzers for embedded devices using the ARM Cortex-M, to create an overview of security testing tools for developers. We will discuss the amount of and quality of documentation for each fuzzer, as most firmware fuzzers available today need extensive manual efforts to configure and analyse the output. Sparse documentation increases the threshold for implementing fuzzing in security testing, and quality guidance material should thus be a priority for developers of these tools. This survey is intended to be a contribution to a not-so-large pool of literature reviews on embedded device fuzzing. It will hopefully highlight some aspects of firmware fuzzing as well as presenting some of the newest research in this field.

S. M. Sørlien · Å. M. Solnør · K. Bernsmed
IIK, NTNU, Trondheim, Norway

K. Bernsmed · M. G. Jaatun (✉)
SINTEF Digital, Trondheim, Norway
e-mail: martin.g.jaatun@sintef.no

M. G. Jaatun et al. (eds.), *Proceedings of the International Conference on Cybersecurity, Situational Awareness and Social Media*, Springer Proceedings in Complexity,
https://doi.org/10.1007/978-981-96-0401-2_18

The rest of the paper is structured as follows. Section 2 explains the survey method, and Sect. 3 documents the results. Section 4 provides a discussion of the implications, and Sect. 5 summarises the results and identifies future work.

2 Method

Our literature review intended to identify, evaluate and narrow down suitable firmware fuzzers. We followed a template for literature reviews including six steps: formulating research questions, performing an initial search for literature, narrowing down the results, assessing the results, extracting interesting information, and lastly analysing the results [28].

We used Google Scholar to identify most up-to-date firmware fuzzers. The main reason for choosing Google Scholar over other databases was that the former also covers grey literature such as reports, blog posts and software documentation. Our search string was "firmware AND fuzzing AND arm cortex M", and gave 3,300 results in Google Scholar. After filtering out papers released before 2020, the results were narrowed down to 1,490. To do a rough elimination of papers not relevant to our research, we read the titles and short summaries available on the Google Scholar search results page. The criteria for this selection was that the paper describes a firmware fuzzer that supports fuzzing ARM Cortex-M CPUs. This left us with approximately 60 possible relevant papers, as illustrated in Fig. 1. When we had identified these, we continued to read abstracts, introductions and conclusions. In addition, we used the *snowball method* where we found firmware fuzzer candidates in other fuzzer papers.

We selected a total of 17 fuzzers we initially found relevant for our research, and read through the papers and their GitHub READMEs to differentiate them. To illustrate and explain the differences between the tools, we included modelled overviews of the architectures of the tools. We adapted them from models from their respective papers, to be able to include them in this paper.

3 Survey Results

Hardware Abstraction Layers (HALs) are abstractions meant to simplify development of firmware implementations. **HALucinator** [4] takes advantage of these to perform High-Level Emulation (HLE), to separate the firmware from the hardware for testing. To be able to do so, a binary analysis is performed to locate library functions of a sample firmware. Following this, generic function implementations can be provided in an emulator for the complete system. They are implemented in an emulator to create an image of the firmware, which can be fuzzed by AFL [4]. The emulator used in HALucinator is based on QEMU [29]. HALucinator supports firmware from ARM Cortex-M architectures that are statically linked into one binary executable.

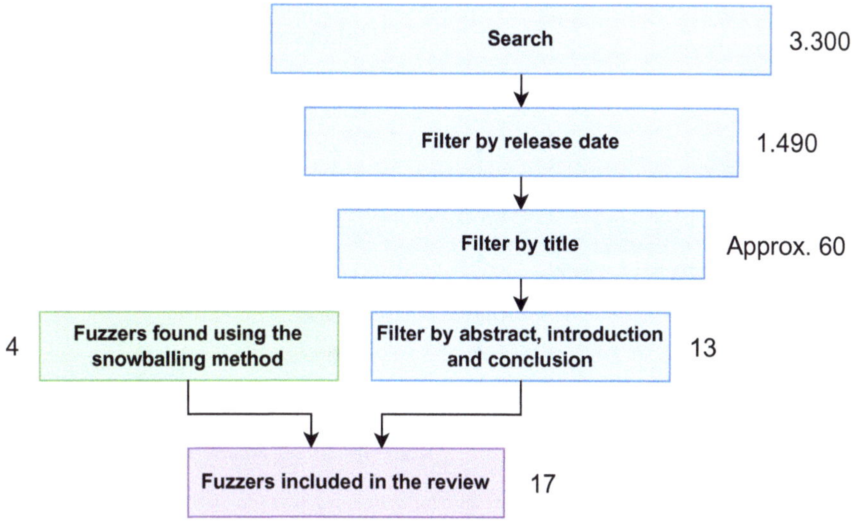

Fig. 1 Search strategy, adapted from [20]

As shown in Fig. 2, HALucinator provides a high-level emulation environment on top of the emulator. The emulator needs to replace the execution of selected function addresses to ensure that the re-hosted firmware is correctly executed. The intercepted functions relates to on-chip or off-chip peripherals of the embedded device. HALucinator simplify this implementation by breaking down the needed implementations per library into *handlers*. These handlers encodes each HAL functions' semantics. The purpose of *peripheral models* is to address common intrinsic elements of what a certain class or kind of peripheral needs to perform. It can be observed in the figure that the IO server enables emulator host interactions. To be able to meaningfully execute the re-hosted firmware it must be able to interact with external devices outside of the CPU. Because of this, each peripheral model defines an interface for the host system to send and receive data and trigger interrupts. An I/O server is then used to make these interfaces available [4]. The GitHub repository was last updated in November of 2022 [3]. It includes documentation about how to use the fuzzer [3]. It also explains the installation steps, a recommended environment, dependencies and basic usage with examples. Lastly, there is a brief explanation on how to analyse the output.

μAFL [25] is a grey-box fuzzing tool that employs hardware-in-the-loop. This implies that to be able to use μAFL, developers need access to the board as well as additional dongles. Dongles are components that can be plugged into the device to provide more functionality to the device. It leverages debugging tools in existing embedded system development to construct an AFL-compatible environment for fuzzing [25]. It uses the debug dongle to bridge the fuzzing environment on the PC and the target firmware on the microcontroller device. These debug dongles from the embedded development environment are used to ensure cooperability between

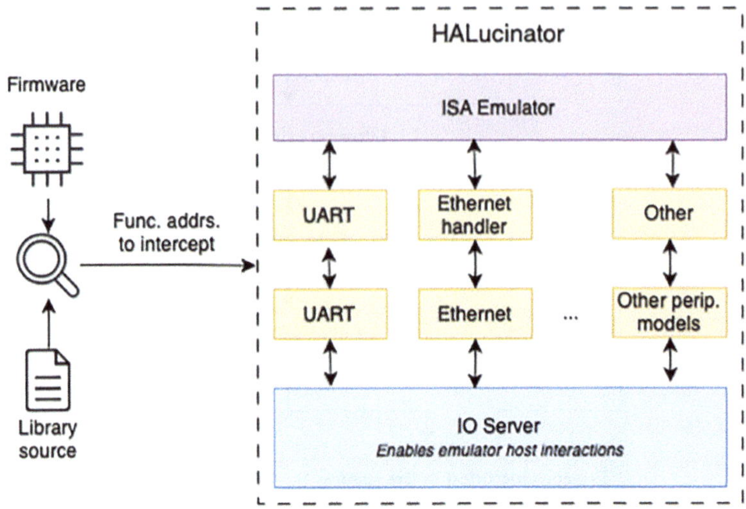

Fig. 2 Overview of HALucinator, adapted from [4]

the engine and AFL [25, 41]. To collect code coverage information without costly code instrumentation, μAFL relies on the ARM Embedded Trace Macrocell (ETM) hardware debugging feature, which transparently collects the instruction trace and streams the results to the PC [25]. The raw ETM data is obscure and needs enormous computing resources to recover the actual instruction trace. μAFL addresses this challenge by using a feedback-driven approach to recover the instruction trace from the raw ETM data. It then uses the recovered instruction trace to guide the fuzzing process. Figure 3 shows the overview of μAFL. The host PC and target board communicate through a debugging dongle. μAFL adopts AFLs genetic algorithm while substituting its process-based execution engine with two essential components, an online trace collector and an offline trace analyser [25]. The host PC feeds the test case through the debug dongle into a reserved memory on the target board. The target board directs the target to begin execution. The host PC sends the command to activate the ETM function. While the firmware is executing, the generated instruction trace is synchronously streamed to the host PC via the debug dongle. The host PC sends command to deactivate ETM. The collected trace information is used to reconstruct the execution paths. The final outcome is checked against the bitmap to see if any new paths have been found. The Github repository was last updated in July of 2022 [24]. The documentation includes a recommended environment. However, there are no instructions on how to actually install it correctly. It explains how to run the fuzzer and a provides list on how to prepare the hardware environment. In addition, there is no documentation on how to analyse the output or how to create a correct configuration file.

FIRM-AFL [43] is a grey-box fuzzer where the main focus is to create a high-throughput emulation to increase efficiency in full system testing [43]. The paper

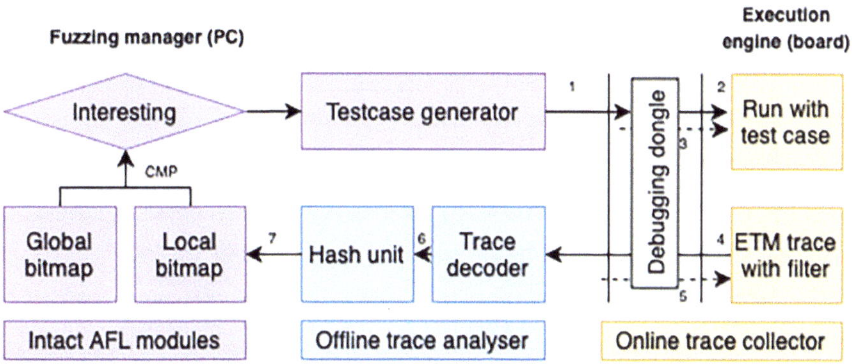

Fig. 3 Overview of μAFL, adapted from [25]

proposes a new technique called *augmented process emulation*, where the idea is to augment process (user-mode) emulation with full system emulation. FIRM-AFL is a prototype to evaluate the feasibility of this technique, and it is built on top of AFL [41] and FIRMADYNE [1, 43]. FIRMADYNE [1] can emulate both ARM and MIPS architectures [16]. The architecture and workflow of FIRM-AFL is shown in Fig. 4. *afl-fuzz* that is the main program that drives the fuzzing. This program picks a seed from the seed queue, performs a random mutation, generates an input, and feeds this input to the target program. To gather code coverage data during the target program's execution, FIRM-AFL uses augmented process emulation to launch the program, instruments the program's branch transitions, and then encodes and stores the code coverage data in a bitmap. To speed up the process of repeatedly executing the target program, which is necessary for fuzzing, FIRM-AFL uses a mechanism called *fork*. This mechanism runs the target program up to a certain point so that the program's code and data have been properly initialized, and then repeatedly forks a child process from it. Because of this, the parent process is called *fork server*. The input is fed into the forked child instance, and then coverage information is collected and stored in a bitmap that is shared between all three instances (afl-fuzz, fork server, and child instance). Lastly, afl-fuzz uses the bitmap from the child instance to compare it with the bitmap obtained from all past executions to determine if this mutated input should be kept as a new seed and stored in the seed queue [43]. The GitHub repository of Firm-AFL has not been updated since October 2020 [42]. In the repository there is information on how to install and use the fuzzer with the commands to run it. It does not mention a recommended environment, configuration, or output analysis.

FIRMCORN [16] is a grey-box fuzzer uses a vulnerable-code search algorithm to determine the entry point for vulnerability-oriented fuzzing and optimized virtual execution technology to improve execution speed, accelerate execution accuracy, and ensure execution stability. The emulator is based on Unicorn Engine [5] and their own fuzzer and crash submodules. An overview of the architecture is modelled in Fig. 5. The framework is separated into five parts: preanalysis, dump, fuzz, hook and crash. These five submodules are all connected to FIRMCORN core. The first part is the

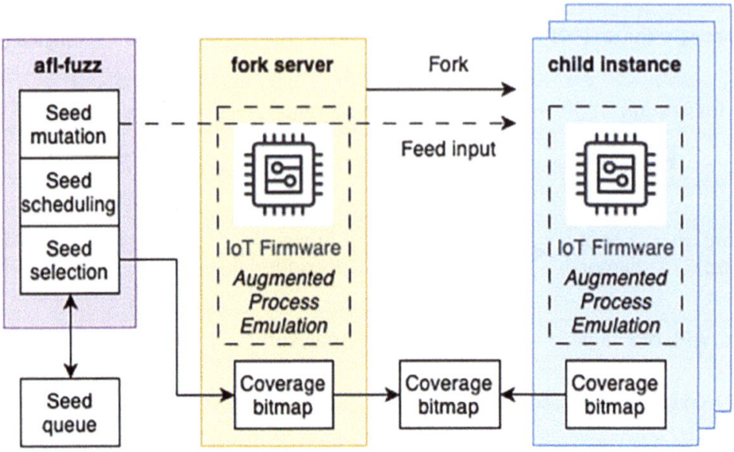

Fig. 4 Overview of FIRM-AFL, adapted from [43]

preanalysis, where an algorithm groups all functions in the firmware based on complexity. The second part is dump, where shell access to the device is obtained in one out of three ways: Telnet/SSH, Device Debug Interface or using the Firmware Update Mechanism. Once shell is obtained, gdbserver is uploaded to the device and this is used to dump context. The hook part is where the Global Offset Table Hook (GOT) information is analyzed. This includes unresolved functions, hardware-specific functions, custom functions and unnecessary functions. The optimized virtual execution part takes place in the core. Here, we can find the CPU emulator, fuzzing policy, the initial environment and a module for heuristic optimization that communicates with the hook part. The fuzz submodule contains the fuzzer, and the crash submodule checks and stores information about crashes [16]. The GitHub repository has not been updated since April 2022 [15]. The repository contains only a link to the technical paper of FIRMCORN. In the paper, there is no information on how to install the fuzzer, run the fuzzer, configure the configuration file, or analyse the output.

EM-fuzz [14] is a grey-box fuzzer that integrates real-time memory checking into fuzzing to improve bug discovery of the fuzzing [14]. It is based on QEMU [29] for emulation and AFL [41] for the fuzzing engine. EM-fuzz aims to reach certain program branches early and execute specific blocks of code more frequently by mutating the input data [14]. It also checks for memory-related vulnerabilities by monitoring the system's memory usage during the execution of the test cases. An overview of the architecture is modelled in Fig. 6. EM-fuzz contains two main modules. The first module is the memory checking instrumentation component which is responsible for recording memory operations that are used in the second main module that performs the guided fuzzing. It uses information about the branches to generate seeds that aid in discovering memory regions that otherwise could be difficult to reach. The parts of the fuzzing module are the fuzzing controller, the seed collector, mutation engine and a process execution component. The controller uses

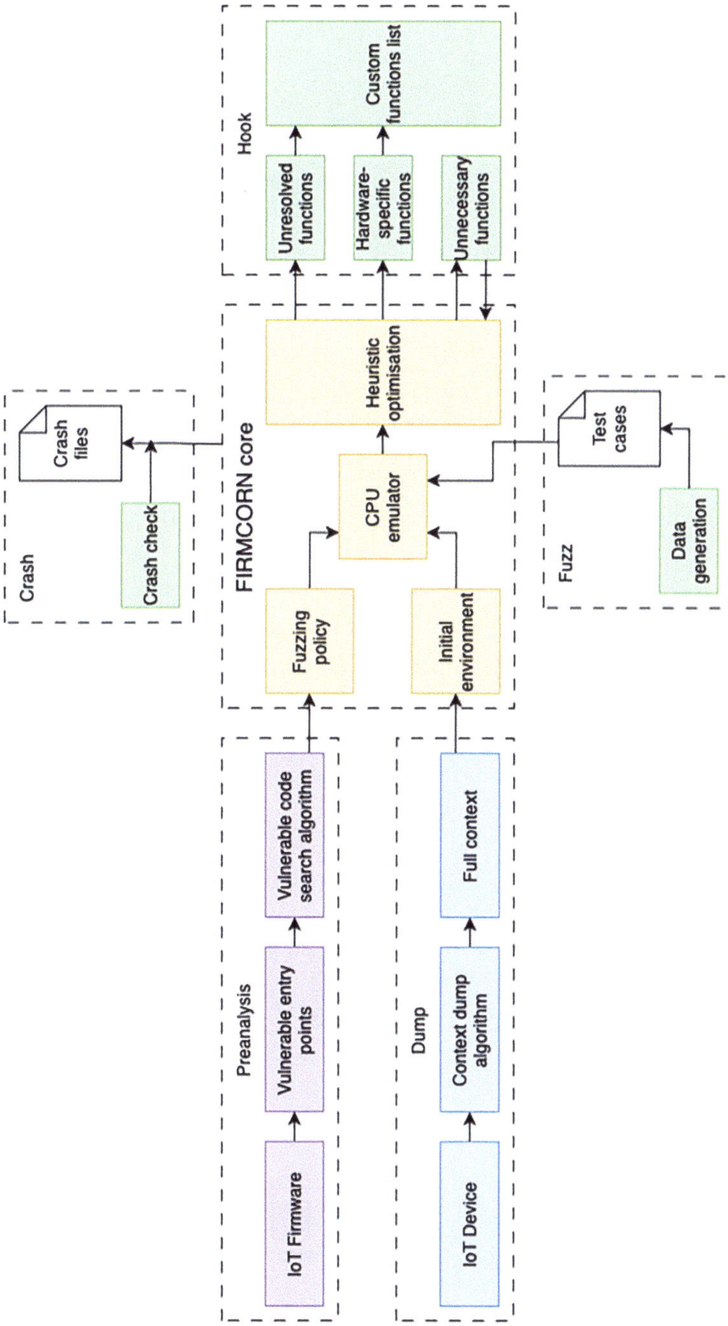

Fig. 5 Overview of FIRMCORN, adapted from [16]

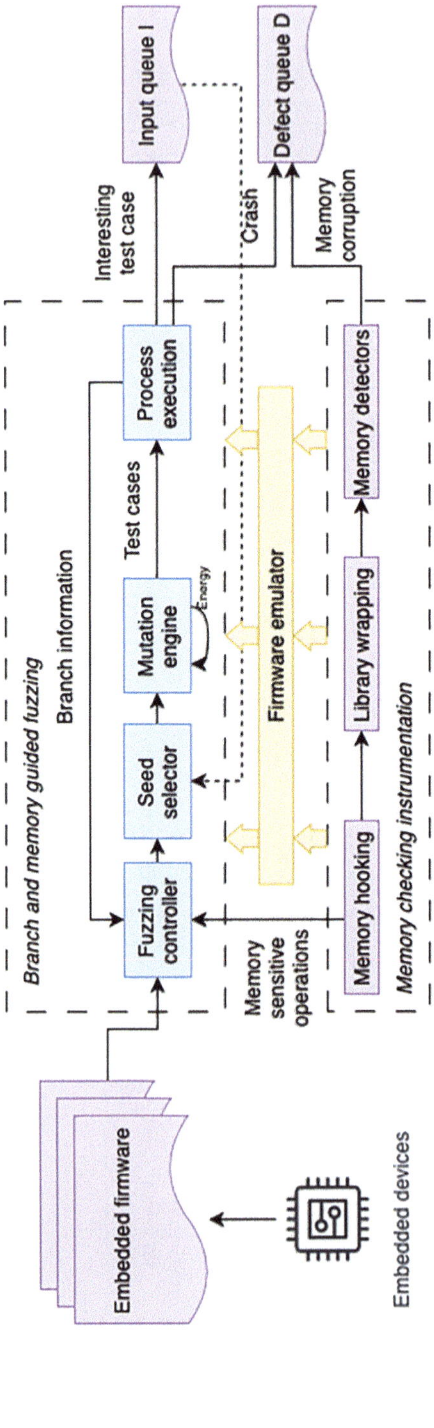

Fig. 6 Overview of EM-fuzz, adapted from [14]

the feedback from earlier executions to have control over which branches have been traversed. The seed selector is responsible for calculating probabilities and adjusting prioritization for the seeds resulting in an execution of a rare branch, ensuring that also rare branches have a high probability of being traversed. The mutator is responsible for maximizing the branch coverage. The process execution component performs the fuzzing, and forwards information on interesting test cases and crashes to their respective queues, and delivers feedback to the controller. The last component is the emulator, which is based on QEMU [14]. As of January 2024, this framework is not open source [40].

FirmHybridFuzzer [38] is a hybrid fuzzing tool that enables testing of MCU firmware without relying on specific peripheral hardware [38]. FirmHybridFuzzer can efficiently test the correctness of a certain portion of the code and maximize code coverage [27]. It is a hybrid fuzzer, which means that it combines the benefit of both traditional fuzzing and symbolic execution [38]. The key solution in this fuzzer is the implementation of a virtual peripheral, allowing it to emulate the behaviour of different peripherals. FirmHybridFuzzer uses AFL [41] as the fuzzer. Figure 7 shows an overview of FirmHybridFuzzer. The input to the fuzzing framework consists of a firmware binary and the output is the potential firmware vulnerabilities. FirmHybridFuzzer works on the firmware as a whole, meaning it tests both the application tasks and the kernel. To do so, the fuzzer uses the symbolic peripheral-based execution technique. That is a technique to be able to execute firmware in the lack of a physical peripheral and hybrid event generation. The fuzzer also uses a mechanism called multiple dimensional coverage feedback based guidance to improve test case generation. To identify common firmware vulnerabilities, FirmHybridFuzzer implements a fault detection mechanism [38]. The last update on the GitHub repository was November 2023 [37]. The documentation contains information on how to install and run the fuzzer. They do not provide all the installation commands, but refer to README files from other repositories. Only brief information on basic usage and configuration is included. Output analysis is not explained [37].

PRETENDER [19] is a fuzzer that observes the interaction between the hardware and firmware to automatically create models of peripherals. What sets this apart from the other fuzzers, is that they are stateful, interactive and transferable. By creating these models of peripherals, the firmware can be executed in a fully virtual environment. The emulator is based on QEMU. It only supports ARM Cortex devices [19]. An overview of the architecture is modelled in Fig. 8. PRETENDER works on unmodified firmware by observing interactions within the hardware and generating models based on the hardware peripherals. This can be observed by the input to the emulated environment in the figure. This is a hardware-in-the-loop technique where the models that are generated are fed into an emulator, in this case QEMU. This is then processed using machine learning and pattern-based analysis, and directed back to the emulated environment [19]. The GitHub repository was last updated in August of 2020 [18]. The README file contains information on how to set up, install and run the fuzzer. Dependencies are mentioned. Commands for running the fuzzer is explained, but there is no information on how to configure the targets or analyse the results [18].

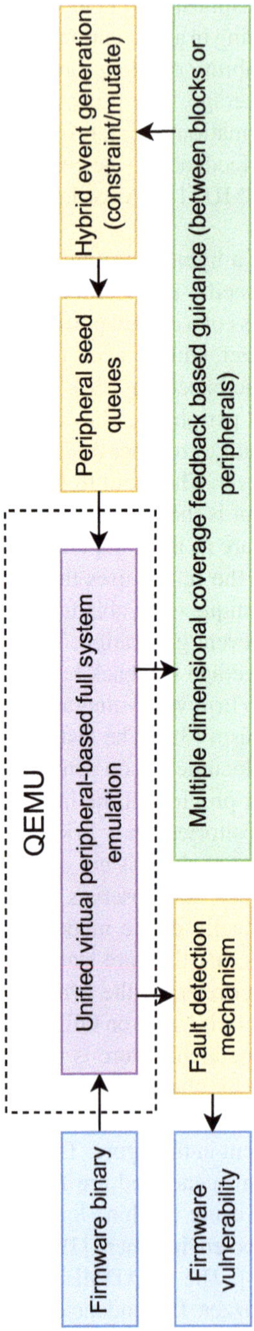

Fig. 7 Overview of FirmHybridFuzzer, adapted from [38]

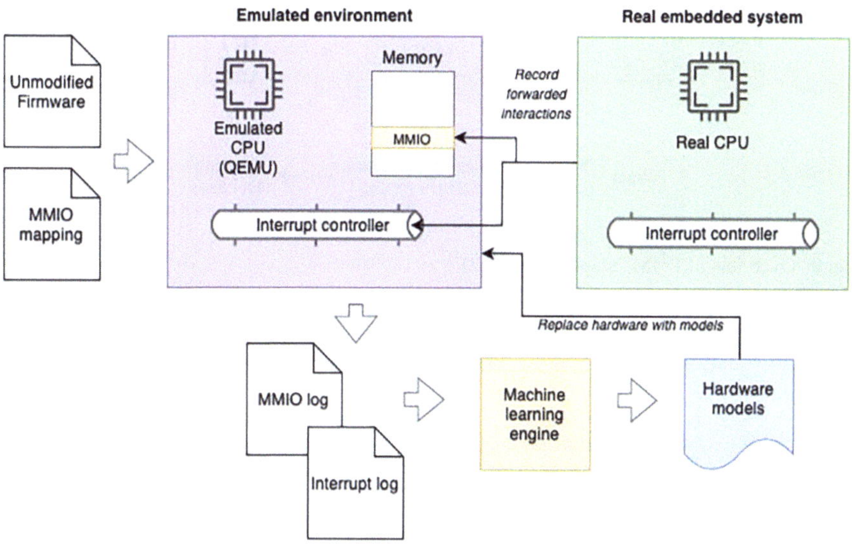

Fig. 8 Overview of PRETENDER, adapted from [19]

P²IM [12] is a fuzzing framework and presents a technique called Processor-Peripheral Interface Modeling, that models I/O accesses for several peripherals. It treats peripheral devices as black boxes. It uses QEMU [29] for emulation and AFL [41] for fuzzing. P²IM uses a technique called explorative execution to automatically infer acceptable state register values during runtime [12]. Handling state register writes is much simpler and the same for all firmware [12]. An overview of the architecture is modelled in Fig. 9. The fuzzer is an unmodified implementation of AFL, and the emulator based on QEMU. The Processor-Peripheral Interface Modeling technique proposed by P²IM is what models the behaviour of the peripherals. These models check the criteria of a property they propose called Processor-Peripheral Interface Equivalence, which is a prerequisite for executing the emulated firmware. The emulated firmware is executed, and feedback is provided back to the fuzzer [12]. The GitHub repository for this fuzzer was last updated in November 2023 [11]. The documentation includes information on how to install the fuzzer, configure the target, perform the fuzzing and analyse the results. Preparing the target, including configuration, is also explained with references to the paper for more information. Its basic usage, including optional customization, statistics and coverage is mentioned, as well as some commands that help in output analysis. Separate README-files for adding an MCU, building QEMU and prepping the firmware for fuzzing is referenced. Thorough explanations for each step are not included, and there is a limited information on how to analyze the results [11].

Fuzzware [31] is a fuzzing tool that self-configures inputs to test a firmware image without knowledge about the underlying architecture [31]. It uses access models to transform fuzzing inputs to meaningful values for the firmware to interpret.

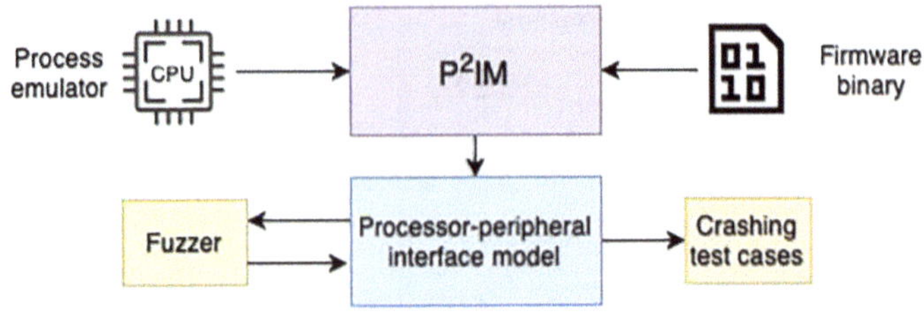

Fig. 9 Overview of P^{2}IM, adapted from [12]

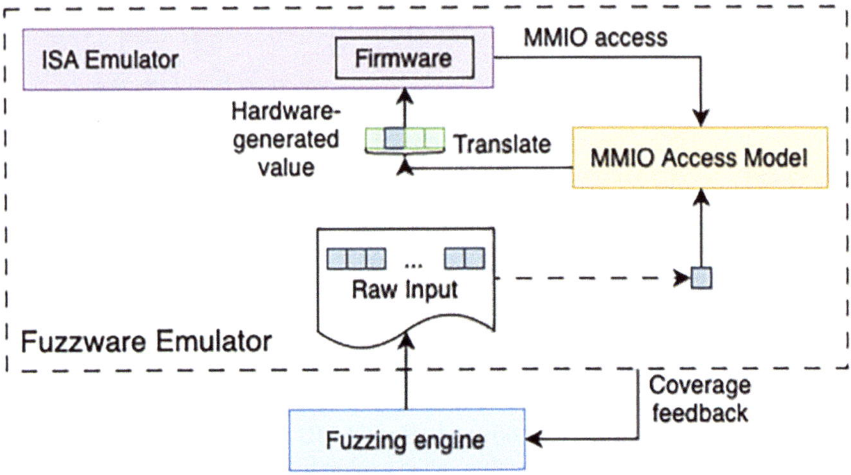

Fig. 10 Overview of Fuzzware, adapted from [31]

It uses AFL [41] as the fuzzing engine and supplements it by providing an emulator and access models to translate firmware images to something that is fuzzable by AFL [31, 41]. The emulator is based on Unicorn Engine. An overview of the architecture is illustrated in Fig. 10. The fuzzer generates input that is provided to the emulator. MMIO accesses triggers parts of the raw input to be consumed by the MMIO access models. Here, they are transformed into hardware-generated values and sent to the emulated target. Then, coverage feedback is forwarded to the fuzzer for further mutation [31]. The GitHub repository was last updated in October of 2023 [33]. The documentation includes information on how to install Fuzzware locally or using Docker, how to configure targets, and how to analyse the results. It also includes thorough explanations for each step, each component and an overall explanation of the fuzzer. It links to separate README-files for each step, where configuration, customization and analysis is explained in-depth [33].

Hoedur [32] attempts to mitigate shortcomings of previous black-box fuzzers by allowing the fuzzer to process feedback from firmware execution and understand inputs in a multi-stream form. This is done by deciding the context of the accesses during execution, which involves the counter, the MMIO register address of the request and the size of the access. Based on these values, Hoedur can generate input streams that are combined into a multi-stream input. The fuzzing engine is mainly based upon libFuzzer, with certain features borrowed from AFL and AFL++ [32, 41]. The emulator is based upon QEMU [29], with some minor modifications. Automated DMA support is not yet implemented in Hoedur, but can be manually handled within the MMIO setting. An overview of the architecture is modelled in Fig. 11. The fuzzer provides a multi-stream input to the emulator. The emulator is what exists within the dotted frame in the figure, to execute the firmware using the multi-stream input. During execution of the target firmware, firmware-specific feedback is recorded. The emulator then provides the extended feedback back to the fuzzer [32]. The GitHub repository was last updated in October of 2023 [34]. The documentation includes information on how to install, fuzz, collect coverage information, and run tracing or hooking scripts. It explains for basic usage, and mentions to use the—help for details on customizing the fuzzer. There is no information on how to configure the target or analyse the results [34].

Ember-IO [6] is a fuzzing framework for firmware that uses model-free memory mapped I/O. This means that the fuzzing framework manages the complexities posed by fuzzing a wide range of memory mapped peripherals without the need to generate register models. It was an initiative to enhance the capabilities of existing fuzzing tools for firmware. Ember-IO uses AFL++ for the fuzzer and QEMU as the emulator to re-host monolithic firmware [6]. This fuzzing framework is focused on monolithic firmware and supports firmware built for ARM Cortex-M microcontrollers. The Ember-IO paper introduces *functional equivalent coverage instrumentation* (FERM-Cov) and *peripheral input playback* (PIP) techniques. These techniques improve the

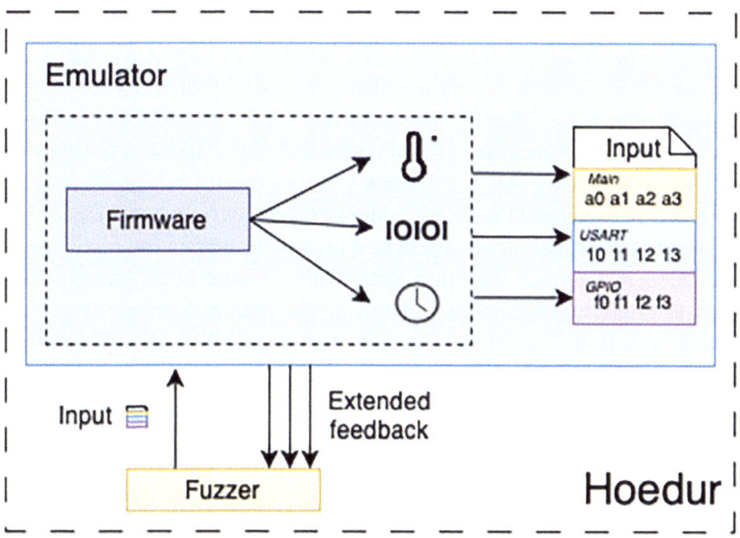

Fig. 11 Overview of Hoedur, adapted from [32]

performance of monolithic firmware fuzzers by enabling a more efficient search through the input-space and thus overcomes the obstacles imposed by MMIO when accessing and reaching deep code paths [6]. The PIP technique provides mutations more likely to reach previously unseen code and FERMCov increases the likelihood of the mutation efforts on inputs to uncover interesting code paths [6]. An overview of the architecture is modelled in Fig. 12. The MMIO manager component is in control of the raw input stream that is sent to the firmware. The FERMCov generator is the component that sends coverage feedback to the fuzzer, and is responsible for the functional equivalent coverage instrumentation. The peripheral memory store within the MMIO manager is responsible for the PIP technique. The interrupt manager sends an interrupt if a sleep instruction is triggered, and also at certain times based on how many blocks are executed [6]. The GitHub repository of Ember-IO was last updated June 2023 [8]. The documentation includes information on installation, fuzzing, configuration and analysis. Configuration is explained. For the analysis part, replaying inputs and debugging crashes is explained. The GNU Project Debugger (GDB) can be used as well by adding some flag to the command. This option has its own file in the repository, where it is explained in depth. The README file also includes some details on Peripheral Input Playback and FERMCov, which can be useful to understand how the fuzzer works [8].

μEMU is a firmware fuzzer that is based on S2E, which is a QEMU-based concolic execution tool [45]. μEMU is an approach to emulate firmware with unknown peripherals. It works by taking the firmware image as input and then symbolically executes it by representing unknown peripherals registers as symbols [45]. Figure 13 shows the architecture of the fuzzer. The fuzzing engine is based on AFL, but is modified to use AFL only for test-case generation and FuzzerHelper for the rest. This includes coverage instrumentation, the fork server, and crash/hang detection. FuzzerHelper is modelled in the Plugins section of Fig. 13. Since S2E 2.0 does not support ARM MCUs, μEMU has made two contributions to include support for ARM architecture: porting ARM dynamic binary translation to the S2E emulation using QEMU, and adding interfaces for QEMU for managing the CPU of ARM Cortex-M through the kernel-based virtual machine (KVM) interface. The GitHub repository was last updated November 2023 [44]. The documentation contains information on installation, including optional methods and recommended environment, fuzzing, configuration and analysis. It includes a subsection on how to update the repository, which is not included in many of the other READMEs. Usage is explained thoroughly, with steps to configure the target and to run the fuzzer. Analysis is explained to some extent, also how to calculate coverage and use GDB to debug. Configuration is explained well in a separate file [45].

SplITS [7] is implemented on top of Fuzzware, and extends it by running a feedback guided search for input-to-state mappings as well as provide a strategy for optimized replacement of input. The framework also applies the FERMCov technique described in Ember-IO [6] to increase the effectiveness of the framework. Embedded devices use input-to-state mappings to bypass problems like infinite loops that can occur if a peripheral does not respond with an expected response in string format (OK). SplITS allows the mapping and replacement of non-contiguous strings

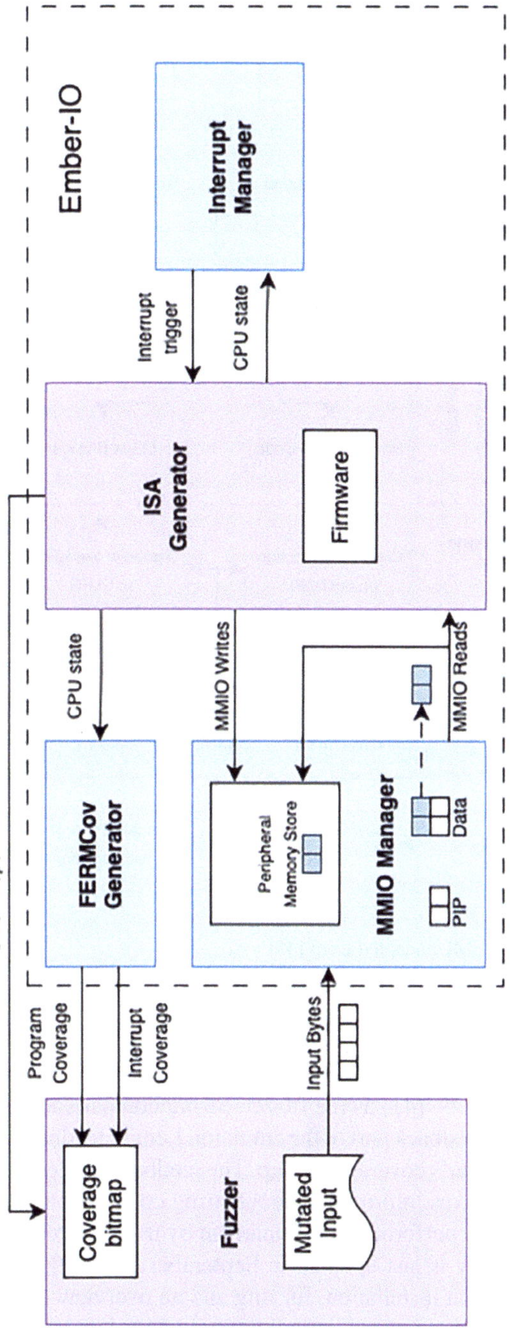

Fig. 12 Overview of Ember-IO, adapted from [6]

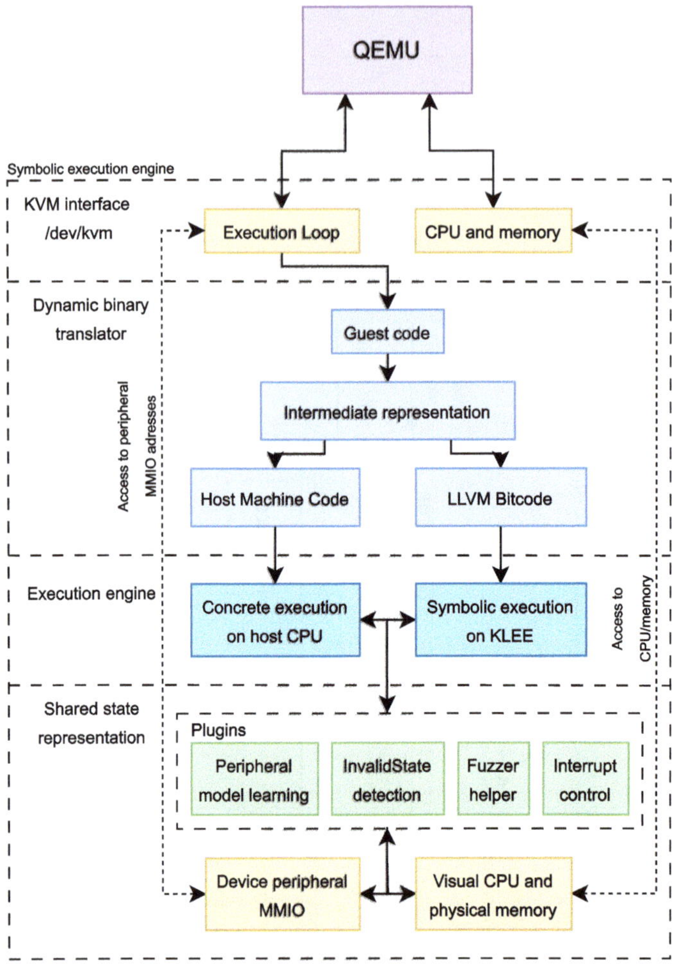

Fig. 13 Overview of μEMU, adapted from [45]

in input-to-state transitions. The overview of SplITS is shown in Fig. 14. SplITS is based on Fuzzware, therefore using Unicorn Engine as the emulator, and extends it by implementing AFL++ [41] as the fuzzer. String comparison is performed in the instrumentation and feedback part of the emulator. Length feedback is also performed here, by using the fuzzer's coverage bitmap. The feedback guided search and replacement component receives information about string comparisons and position of the mutated bytes. Then it performs the comparison by using incremental feedback [7]. The GitHub repository is last updated in September 2023 [9]. The documentation includes information on installation, fuzzing and an overview of the architecture of the fuzzer. It mentions a recommended environment, dependencies, and explains basic usage. It does not contain any further information, but since the fuzzer is built

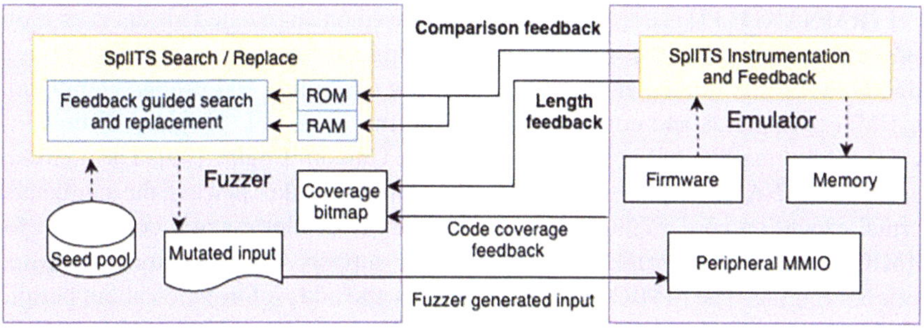

Fig. 14 Overview of SplITS, adapted from [7]

on top of Fuzzware, the documentation from Fuzzware is also included in the repository. Like explained earlier in this review, Fuzzware is thoroughly documented, and includes information of configuration, customization and analysis.

Jetset [22] uses guided symbolic execution to generate a model of the device's peripherals and passes it on to QEMU [29] to emulate the hardware. AFL is used as the fuzzing engine. To run Jetset, developers need access to the executable code, the memory layout, the entry point address and the program goal address. Jetset has no support for devices using DMA because these devices gains access to the memory without consulting the CPU. Therefore the firmware cannot observe these accesses [22]. The architecture of Jetset is modelled in Fig. 15. Jetset uses the firmware binary, its entry point and memory map as input. Directed symbolic execution is performed in the device interference stage to discover interesting locations in the firmware for further analysis. In the next step, the device synthesis, constraints discovered in the previous step is used to construct a model of the device. Lastly, this model is used in QEMU to emulate the firmware. AFL is used for fuzzing [22]. The GitHub repository of this fuzzer was last updated March 2022 [21]. The documentation contains information on installation, fuzzing, reproducing the crashes as well as links to related repositories and FAQs. It mentions dependencies and shows the basic usage with explanations. It also includes information on how to reproduce the results, but does not contain any additional documentation on how to analyze and interpret the results. It does not include any background information or information on how to configure or customize. It also does not recommend a specific environment [21].

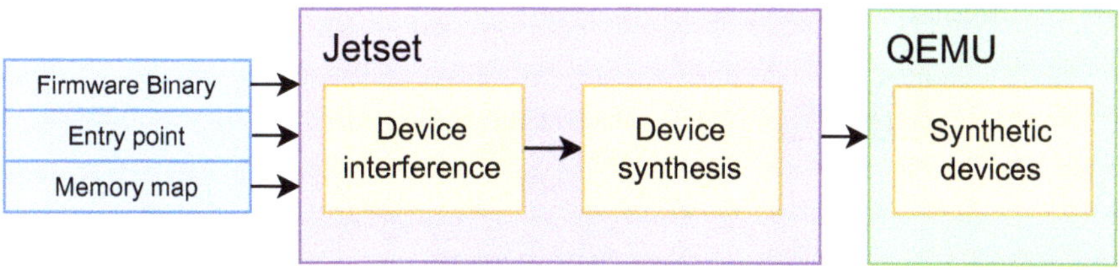

Fig. 15 Overview of Jetset, adapted from [22]

FIRMNANO [17] is a fuzzing framework based on augmented virtual execution, similarly to Firm-AFL (3), to implement fuzzing for firmware. FIRMNANO uses MMIO modeling, implements a virtual interrupt controller, and supports emulation of DMA peripherals and controllers [17]. The fuzzer is AFL [41] and afl-unicorn, which allows fuzzing of binaries emulated with Unicorn Engine [39]. The architecture of FIRMNANO is shown in Fig. 16. The emulation takes place in the augmented virtual execution core. Within the core is the CPU emulator and a component for MMIO modelling, interrupt handles and DMA support. The emulator is based on Unicorn Engine. The MMIO model part records and adds information about peripherals to a *peripheral_list*. During the pre-execution, this list is traversed to map the regions for MMIO. During execution, these are provided to the Unicorn Engine API. The interrupt handle part involves a virtual interrupt controller that streamlines the emulation of the interrupts. The DMA support part uses information about firmware symbols to emulate DMA-related functions. After the firmware has been emulated in the core, it is fuzzed by AFL. The outputs are mutated and directed back towards the core for execution on the emulated firmware [17]. As of March 2024, we haven't received a response from the paper's authors regarding the availability of the tool as open source, despite our attempts to contact them.

HD-Fuzz [23] is a firmware fuzzer developed to be aware of the underlying hardware dependencies by utilizing a hybrid MMIO modeling scheme. This is a technique where fuzzing and symbolic execution is adapted to the firmware. It is implemented on top of a QEMU-based [29] platform for symbolic execution, S2E 2.0 [2], and uses AFL [41] for fuzzing. HD-Fuzz is unable to observe DMA accesses, and therefore cannot emulate these responses [23]. Another limitation of HD-Fuzz is per-firmware manual effort. As depicted in Fig. 17, the workflow of HD-Fuzz is divided into 4 parts: inputs, rehosting, hybrid MMIO modeling and outputs. The first step is to configure and input the firmware to the rehosted system. Here, the firmware is separated into *system-side firmware code* and *user-side firmware code* to execute

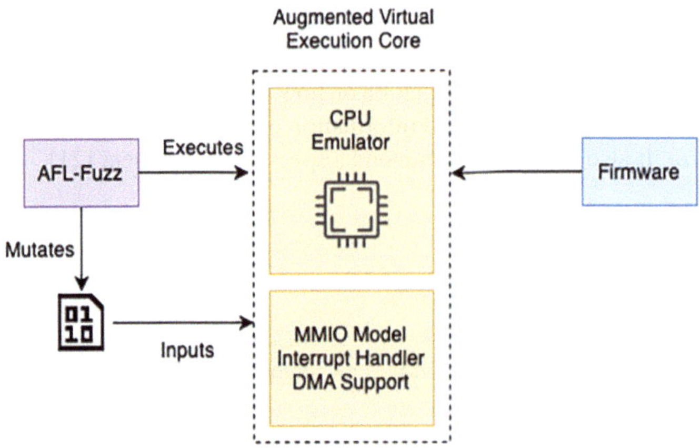

Fig. 16 Overview of FIRMNANO, adapted from [17]

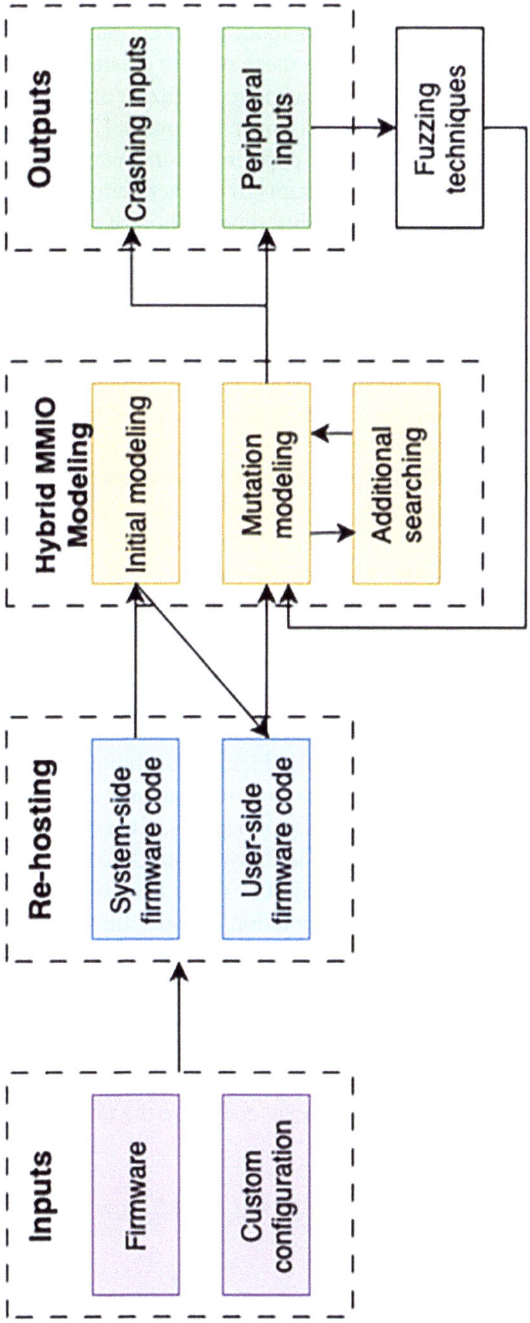

Fig. 17 Overview of HD-Fuzz, adapted from [23]

them sequentially. To include MMIO modeling, a hybrid solution is implemented. The system-side code is used for initial modelling, then the user-side code is used for mutational modeling using the hybrid method. The additional search and is used to deduce MMIO responses. HD-Fuzz uses these steps to create inputs: crashing inputs and peripheral inputs. Lastly, these are passed to the fuzzer and the results re-iterated through the hybrid MMIO method to discover new paths [23]. As of March 2024, we haven't received a response from the paper's authors regarding the availability of the tool as open source, despite our attempts to contact them.

SAFIREFUZZ [36] implements a different method of rehosting than many of the competitors. It is a technique they call *near-native rehosting*. They have based the peripheral interaction emulation on the approach used in HALucinator (Sect. 3), using HAL-based hooks [36]. Instead of rehosting a virtual image of the firmware, they adapt binaries to be rehosted in an ARM Cortex-A system that is more powerful [36]. It uses LibAFL as the fuzzing engine. The engine can only run on ARMv7-A cores, e.g. to be found in Cortex-A cores in Raspberry Pis. SAFIREFUZZ does not rely on an emulator and is thus not emulating a whole architecture [36]. This is to avoid the performance penalty induced by emulators such as QEMU. Instead, it uses a light-weight dynamic binary rewriting approach called near-native rehosting. This technique exploits that some ARMv8-A cores provide userspace compatibility with the AArch32 and Thumb instruction set variants [36]. This is done instead of emulating the firmware through lifting and recompilation. Like modelled in Fig. 18, SAFIREFUZZ implements a fuzzing engine, LibAFL, that executes the target, is responsible for rehosting and instrumentation. High-Level Emulation using HALs is used to handle unknown peripherals, by implementing a harness that is tailored to the firmware. This is responsible for memory initialization and records HAL-hooks, and does this by modifying the initial basic block. Translation of these blocks is performed by the engine, while at the same time instrumenting the target, moving the execution of the correct hooks and estimate interrupts. To ensure efficiency, only one basic block is rewritten at a time. This is to avoid multiple rewrites to blocks, and reduce overhead in instrumentation [36]. The GitHub repository was last updated in September of 2023 [35]. The documentation includes information about installation, usage, harnessing and performance. It specifies recommended environment, as well as packages and libraries. Commands to run the fuzzer are provided and briefly explained. There is no documentation on analysis and interpretation of the results. Overall, the documentation for this fuzzer is short and lacks explanations and options for customization. It mentions where to specify the configuration file, but does not contain any information on how to properly configure the target [35].

4 Discussion

After having examined the candidate tools, we will now discuss the differences and similarities between them. Apart from the technical differences of the fuzzers, it is important to consider ease-of-use to evaluate the different fuzzers' applicability in

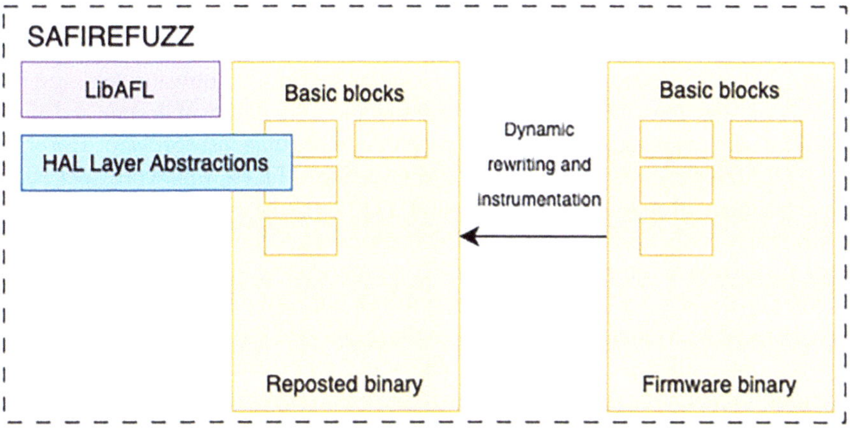

Fig. 18 Architecture of SAFIREFUZZ, adapted from [36]

real-word firmware security testing. Therefore, we will begin by discussing the non-technical details of the fuzzers, and then proceed with the architectural and functional differences.

The most important aspect of the ease-of-use of a fuzzer is the documentation, both for installation and further usage. Most fuzzers in this review included information on how to install the tool, but there are few fully developed implementations for firmware developers to use out-of-the-box. If there isn't enough guidance on how to use fuzzing in different situations or on how to understand its results, it will be more difficult to adopt fuzzing as a new method for security testing.

The only fuzzers in this review that have thorough and descriptive documentation for installation, usage and analysis, are Fuzzware, Ember-IO and μEMU. SplITS indirectly includes documentation as it uses Fuzzware as a submodule, and therefore uses the same commands. Because of this, Fuzzware's documentation can be used for fuzzing with SplITS.

Another important aspect when it comes to comparing the fuzzers is whether it is actively maintained and up-to-date with architectures and technologies. Firmware fuzzing is a relatively new concept, and a large portion of the research that exists on the topic has been conducted in the last couple of years. Therefore, we expect new methods or improvements to existing methods to be discovered rapidly. Choosing a tool that is actively maintained could ensure that the latest technologies or methods are being used. It also makes it more likely that any issues will be replied to and possibly fixed.

A third important criterion to ensure ease-of-use in commercial devices, is for the licensing to allow the tool to be used in a commercial environment. Developers need be allowed to run the code and use the tool. While working on this review, we contacted the developers of fuzzers that were missing appropriate licensing, and licences were added to their repositories. We did not receive a response from FIRMCORN, which is still missing a licence.

AFL has grown to become the industry standard for fuzzing software. Amongst the reasons why, are low-effort testing due to automation, simplicity of design, and compatibility with emulators to expand the support to more than just software. The original AFL fuzzer has been bypassed by a new version, AFL++, which provides additional features and options for customisation [13]. Out of the fuzzers we have included in this review, only SplITS and Ember-IO implements AFL++ directly. Some of the other fuzzers have the option to use AFL++, however it must be set specifically. Another difference is whether the fuzzers use an out-of-the-box implementation of AFL, or use it with modifications. Some of the fuzzers (μEMU, P2IM and Fuzzware) use AFL unmodified, as a component solution in their products. Most of the remaining fuzzers modify the fuzzing engine to fit their use cases [32]. As discussed above, Hoedur implement their own fuzzing engine based on libFuzzer [32].

When it comes to emulating the firmware, almost all the fuzzers in this review use QEMU directly or a QEMU-based implementation. Fuzzware, FIRMCORN, SplITS, and FIRMNANO use Unicorn as the emulator, and Unicorn is based on QEMU. A fuzzer that stands out from the rest is SAFIREFUZZ, as it does not use an emulator, but the near-native rehosting technique.

As of 2024, complete emulation of the varieties of MCUs is not yet possible. There are two techniques for testing firmware: hardware-in-the-loop and emulation for virtually rehosting the hardware. Out of all the candidate fuzzers, only μAFL uses hardware-in-the-loop. By using this method, it bypasses the need for abstractions by forwarding hardware accesses to a physical device during emulation [25]. Although this makes it possible to analyse firmware dynamically, this method is not very useful for fuzz testing. It requires one instance of the target hardware per fuzzing thread since the hardware state and fuzzer must be maintained consistent, in addition to the fact that forwarding is typically a bottleneck for most of these systems [31].

The remaining fuzzers discussed in this review apply different techniques to emulate the hardware. Some of them replace the interactions in the Hardware Abstraction Layer, including HALucinator and SAFIREFUZZ. μEmu and P^2IM use heuristic models for automating the same process. A third technique is to perform the fuzzing on all the registers and use other methods to reduce overhead. This is done in Fuzzware and Ember-IO. Lastly, techniques such as symbolic execution and coverage-guidance is used in several of the fuzzers to make the process more efficient [7]. An alternative to symbolic execution is developed in Ember-IO to create a fuzzer that does not generate models as part of the emulation. This is done to avoid the increase in manual labor to avoid failures caused by misrepresentations in heuristic-based models. It can also include testing of error handlers, which is known to cause a significant amount of bugs. This is the FERMCov technique described in the review. This technique is applied in some of the other fuzzers as well: SplITS and experimentally on Fuzzware [7].

HALucinator and SAFIREFUZZ abstracts away the hardware interface using a technique that enables the use of fuzzers like AFL [41] without having to perform any peripheral functions. While these HAL-based fuzzers are unable to operate on the driver level, μAFL targets these specifically. This is different from the other approaches, as drivers are peripherals that interact with the outside world and should be tested accordingly [25].

While reviewing the papers, we discovered that several of the authors are involved in more than one of the fuzzers in this review. This could be a reason as to why many of the fuzzers have similar techniques, components or even build directly on top of each other. In the following we will discuss some similarities between the fuzzers.

Fuzzware and Hoedur are created by almost the same research team, where Hoedur is a strictly improved version of Fuzzware. Both are built using elements from AFL/AFL++ as their fuzzing engine. One major difference in their architecture is the emulator. Fuzzware uses Unicorn and Hoedur uses QEMU. The reason for Hoedur not using the same emulator as Fuzzware is that QEMU makes it easier to support hardware features [30]. However, they are also similar because Hoedur includes the Fuzzware modeling, in addition to use the multi-stream inputs.

Ember-IO and SplITS are created by the some of the same developers, and was intended to target two different types of firmware. Ember-IO was developed to cover the repetitive access pattern that is the case for many peripherals. Therefore it is beneficial for fuzzing firmware that performs a large amount of input and output operations. SplITS is intended for firmware that performs string comparisons, and applies Input To State (ITS) mappings to do so. Like mentioned in the review, Ember-IO is built using AFL++ and QEMU while SplITS is built on top of Fuzzware which uses AFL and Unicorn. This is because the Peripheral Input Playback (PIP) technique applied in Ember-IO led to a slight decrease of consistency in performance - some data could be repeated by the PIP technique, meaning that a byte in the firmware does not necessarily correspond to a single byte in the fuzzer generated input [10].

Another difference is that SplITS depends on a one-to-one mapping between bytes from the observed firmware buffer and the fuzzer generated input bytes for input-to-state mapping and replacement. If PIP is applied to the data values that are loaded into the firmware buffer, this can create a one-to-many mapping, preventing replacement with a single suitable value. While SplITS' techniques can be applied to Ember-IO, the developers of SplITS found the aforementioned issue would occur in some fuzzer generated inputs, preventing SplITS from being applied successfully to those inputs, reducing SplITS' overall consistency on Ember-IO compared to Fuzzware [10].

Pretender [19] and P^2IM both use automation of the emulation procedure by using direct MMIO modeling. They do this in two different ways: Pretender uses recordings of real activity in the MMIO region of the device to model the peripherals, while P^2IM uses *blind fuzzing*, which is corresponding to black-box fuzzing. Both methods have advantages and disadvantages, the method used in Pretender requires access to the physical device while the method used in P^2IM is less generic and will not produce an accurate model in many devices [4].

One of the most difficult aspects of analyzing the fuzzed firmware is the false positive rate. A false positive is a "false alarm" and occurs when the fuzzer incorrectly identifies parts of the firmware as vulnerable. A high false positive rate diminishes the effectiveness of the fuzzing tool, as it means spending more time sorting through false alarms rather than focusing on actual vulnerabilities, which can be frustrating and time-consuming. Thus, triggering crashes does not need to mean that it represents a security issue.

An aspect to consider when discussing false positives found while fuzzing a firmware image of an embedded device, is the hardware constraints. As discussed in the background section, developers have to consider if vulnerabilities are likely to be exploitable if the code is not reachable on the physical device. Even though the optimal solution is to have the firmware secure in all environments, this would be a question of prioritization for the development team. Ember-IO mentions that two out of the 6 new bugs discovered by the fuzzer could be vulnerabilities that would not be reachable on the physical device. Even though this does not count as a false positive, the development team of the device might not expect to trigger them, unless the embedded device is defective.

The issue of false positives arises partly because inputs for a particular register is defined by unknown hardware [6]. This makes it challenging to deduce directly from the firmware. For fuzzers like P^2IM and μEmu, this is unlikely to happen because in these approaches they enforce register models based on common access patterns. In these fuzzers some registers can only contain values written by firmware and thus prevents unexpected inputs. Other fuzzers, such as Fuzzware and Ember-IO, have a different approach where all registers are considered as valid sources to insert the fuzzing input.

Some false positives can be linked to the lack of DMA support in several of the fuzzers. DMA enables the RAM to be accessed by the peripherals directly without running it through the processor. Therefore, using symbolic execution is not useful in discovering information about the DMA accesses [45]. We will discuss this further below.

DMA is mentioned and explained in several of the papers, some including options on how to incorporate support for DMA input generation. Amongst these are manual configuration of MMIO settings and installing separate components on top of the fuzzer. As explained, not including DMA support can cause an increase in the amount of false positives [23], but none of the fuzzers in our study include automated DMA support at the moment.

The research team behind Ember-IO and SplITS have followed up with a supplementary component, DICE, to better accommodate fuzzing of firmware that utilizes DMA [12]. The authors mention the option to implement DICE on top of existing fuzzers to include automated DMA support [6]. DICE [26] is a component to extend existing firmware analyzers which enables them to create or modify DMA input channels (from peripherals to firmware) [26]. As soon as the firmware inserts the source and destination DMA transfer pointers into the DMA controller, DICE recognises DMA input channels [26]. Subsequently, DICE modifies the data transmitted via DMA on behalf of the firmware analyzer.

The only fuzzer included in this review that already includes support for DMA is HALucinator. This is due to the HLE techniques applied in the emulation of the firmware. HALucinator handles DMA by using the same HALs as the developers use to perform DMA [4]. Here, the DMA accesses are removed from the program. P^2IM however, only considers MMIO interaction sequences as input, so if a crash is found it must be mapped back to the corresponding input. In addition, the inputs are replayable against virtualized and actual targets [4].

For the fuzzers without DMA support, a possible way of incorporating support for DMA is by configuring a simple DMA overlay in the memory maps [30]. This is done by manually setting an MMIO overlay region to allow manually specifying DMA buffers in case they are located in RAM. The resulting behavior is that fuzzing input is provided each time the specified region is accessed.

5 Conclusion

Even though the research field on fuzzing embedded devices is growing, firmware fuzzing is still so complicated that there is a long way to go until we can expect adaptation in most security testing routines of embedded devices. An important step in the right direction is ensuring thorough and descriptive documentation, which is why we have decided to emphasise documentation in this review. If this ensures adaptation in real-world projects, this could lead to more feedback that the developers of the fuzzers can use to improve the tools. Several of the papers in this review discuss the possibility of future support for DMA. After communicating with the developers of the different fuzzers, our impression is that DMA support is a priority in future work for most of them.

Table 1 shows a summary of all the fuzzers, visualizing the similarities and differences between them. The table shows which fuzzing engine, emulator and technique each of the firmware fuzzers use.

Table 1 Overview of the firmware fuzzing tools mentioned in this review

	Fuzzer	Emulator	Technique
HALucinator [4]	AFL	QEMU	HAL-based
μAFL [25]	AFL	QEMU	Hardware-in-the-loop
FIRM-AFL [1]	AFL	FIRMADYNE	Coverage-feedback [40]
FIRMCORN [16]	Their own implementation	Unicorn	Pattern-based
EM-Fuzz [14]	AFL	QEMU	Coverage-feedback [40]
FirmHybridFuzzer [38]	AFL	QEMU	Symbolic execution
PRETENDER [19]	Their own implementation	QEMU	Pattern-based
P²IM [12]	AFL	QEMU	Pattern-based
Fuzzware [31]	AFL	Unicorn	Symbolic execution
Hoedur [32]	libFuzzer	QEMU	Symbolic execution with multi-stream input
Ember-IO [6]	AFL++	QEMU	Coverage-feedback (FERMCov)
μEMU [45]	AFL	QEMU (S2E)	Symbolic execution
SplITS [7]	AFL++	Unicorn	Symbolic execution with FERMCov [6]
FIRMNANO [17]	AFL	Unicorn	Coverage-feedback
HD-Fuzz [23]	AFL	QEMU (S2E)	Symbolic execution
Jetset [23]	AFL	QEMU (S2E)	Symbolic execution
SAFIREFUZZ [36]	LibAFL	Near-native rehosting[a]	HAL-based

[a] Safirefuzz does not rely on an emulator

Acknowledgements The research in this paper was co-funded by the European Union through the Horizon Europe NEMECYS project, grant number 101119747.

References

1. Chen, D.D., Woo, M., Brumley, D., Egele, M.: Towards automated dynamic analysis for linux-based embedded firmware. In: NDSS (2016)
2. Chipounov, V., Kuznetsov, V., Candea, G.: S2E: a platform for in-vivo multi-path analysis of software systems. ACM Sigplan Notices **46**(3), 265–278 (2011)
3. Clements, A.A., Gustafson, E., Scharnowski, T., Grosen, P., Fritz, D., Kruegel, C., Vigna, G., Bagchi, S., Payer, M.: Halucinator. https://github.com/sandialabs/halucinator
4. Clements, A.A., Gustafson, E., Scharnowski, T., Grosen, P., Fritz, D., Kruegel, C., Vigna, G., Bagchi, S., Payer, M.: HALucinator: Firmware re-hosting through abstraction layer emulation.

In: 29th USENIX Security Symposium (USENIX Security 20), pp. 1201–1218. USENIX Association (2020). https://www.usenix.org/conference/usenixsecurity20/presentation/clements
5. Engine, U.: The ultimate CPU emulator. https://www.unicorn-engine.org
6. Farrelly, G., Chesser, M., Ranasinghe, D.C.: Ember-IO: Effective firmware fuzzing with model-free memory mapped IO. In: Proceedings of the 2023 ACM Asia Conference on Computer and Communications Security, ASIA CCS'23, pp. 401–414. Association for Computing Machinery, New York, NY, USA (2023). https://doi.org/10.1145/3579856.3582840
7. Farrelly, G., Quirk, P., Kanhere, S., Camtepe, S., Ranasinghe, D.: SplITS: split input-to-state mapping for effective firmware fuzzing. In: European Symposium on Research in Computer Security, pp. 290–310 (2023). https://www.researchgate.net/publication/373142220_SplITS_Split_Input-to-State_Mapping_for_Effective_Firmware_Fuzzing
8. Farrelly, G., Quirk, P., Kanhere, S.S., Camtepe, S., Ranasinghe, D.C.: Ember-IO. https://github.com/Ember-IO/Ember-IO-Fuzzing/tree/main
9. Farrelly, G., Quirk, P., Kanhere, S.S., Camtepe, S., Ranasinghe, D.C.: SplITS. https://github.com/SplITS-Fuzzer/SplITS
10. Farrelly, G.D.: Personal communication
11. Feng, B., Mera, A.: P2IM. https://github.com/RiS3-Lab/p2im
12. Feng, B., Mera, A., Lu, L.: P^{2}IM: Scalable and hardware-independent firmware testing via automatic peripheral interface modeling. In: 29th USENIX Security Symposium (USENIX Security 20), pp. 1237–1254 (2020)
13. Fioraldi, A., Mantovani, A., Maier, D., Balzarotti, D.: Dissecting American fuzzy lop: a FuzzBench evaluation. ACM Trans. Softw. Eng. Methodol. **32**(2) (2023). https://doi.org/10.1145/3580596
14. Gao, J., Xu, Y., Jiang, Y., Liu, Z., Chang, W., Jiao, X., Sun, J.: EM-Fuzz: augmented firmware fuzzing via memory checking. IEEE Trans. Comput.-Aided Des. Integr. Circuits Syst. **39**(11), 3420–3432 (2020). https://doi.org/10.1109/TCAD.2020.3013046
15. Gui, Z., Shu, H., Kang, F., Xiong, X.: FIRMCORN V2. https://github.com/S4CH/FIRMCORN-V2
16. Gui, Z., Shu, H., Kang, F., Xiong, X.: FIRMCORN: vulnerability-oriented fuzzing of IoT firmware via optimized virtual execution. IEEE Access **8**, 29826–29841 (2020). https://doi.org/10.1109/ACCESS.2020.2973043
17. Gui, Z., Shu, H., Yang, J.: FIRMNANO: toward IoT firmware fuzzing through augmented virtual execution. In: 2020 IEEE 11th International Conference on Software Engineering and Service Science (ICSESS), pp. 290–294 (2020). https://doi.org/10.1109/ICSESS49938.2020.9237719
18. Gustafson, E., Muench, M., Spensky, C., Redini, N., Machiry, A., Fratantonio, Y., Balzarotti, D., Francillon, A., Choe, Y.R., Kruegel, C., Vigna, G.: Pretender: automatically emulating hardware. https://github.com/ucsb-seclab/pretender
19. Gustafson, E., Muench, M., Spensky, C., Redini, N., Machiry, A., Fratantonio, Y., Balzarotti, D., Francillon, A., Choe, Y.R., Kruegel, C., Vigna, G.: Toward the analysis of embedded firmware through automated re-hosting. In: 22nd International Symposium on Research in Attacks, Intrusions and Defenses (RAID 2019), pp. 135–150. USENIX Association, Chaoyang District, Beijing (2019). https://www.usenix.org/conference/raid2019/presentation/gustafson
20. Jaatun, M.G., Sæle, H.: A checklist for supply chain security for critical infrastructure operators. In: The International Conference on Cybersecurity, Situational Awareness and Social Media, pp. 235–249. Springer (2023)
21. Johnson, E., Bland, M., Zhu, Y., Mason, J., Checkoway, S., Savage, S., Levchenko, K.: Jetset: targeted firmware rehosting for embedded systems. https://github.com/aerosec/jetset
22. Johnson, E., Bland, M., Zhu, Y., Mason, J., Checkoway, S., Savage, S., Levchenko, K.: Jetset: targeted firmware rehosting for embedded systems. In: 30th USENIX Security Symposium (USENIX Security 21), pp. 321–338. USENIX Association (2021). https://www.usenix.org/conference/usenixsecurity21/presentation/johnson
23. Kim, J., Yu, J., Lee, Y., Kim, D.D., Yun, J.: Hd-fuzz: hardware dependency-aware firmware fuzzing via hybrid MMIO modeling. J. Netw. Comput. Appl. (2024). Available at SSRN 4493040

24. Li, W., Shi, J., Li, F., Lin, J., Wang, W., Guan, L.: microAFL. https://github.com/MCUSec/microAFL
25. Li, W., Shi, J., Li, F., Lin, J., Wang, W., Guan, L.: μAFL: non-intrusive feedback-driven fuzzing for microcontroller firmware. In: Proceedings of the 44th International Conference on Software Engineering, ICSE'22, pp. 1-12. Association for Computing Machinery, New York, NY, USA (2022). https://doi.org/10.1145/3510003.3510208
26. Mera, A., Feng, B., Lu, L., Kirda, E.: Dice: Automatic emulation of DMA input channels for dynamic firmware analysis. In: Proceedings of the 42nd IEEE Symposium on Security and Privacy, S&P/Oakland'21 (2021)
27. Pak, B.S.: Hybrid Fuzz Testing: Discovering Software Bugs via Fuzzing and Symbolic Execution. School of Computer Science Carnegie Mellon University (2012)
28. Pare, G.K.S.: Handbook of eHealth Evaluation: An Evidence-Based Approach [Internet]. University of Victoria, Victoria (BC) (2017). https://www.ncbi.nlm.nih.gov/books/NBK481583/
29. QEMU: QEMU wiki. https://wiki.qemu.org/Main_Page
30. Scharnowski, T.: Personal communication
31. Scharnowski, T., Bars, N., Schloegel, M., Gustafson, E., Muench, M., Vigna, G., Kruegel, C., Holz, T., Abbasi, A.: Fuzzware: using precise MMIO modeling for effective firmware fuzzing. In: 31st USENIX Security Symposium (USENIX Security 22), pp. 1239–1256. USENIX Association, Boston, MA (2022). https://www.usenix.org/conference/usenixsecurity22/presentation/scharnowski
32. Scharnowski, T., Wörner, S., Buchmann, F., Bars, N., Schloegel, M., Holz, T.: Hoedur: embedded firmware fuzzing using multi-stream inputs. In: USENIX Security Symposium (USENIX Sec) (2023)
33. Scharnowski, T., et al.: Fuzzware repository. https://github.com/fuzzware-fuzzer/fuzzware
34. Scharnowski, T., et al.: Hoedur repository. https://github.com/fuzzware-fuzzer/hoedur
35. Seidel, L., Maier, D., Muench, M.: Safirefuzz. https://github.com/pr0me/SAFIREFUZZ
36. Seidel, L., Maier, D., Muench, M.: Forming faster firmware fuzzers. In: Proceedings of the 32nd USENIX Conference on Security Symposium, pp. 2903–2920 (2023)
37. Situ, L., Zhang, C., Guan, L., Zuo, Z., Wang, L., Li, X., Liu, P., Shi, J.: FirmHybridFuzz. https://github.com/stuartly/FirmHybridFuzz
38. Situ, L., Zhang, C., Guan, L., Zuo, Z., Wang, L., Li, X., Liu, P., Shi, J.: Physical devices-agnostic hybrid fuzzing of IoT firmware. IEEE Internet Things J. **10**(23), 20718–20734 (2023). https://doi.org/10.1109/JIOT.2023.3303780
39. Voss, N.: afl-unicorn. https://github.com/Battelle/afl-unicorn
40. Yun, J., Rustamov, F., Kim, J., Shin, Y.: Fuzzing of embedded systems: a survey. ACM Comput. Surv. **55**(7), 1–33 (2022)
41. Zalewski, M.: American fuzzy lop (AFL) (2023). http://lcamtuf.coredump.cx/afl/. Accessed on September 4, 2023
42. Zheng, Y., Davanian, A., Yin, H., Song, C., Zhu, H., Sun, L.: FIRM-AFL. https://github.com/zyw-200/FirmAFL
43. Zheng, Y., Davanian, A., Yin, H., Song, C., Zhu, H., Sun, L.: FIRM-AFL: high-throughput greybox fuzzing of IoT firmware via augmented process emulation. In: 28th USENIX Security Symposium (USENIX Security 19), pp. 1099–1114. USENIX Association, Santa Clara, CA (2019). https://www.usenix.org/conference/usenixsecurity19/presentation/zheng
44. Zhou, W., Guan, L., Liu, P., Zhang, Y.: μEmu. https://github.com/MCUSec/uEmu
45. Zhou, W., Guan, L., Liu, P., Zhang, Y.: Automatic firmware emulation through invalidity-guided knowledge inference. In: 30th USENIX Security Symposium (USENIX Security 21). USENIX Association (2021). https://www.usenix.org/conference/usenixsecurity21/presentation/zhou

Towards a Secure Manufacturing Framework for Single Malt Whisky in Industry 4.0

Jacob Connell, Dimitrios Kasimatis, Pavlos Papadopoulos, William J. Buchanan, Panagiotis Sarigiannidis, and Nikolaos Pitropakis

Abstract Recent instances have brought to light the inherent vulnerability of the manufacturing sector to cyber security threats, particularly within the food and beverage industries. With the rapid adoption of Industry 4.0, supply chains are exposed to an expanding surface area of potential attacks. This paper delves into the paramount issue of supply chain security and explores the utilisation of distributed ledger technology to enhance the reliability and security of the whisky manufacturing process. We introduce an experimental framework that is focused on performance enhancements and security optimisations. Furthermore, a review of the implemented testbed evaluates the robustness of the business logic, emphasising additional benefits derived from implemented features. This analysis serves to showcase the advancements made in response to recent disruptions within similar supply chains.

J. Connell · D. Kasimatis (✉) · P. Papadopoulos · W. J. Buchanan · N. Pitropakis
Edinburgh Napier University, Edinburgh, UK
e-mail: d.kasimatis@napier.ac.uk

P. Papadopoulos
e-mail: p.papadopoulos@napier.ac.uk

W. J. Buchanan
e-mail: b.buchanan@napier.ac.uk

N. Pitropakis
e-mail: n.pitropakis@napier.ac.uk

P. Sarigiannidis
University of Western Macedonia, Kozani, Greece
e-mail: psarigiannidis@uowm.gr

M. G. Jaatun et al. (eds.), *Proceedings of the International Conference on Cybersecurity, Situational Awareness and Social Media*, Springer Proceedings in Complexity,
https://doi.org/10.1007/978-981-96-0401-2_19

1 Introduction

Manufacturing, evolving through various Industrial Revolutions, is now embracing the fourth era with Industry 4.0. This evolution addresses the complexities and vulnerabilities of supply chains, as highlighted by recent global challenges like pandemics and geopolitical tensions [25]. Industry 4.0 is reshaping manufacturing by integrating Internet of Things (IoT) technologies to enhance connectivity, efficiency, and transparency [18].

Single Malt Whisky is a spirit exclusively made from malts processed by a single distillery. Although originating from a single source, these whiskies may be a blend of various casks and ages to enhance their flavour profile [22]. The journey of these raw assets, from the initial harvesting phase to the final bottling process, is crucial. This procedure implies a complex relationship between numerous assets and the final product, necessitating a sophisticated data model to reflect this accurately. Cybersecurity, however, is a growing concern, particularly in the food and agriculture sectors, as evidenced by the UK Government's 2022 Cyber Security Survey [10]. These attacks reveal the vulnerability of these sectors to cyber threats. The industry faces challenges like inadequate investment in security, lack of skills, and reliance on equipment vendors for protection [16].

Furthermore, confidentiality in certain data aspects, such as financial transactions and sensitive company information, is paramount. This is especially pertinent in industries with strict legislation regarding supply chain transparency. In the context of single malt whisky, a unique blend of various batches, distillers are prohibited from disclosing the age of each whisky in the blend to consumers. The law mandates only the age of the youngest whisky to be disclosed, rounded to the nearest whole year [15]. This legislative requirement inherently limits supply chain transparency. The proposed testbed will address this by safeguarding specific cask information from downstream partners and consumers. Additionally, the traceability system will verify the whisky's age while complying with legal restrictions on information disclosure.

This paper proposes a secure framework for Industry 4.0 manufacturing using a case study on whisky production, that can also be applied to other industry supply chains. The 'SupplyFlow' system, leveraging blockchain technology, offers a secure, transparent process for data sharing and fraud protection among various participants. This approach is visualised in Fig. 1, illustrating a distributed network ensuring data integrity and robustness in production.

The contributions of this work can be outlined as follows:

- The design and technical development of SupplyFlow, an Industry 4.0 data solution tailored for the manufacturing of single malt Scotch whisky, that upholds security standards, preserves privacy and aligns seamlessly with prevailing legislation.
- The extensive performance examination exhibited by SupplyFlow across various evaluation scenarios, providing valuable insights into its operational efficiency and effectiveness.

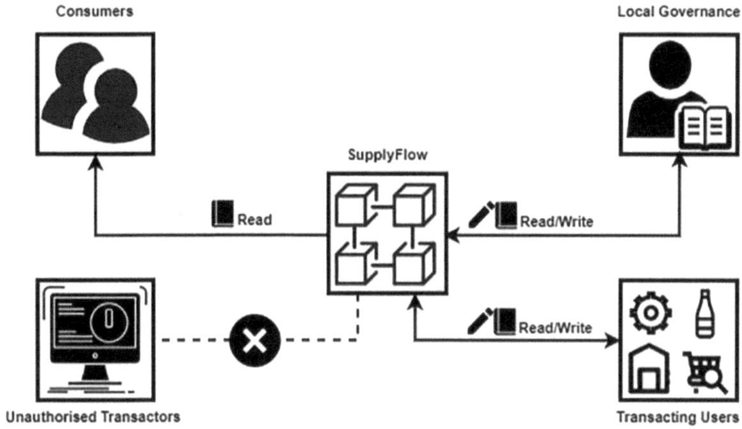

Fig. 1 SupplyFlow overview

- The comprehensive evaluation and demonstration of the security attributes inherent in SupplyFlow, by investigating the system's resilience against prevalent threats, drawing connections to recent and noteworthy examples.

The rest of the paper is structured as follows. Section 2 briefly outlines the background of the technological foundations of this paper while examining related works. Section 3 provides an overview of the testbed and network configuration, while Sect. 4 empirically evaluates the proposed solution. Finally, Sect. 5 draws the conclusions while giving some suggestions for future work.

2 Background Knowledge

2.1 Industry 4.0

Industry 4.0 transitions from basic automation to sophisticated Cyber-physical systems, enhancing global collaboration, real-time analytics, and social trust. This advancement, however, brings challenges like securing smart sensor data and protecting cloud systems from malware [12]. With interconnected supply chains, traditional security perimeters become inadequate, prompting a shift towards Zero Trust architecture, focusing on rigorous access control and secure authentication [8]. Cloud computing is prevalent in small-medium enterprises for operational management, but its centralised nature poses security challenges. Distributed Ledger Technologies (DLT) emerge as a solution, offering secure, collaborative environments for improved delivery times and quality control in IoT systems [24].

2.2 Distributed Ledger Technology

DLTs represent a paradigm shift from traditional centralised databases to distributed systems. In DLTs, each network member independently stores data, enhancing query performance, reliability, and security. Unlike older systems where an administrator maintains database consistency, modern DLTs require consensus on data integrity, presenting challenges and opportunities [33]. Also, DLTs align with ISO 27000 standards and, can address security concerns by design [35]. Blockchain, the most recognised DLT, has revolutionised various sectors since its introduction, but also faces confidentiality challenges due to its consensus methods, necessitating robust privacy controls [3]. Blockchain applications are diverse. In manufacturing, blockchain enables IoT devices to transact on the ledger securely. Awwad et al. (2018) highlight its use in supply chains, citing advantages like enhanced security and scalability [5].

2.3 Hyperledger Fabric

Hyperledger Fabric (HLF), a notable DLT framework within the Linux Foundation's Hyperledger projects, is designed for collaboration and data sharing. It offers a consortium network structure, allowing for decentralised transaction control and efficient private data protection, making it ideal for various enterprise and business-to-business applications [29]. HLF's architecture includes several key components such as peers managing transactions and maintaining ledgers, ordering nodes for consensus building, and a pluggable consensus mechanism for flexibility in transaction validation [3]. Its consensus process comprises endorsement, ordering, and validation stages, ensuring secure and synchronised data updates across the network while reducing computational overhead [29]. Furthermore, its support for modularity in storage methods and scalability aligns well with varied enterprise requirements. Private data collections within HLF enable organisations to maintain confidentiality while participating in shared processes. The network's robust identity management, facilitated by membership service providers and using tools like Idemix, enhances privacy and security [32], thus making it particularly effective in supply chain environments.

2.4 Related Works

Supply chain management systems require traceability, transparency, and accountability, especially in an Industry 4.0 environment with cloud computing, IoT integration, and cybersecurity concerns [13]. The whisky industry, a sector with significant investments and a complex supply chain, exemplifies these needs. It requires protection against counterfeiting and efficient tracking of duty payments and exports.

Recent studies indicate a growing interest in applying blockchain technology to this domain, although direct applications in aged spirit production are sparse [28].

Though not directly applicable to manufacturing, the UnicalCoin model [19] demonstrates the challenges of using public and permissionless blockchains in supply chains. Similarly, a study on the agricultural sector utilises the Ethereum network and InterPlanetary File System (IPFS) for data verification yet needs more measures for private data communication [31]. The work of Marchese and Tomarchio [21] suggests a supply chain system without private data collections or an effective crash tolerance mechanism, posing risks in confidentiality and network resilience. Malik et al. [20] explores a centrally governed food traceability network using Python, highlighting the challenges of global processing and transaction validation in shared systems.

The study of Tian [34] addresses food safety in China, suggesting a blockchain-based agri-food traceability system incorporating Radio-frequency identification (RFID) technology, a prevalent method in manufacturing. Cañada [7], an industry expert, proposed an Ethereum-based supply chain method, notable for its traceability but lacking in data privacy and role management. Ernst & Young (EY) Ops Chain's collaboration with Italian beer brand Peroni demonstrates the application of the Ethereum network for product traceability, utilising Zero-Knowledge Proofs (ZKP) for private data sharing [14]. However, it faces scalability and performance issues.

A comparative analysis of these studies, as displayed in Table 1 reveals varied approaches to blockchain application in supply chains, measured in the following categories:
(A) Data privacy allowing the selective sharing and storage of information.
(B) Security analysis of known technological threats and measures implemented.
(C) Performance analysis to assess scalability.
(D) Performance analysis under malicious behaviours.

Table 1 Comparison of Related Works

Method	Technology	Access type	A	B	C	D	E	F
Longo et al. (2019) [19]	Unicoin	Private	✗	✗	✗	✗	✗	✗
Shahid et al. (2020) [31]	Ethereum	Public	✓	✓	✓	✗	✗	✓
Marchese et al. (2021) [21]	HLF v1	Consortium	✗	✗	✗	✗	✓	✗
Cañada (2019) [7]	Ethereum	Public	✗	✗	✗	✗	✗	✓
Malik et al. (2019) [20]	ProductChain	Private	✗	✓	✓	✗	✓	✓
SupplyFlow (this work)	HLF v2	Consortium	✓	✓	✓	✓	✓	✓

(E) Consideration for local governance and legislation.
(F) Network crash tolerance.

The studies are also categorised by blockchain access type into the three main types of public, private and consortium-based:

Public Blockchains: Open and decentralised, exemplified by Bitcoin and Ethereum, these blockchains permit unrestricted participation in transaction verification.

Private Blockchains: Governed by a single entity with limited access, these blockchains are engineered for enhanced transaction efficiency and reduced energy use, sacrificing decentralisation in the process.

Consortium Blockchains: Managed jointly by several organisations, these blockchains blend limited decentralisation with controlled access.

The main challenges include ensuring data privacy and network resilience while utilising public blockchains. This work addresses privacy and permission control issues in such networks by employing a Hyperledger Fabric-based system for secure data sharing in supply chains, enhancing agility and scalability in whisky manufacturing. It ensures compliance and legislative coherence, offering a fault-tolerant, flexible network with controlled access. It securely maintains financial records between authorised organisations, with provisions for record expiration and validation. The study also examines the network's resilience against malicious attacks, assessing its impact on availability and performance.

3 Experimental Setup

A local implementation of the testbed system has been developed, along with the chaincode, that processes the business logic for this specific use case. This testbed was deployed on a 2.6 Hz 6-Core Intel i7 Macbook Pro with 16GB 2400 MHz DDR4 RAM. The architecture of this local implementation is depicted in Fig. 2. The system uses Hyperledger Fabric images within Docker containers to establish the network structure, using Docker engine 19.03.13 and Hyperledger Fabric v2.3.0 images.

For testing, Ansible scripts were utilised to construct the testbed to allow straighforward reproduction from configuration files. MiniFabric, part of the Hyperledger Lab, includes Bash scripts that create channel artefacts for the designed network [17]. These tools enable each peer to interact with the chaincode. The Ansible scripts also allow multiple actions across different peers to be executed through single function calls, enabling the reproduction of the testbed with a basic understanding of the Hyperledger Fabric underlying architecture. Since data records are not stored in a traditional database but are managed by chaincode, this chaincode is responsible for maintaining their integrity and access control. Hence, the chaincode manages access to individual and collective data, accounting for scenarios where a user may possess write but not read permissions, as seen in the case of IoT sensors [26].

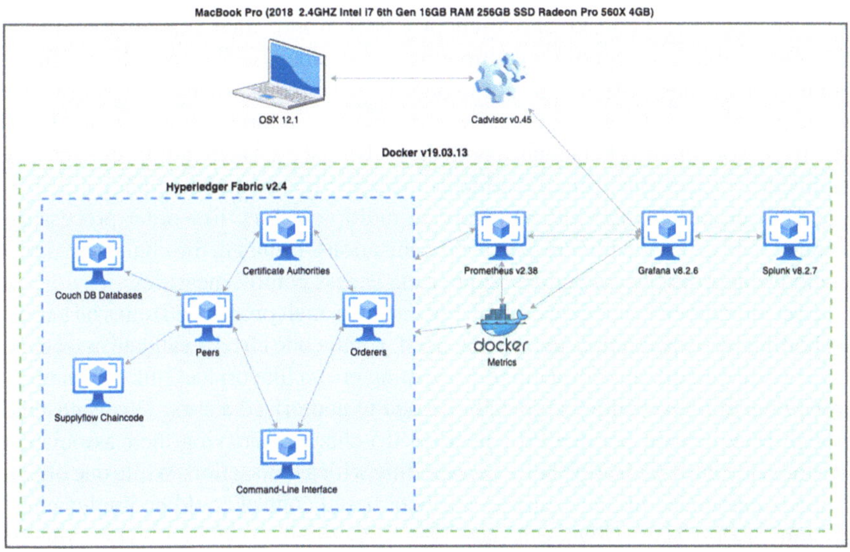

Fig. 2 Testbed architecture

Private data collections help maintain the integrity and confidentiality of sensitive information within the blockchain framework. This is critical for supply transactions between retailers, processors, and producers, where certain information must be kept private from competitors. Regulations in specific industries dictate sharing information about whisky casks with retailers and consumers [15]. Local governing bodies such as His Majesty's Revenue and Customs (HMRC), process duty payments confidentially while verifying these transactions on the main ledger. These private data collections are stored on the peers of member organisations with the authority to hold this data in their state database. Peers can be connected to multiple channels and chaincodes simultaneously, maintaining a separate state [9].

HLF is a private blockchain network where authorisation is a prerequisite for the supply chain network participation. This setup allows peers to be part of several channels, thus enabling businesses to engage in various supply chains or networks. In the described testbed, changing a channel transaction requires endorsement by most of the organisation's administrators [1].

Endorsement policies can be specific, such as requiring a distillery to endorse changes in its supply chain along with other organisations. No single party should exert complete control except for specific business needs in a consortium channel. For example, these chains could exclude a distillery linked to multiple retail chains from the supply chain. A rigid, hardcoded endorsement policy may lead to security vulnerabilities. HLF advocates a majority endorsement process, which adapts as network membership changes. A hardcoded policy, if not updated, might become ineffective, representing only a minority of network members [4].

Chaincode record checks during endorsement ensure transactional compliance with the required asset conditions, enforcing consensus-based business logic. Transactions failing these checks are not endorsed, with peers strictly adhering to the chaincode. Chaincode uses the local world state for endorsing transactions. However, if the world state is not updated until the end of the transaction lifecycle, security and integrity issues can occur. Developers must be vigilant in their coding practices to avoid such pitfalls. Operations involving multiple actors, like order processing, maintain data privacy within respective organisations. Filters in the chaincode direct data to the appropriate collections, supporting access control measures.

Asset interdependencies are critical. For instance, malt production relies on barley, and bottling whisky requires the whisky itself. Chaincode checks validate asset existence before processing transactions, preventing errors like orphan links or manipulation. Function invocation is typically limited to authorised actors. This is ensured through Membership Service Providers (MSP) checks, verifying their association with the relevant organisation before proceeding with a transaction. While one organisation is hardcoded in this example, attribute-based control could be implemented for more flexible access management.

In whisky manufacturing, the cask maturation date is significant. Given the investment interest in whisky maturation, the accuracy and immutability of records are paramount. However, the precision needed for validating timestamps during endorsements can cause problems. Mitra [23] proposes a time variance allowance from a local peer's trusted time source since whisky bottles are usually labelled to a year's accuracy [15], verifying the transaction date within a day is considered acceptable. Each endorsing peer in this testbed verifies the transaction date against its trusted time source, ensuring proof of maturation dates with chaincode disallowing any external data input.

The endorsement process involves peers running the chaincode independently and verifying transactions. Each peer signs the transaction with its private key before returning it to the proposing peer. To mitigate risks, the endorsement policy should balance robustness against performance impact and consider endorser diversity. The policy might require a majority of organisational admin consensus for channel updates and lifecycle endorsements.

Effective network monitoring is crucial for optimal performance and security in blockchain platforms. This setup involves utilising multiple data sources, configuring node metrics via Prometheus, and employing Splunk security information and event management (SIEM) for comprehensive network analysis [30]. The Fabric logger, key for transaction logs, facilitates peer connections for data access, offering vital chaincode and ledger details. Docker logs enhance container analytics, augmenting Splunk's capabilities. These, alongside client application logs tracking latency and consensus, ensure thorough transaction, performance, and security surveillance. Typically, real-world applications deploy more extensive resources, with Splunk independently connected through a forwarder.

In a distributed network, ordering nodes are key as they sequence and distribute blocks to peers. If these nodes fail, channel publishing halts. The network uses the ETCD's RAFT algorithm [11] for crash fault tolerance (CFT). Through majority elections, RAFT ensures functionality if enough nodes remain for a vote. A node can be self-nominated as a leader, and if no leader currently exists, it becomes the leader after majority approval, starting an Election Term. This term ends when a follower, not receiving heartbeats, becomes a candidate. Repeated election cycles can cause network latency. Having an odd number of ordering nodes helps prevent vote splits. In some networks, organisations control these nodes for reliability [36].

Private data collections may have tailored endorsement policies. These policies should be as inclusive as any specific endorsement policy [2]. For example, a policy might require endorsements from both HMRC and a distillery, ensuring secure validation of transactions. In conclusion, HLF's design requires careful consideration of endorsement policies, asset interdependencies, and transaction verification to maintain network integrity, especially in industries like whisky manufacturing with significant business and legal implications.

4 Evaluation

Several benchmark tests were conducted to evaluate the performance of the experimental setup. The conducted experiments included:

- Monitoring throughput and latency at varying transactions per second (TPS) of 20, 30, 40, 60, 80, 100, 120, and 140.
- Assessing the impact on CPU and memory usage between 1,000 and 10,000 invoked transactions.

To gauge the impact of private data on transaction performance, throughput and latency were assessed under comparable conditions involving private data. This study is critical, given sparse data on the effects of such overheads. Anticipated performance impacts, possibly from added computational needs in data validation, will inform future architectural designs.

4.1 Throughput and Latency

Throughput and latency results were collected using Caliper to measure the network tolerance to various transaction inputs per second. Throughput testing shows a visible linear relationship between the input TPS and the throughput TPS, as displayed in Fig. 3. After input of 40 TPS, the relationship declines as the throughput appears to reach a threshold limitation. The high load metrics graph displayed in Fig. 4 also indicates the limitations of the configuration of this network. Transaction loads over 1,000 see an increase in latency, with a reduction in the network's overall efficiency. These readings were taken at the invocation of transactions during the resource analysis stage.

Fig. 3 Throughput

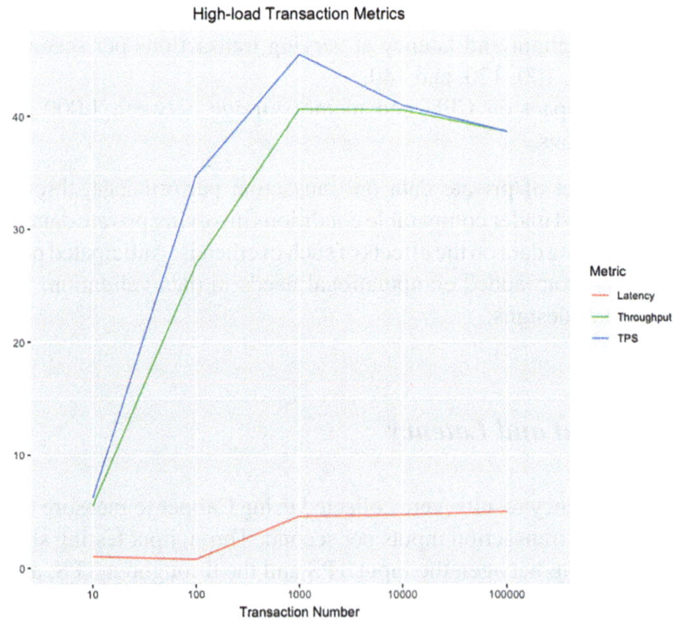

Fig. 4 High-load metrics

Using Hyperledger Caliper [6] to measure the latency and throughput to compare with private data collection performance, it's clear there appears to be a significant difference in performance at high transaction rates as displayed in Fig. 5. The

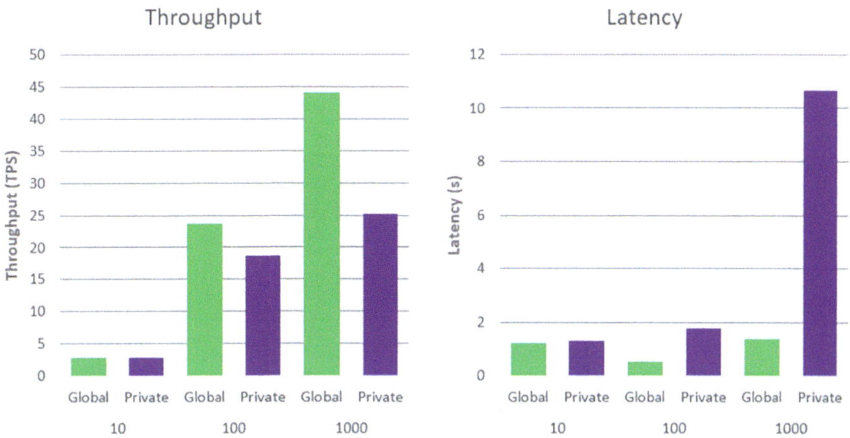

Fig. 5 Private data collection performance metrics

throughput is higher for global collections, with a greater difference as the number of transactions increases. The latency at higher levels shows that the private collections are affecting the performance significantly. Repeated tests, confirmed the latency over 10 s was not an anomaly.

4.2 System Resources

CPU and memory resource utilisation results for different levels of transaction invocation were collected from Prometheus and container statistics. On 1,000 transactions (Tx), there is a spike in CPU usage from the involved nodes with a minutely longer transaction time, as portrayed in Fig. 6. An increase in memory consumption can be seen during the CPU peak as nodes process this data and validate it for their state databases, as shown in Fig. 7.

On the submission of 10,000 Tx, the CPU resources for those endorsing the operation varied significantly compared to lower levels. A tenfold increase in the transaction count saw a fourfold increase in the overall computational time, as shown in Fig. 8. At the same time, the memory resources increased to keep up with the buffer of the limited batch size, as displayed in Fig. 9.

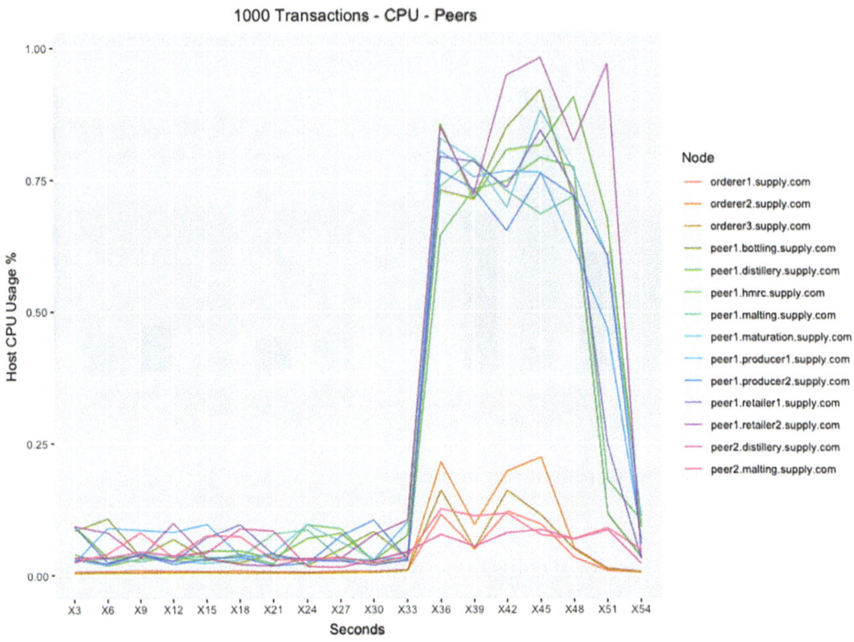

Fig. 6 1,000 Tx CPU usage

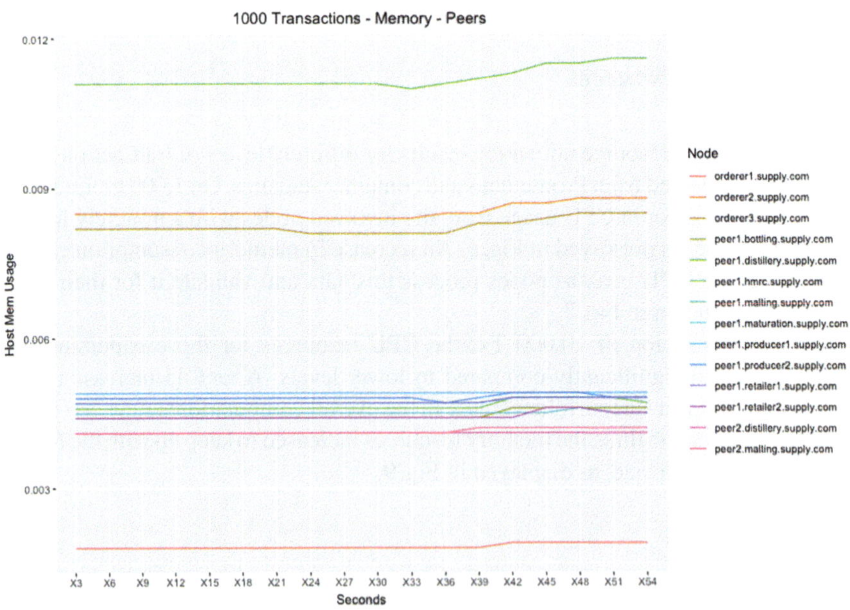

Fig. 7 1,000 Tx memory usage

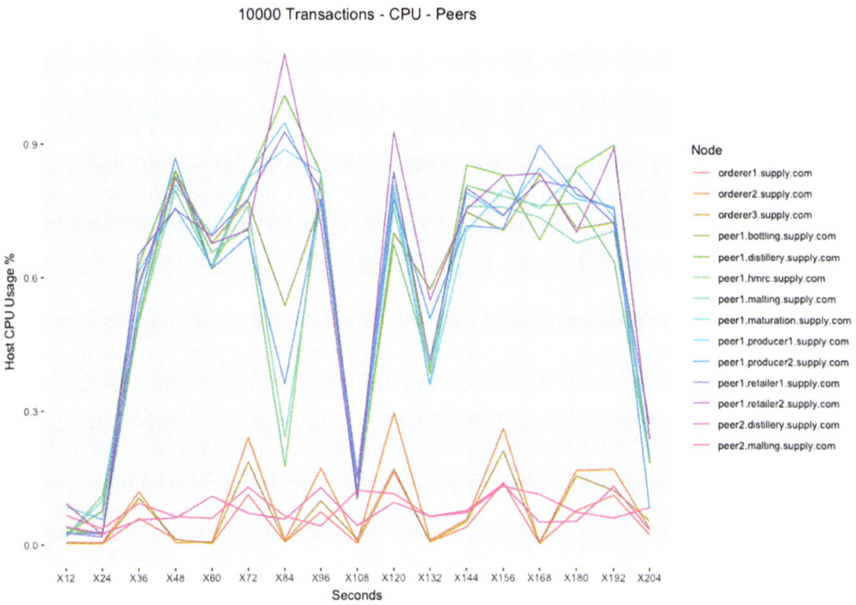

Fig. 8 10,000 Tx CPU usage

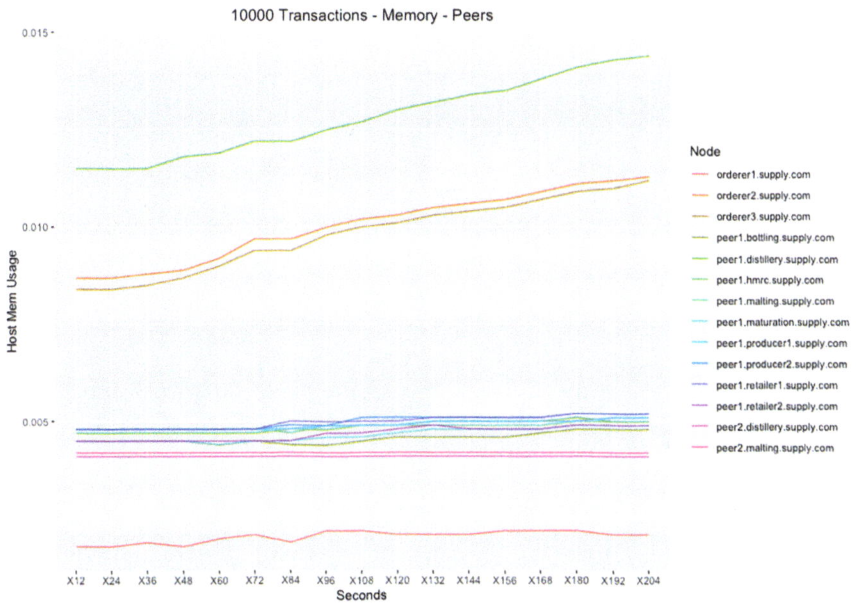

Fig. 9 10,000 Tx memory usage

4.3 Network Resilience

Several manufacturing companies have been ransomware victims, impacting their operations and sometimes compelling them to pay the ransom [27]. A resilient solution is essential, highlighting the advantage of the Fabric blockchain network. Its distributed data structure means a breach in one peer or database minimally affects the network and recovery process. For complete network compromise, an attacker must breach each peer within and across multiple organisations, each with unique security setups.

This experiment demonstrates the network's resilience to database loss from ransomware. Transaction tests are conducted in three phases: regular operation, during an attack, and post-recovery. In normal conditions, 52 transactions succeeded, as shown in Fig. 10. When simulating a database loss, the peer's immediate disconnection is noticeable. Alternative peers are engaged to maintain service, with 52 successful transactions despite the outage, as Illustrated in Fig. 11. In Fig. 12, recovery involves refreshing the affected peer, leading to a fully functional network with 52 successful transactions post-recovery.

System performance metrics, including throughput and latency, remained stable throughout the experiment, as depicted in Fig. 13. The experiment confirms that network design can sustain node loss without performance degradation, provided there is a fallback mechanism for the calling client or application. Load-balancing

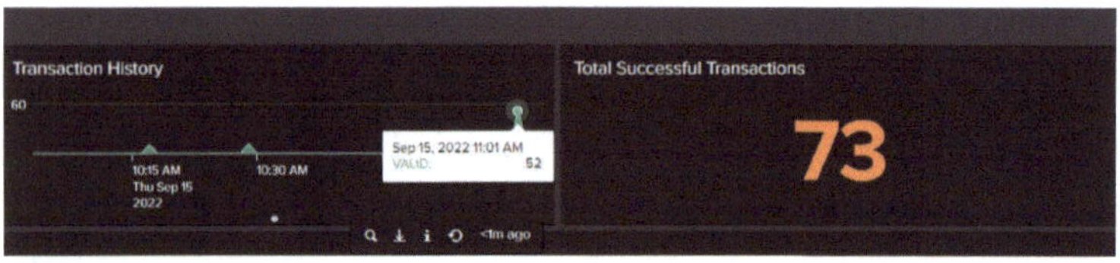

Fig. 10 Transactions prior to loss

Fig. 11 Transactions during loss

Fig. 12 Transactions post recovery

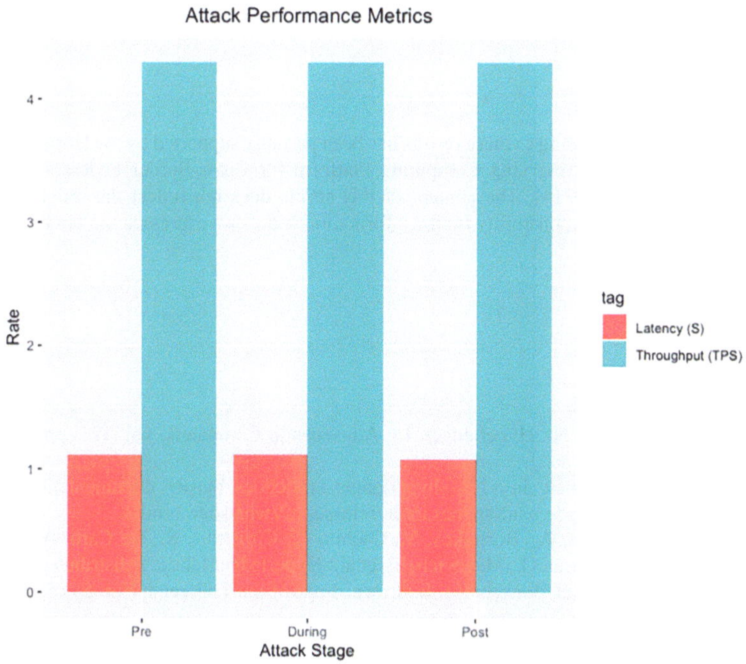

Fig. 13 Database attack performance metrics

systems in organisational networks could effectively monitor peers and distribute workloads.

5 Conclusion

Our work investigated the implementation of DLT to enhance whisky manufacturing in the context of Industry 4.0. This project showcased the promising potential of HLF in whisky manufacturing and emphasised its adaptability and security advantages when properly deployed. Additionally, this work highlighted the performance efficiency of the developed proof of concept (PoC). Despite the limitations imposed by running the network on a single testbed, the study highlighted the indispensable role of a secure front-end system in managing client interactions and transaction processes.

HLF future research can focus on advancing Byzantine Fault Tolerance (BFT) and cryptographic methods and addressing internal threats like DOS attacks for network resilience. Idemix enhancements can be explored for discreet operations in competitive industries, emphasising the need for standardised HLF security frameworks. There is also a need to extend business logic with carbon tracking and IoT integration while ensuring General Data Protection Regulation (GDPR) compliance and data integrity.

Funding The research leading to these results has been partially supported by the Horizon Europe Project Trust & Privacy Preserving Computing Platform for Cross-Border Federation of Data (TRUSTEE), (GA 101070214). The content of this article does not reflect the official opinion of the European Union. Responsibility for the information and views expressed therein lies entirely with the authors.

References

1. Aggarwal, S., Kumar, N.: Hyperledger. In: Advances in Computers, vol. 121, pp. 323–343. Elsevier (2021)
2. Alewine, J.: What's new in v1.2—hyperledger-fabricdocs master documentation (2018). https://hyperledger-fabric.readthedocs.io/en/release-1.2/whatsnew.html
3. Androulaki, E., Barger, A., Bortnikov, V., Cachin, C., Christidis, K., De Caro, A., Enyeart, D., Ferris, C., Laventman, G., Manevich, Y., et al.: Hyperledger fabric: a distributed operating system for permissioned blockchains. In: Proceedings of the Thirteenth EuroSys Conference, pp. 1–15 (2018)
4. Androulaki, E., De Caro, A., Neugschwandtner, M., Sorniotti, A.: Endorsement in hyperledger fabric. In: 2019 IEEE International Conference on Blockchain (Blockchain), pp. 510–519 (2019). https://doi.org/10.1109/Blockchain.2019.00077
5. Awwad, M., Kalluru, S.R., Airpulli, V.K., Zambre, M.S., Marathe, A., Jain, P.: Blockchain technology for efficient management of supply chain. In: Proceedings of the International Conference on Industrial Engineering and Operations Management, pp. 440–449 (2018)
6. Caliper: Hyperledger caliper (2022). https://hyperledger.github.io/caliper/
7. Cañada, A.C.: Implementing a supply chain on the ethereum blockchain. HackerNoon.com (2019). https://medium.com/hackernoon/implementing-a-supply-chain-in-the-ethereum-blockchain-dcc91ea718ab
8. Collier, Z.A., Sarkis, J.: The zero trust supply chain: managing supply chain risk in the absence of trust. Int. J. Prod. Res. **59**(11), 3430–3445 (2021)
9. Dreyer, J., Fischer, M., Tonjes, R.: Towards configuring Hyperledger Fabric 2.0 blockchain platform for Industry 4.0 applications. In: Proceedings—2021 IEEE International Conference on Industry 4.0, Artificial Intelligence, and Communications Technology, IAICT 2021 pp. 241–247 (2021). https://doi.org/10.1109/IAICT52856.2021.9532547
10. Ell, M.: Cyber security breaches survey 2022 (2022). https://www.gov.uk/government/statistics/cyber-security-breaches-survey-2022/cyber-security-breaches-survey-2022
11. etcd: etcd/raft (2021). https://github.com/etcd-io/etcd/tree/main/raft
12. Flatt, H., Schriegel, S., Jasperneite, J., Trsek, H., Adamczyk, H.: Analysis of the cyber-security of industry 4.0 technologies based on rami 4.0 and identification of requirements. In: 2016 IEEE 21st International Conference on Emerging Technologies and Factory Automation (ETFA), pp. 1–4 (2016). https://doi.org/10.1109/ETFA.2016.7733634
13. Ghadge, A., Kara, M.E., Moradlou, H., Goswami, M.: The impact of industry 4.0 implementation on supply chains: Ims. J. Manuf. Technol. Manag. **31**(4), 669–686 (2020)
14. Global, E.Y.: Birra Peroni is the first industrial organization to mint unique non-fungible tokens using EY OpsChain traceability (2021). https://www.ey.com/en_gl/news/2021/05/birra-peroni-is-the-first-industrial-organization-to-mint-unique-non-fungible-tokens-using-ey-opschain-traceability
15. Government, S.: The Scotch Whisky Regulations 2009, vol. 2890 (2009). https://www.legislation.gov.uk/uksi/2009/2890/made

16. Gupta, R., Phanden, R.K., Sharma, S., Srivastava, P., Chaturvedi, P.: Security in manufacturing systems in the age of industry 4.0: Pitfalls and possibilities. In: Advances in Industrial and Production Engineering, pp. 105–113. Springer (2021)
17. Hyperledger: Minifabric (2020). https://github.com/hyperledger-labs/minifabric
18. Liao, Y., Deschamps, F., Loures, E.d.F.R., Ramos, L.F.P.: Past, present and future of industry 4.0—a systematic literature review and research agenda proposal. Int. J. Prod. Res. **55**(12), 3609–3629 (2017)
19. Longo, F., Nicoletti, L., Padovano, A., d'Atri, G., Forte, M.: Blockchain-enabled supply chain: an experimental study. Comput. Ind. Eng. **136**, 57–69 (2019). https://doi.org/10.1016/j.cie.2019.07.026
20. Malik, S., Kanhere, S.S., Jurdak, R.: ProductChain: scalable blockchain framework to support provenance in supply chains. In: 2018 IEEE 17th International Symposium on Network Computing and Applications (NCA), pp. 1–10. IEEE (2018). https://doi.org/10.1109/NCA.2018.8548322
21. Marchese, A., Tomarchio, O.: An agri-food supply chain traceability management system based on hyperledger fabric blockchain. In: Proceedings of the 23rd International Conference on Enterprise Information Systems, pp. 648–658. SCITE—Science and Technology Publications (2021). https://doi.org/10.5220/0010447606480658
22. Miller, G.H.: Whisky Science: A Condensed Distillation. Springer (2019)
23. Mitra, A.: TimeFabric: Trusted Time for Hyperledger Fabric (2021). https://uwspace.uwaterloo.ca/handle/10012/16784
24. Moeuf, A., Pellerin, R., Lamouri, S., Tamayo-Giraldo, S., Barbaray, R.: The industrial management of SMES in the era of industry 4.0. Int. J. Prod. Rese. **56**(3), 1118–1136 (2018)
25. Mussomeli, A., Delesalle, P., Kilpatrick, J.: The new supply chain equilibrium. Deloitte (2022). https://www2.deloitte.com/uk/en/insights/focus/supply-chain/supply-chain-agility-efficiency.html
26. Papadopoulos, P., Buchanan, W.J., Sayeed, S., Pitropakis, N.: Towards the creation of the future fish farm. arXiv preprint arXiv:2301.01618 (2023)
27. Ryan, M.: Ransomware Revolution: The Rise of a Prodigious Cyber Threat. Springer (2021)
28. Sajja, P.: Food Traceability on Blockchain (2022). https://www.farmtoplate.io/
29. Saraf, C., Sabadra, S.: Blockchain platforms: a compendium. In: 2018 IEEE International Conference on Innovative Research and Development (ICIRD), pp. 1–6. IEEE (2018)
30. Senthil, N.: Metrics reference—hyperledger-fabricdocs main documentation (2020). https://hyperledger-fabric.readthedocs.io/en/release-2.2/metrics_reference.html
31. Shahid, A., Almogren, A., Javaid, N., Al-Zahrani, F.A., Zuair, M., Alam, M.: Blockchain-based agri-food supply chain: a complete solution. IEEE Access **8**, 69230–69243 (2020). https://doi.org/10.1109/ACCESS.2020.2986257
32. Stamatellis, C., Papadopoulos, P., Pitropakis, N., Katsikas, S., Buchanan, W.J.: A privacy-preserving healthcare framework using hyperledger fabric. Sensors **20**(22), 6587 (2020)
33. Sunyaev, A.: Distributed ledger technology. In: Internet Computing, pp. 265–299. Springer (2020)
34. Tian, F.: A supply chain traceability system for food safety based on HACCP, blockchain & internet of things. In: 2017 International Conference on Service Systems and Service Management, pp. 1–6. IEEE (2017)
35. Union, I.T.: Security Aspects of Distributed Ledger Technologies (2021). https://figi.itu.int/wp-content/uploads/2021/04/Security-Aspects-of-Distributed-Ledger-Technologies-1.pdf
36. Yang, G., Lee, K., Lee, K., Yoo, Y., Lee, H., Yoo, C.: Resource analysis of blockchain consensus algorithms in hyperledger fabric. IEEE Access **10**, 74902–74920 (2022)

Implementation of Zones and Conduits in Industrial Control and Automation Systems

Lars Halvdan Flå, Mary Ann Lundteigen, Fredrik Gratte, and Martin Gilje Jaatun

Abstract Despite being established concepts in standards we argue that zones and particularly conduits can benefit from more detailed discussions of their architecture and implementation. In this paper we make three contributions towards this. Firstly, we describe detailed principles for implementing conduits. Secondly, we outline a process for connecting zones with potentially different Security Levels (SLs), expressed in the form of a flow chart. Thirdly, we discuss a few highlighted challenges related to the application of zones and conduits in practice.

1 Introduction

The IEC 62443 standards series describes cybersecurity of industrial automation and control systems (IACS), and based on our earlier survey [1], the IEC 62443 standard is the only quantifiable approach currently in use for determining and validating SLs in IACS.

Two key concepts introduced by IEC 62443 are zones and conduits. Establishing zones and conduits, i.e., grouping assets based on their security requirements, is a key activity when securing industrial control systems in accordance with IEC 62443. However, the details surrounding the architecture of these concepts seem to have

L. H. Flå (✉) · M. G. Jaatun
SINTEF Digital, Trondheim, Norway
e-mail: lars.flaa@sintef.no

M. G. Jaatun
e-mail: martin.g.jaatun@sintef.no

M. A. Lundteigen
Department of Engineering Cybernetics, Norwegian University of Science and Technology, Trondheim, Norway
e-mail: mary.a.lundteigen@ntnu.no

F. Gratte
Atina, Athens, Greece
e-mail: fredrik.gratte@atina.no

M. G. Jaatun et al. (eds.), *Proceedings of the International Conference on Cybersecurity, Situational Awareness and Social Media*, Springer Proceedings in Complexity,
https://doi.org/10.1007/978-981-96-0401-2_20

received less attention than one could imagine, given how established these concepts are. This holds particularly true in the case of conduits.

Inspired by interactions with the industry, the overall objective of this paper is to highlight challenges in using the zones and conduits paradigm. Furthermore, we aim to sketch a practical approach to defining the architecture of zones and conduits. More specifically, in this article we:

1. suggest principles for zone and conduit architecture
2. suggest principles for connecting zones with different SLs
3. highlight challenges with using zones and conduits in practice

The remainder of this article is structured as follows: Sect. 2 presents related work. Section 3 presents relevant concepts from the IEC 62443 standard. Section 4 describes proposed architectural principles for implementing conduits. Section 5 presents a process for the secure connection of zones with different SLs. Section 6 discusses practical challenges of implementing zones and conduits in practice. Section 7 describes further work, while Sect. 8 concludes the paper.

2 Related Work

Arguably the contribution that has had the most influence on IACS architecture for cybersecurity is the Purdue model. This model originated from Purdue University in the 1990s, and despite not being created with cybersecurity in mind, it has seen wide adoption within cybersecurity for industrial control systems. The model assigns systems and components into vertical layers. In some cases, it also indicates how vertical layers can be further horizontally segmented, but this is often limited to instances where one enterprise has several factories or plants (see for instance Fig. 18 in IEC 62443-1-1:2009 [2]).

In addition to the Purdue model, there are also several whitepapers and product supplier reference guides for IACS cybersecurity design. An example of a product supplier reference guide is the Industrial Automation Security Design Guide 2.0 from Cisco, which has a chapter on segmenting the network into smaller trust zones [3]. This document (which refers to the IEC 62443 standard) defines a zone as "a collection of physical and functionally united assets that have similar security requirements". In addition to zone characteristics already known from IEC 62443, they additionally specify that the border is used to define access with another zone or outside system. A conduit is described as something that supports and defines allowed communication between two or more zones, with attributes that define which zones are interconnected by the conduit, type of data flow allowed, and security policies and procedures. With regards to the segmentation of OT networks, the document argues that area zones provide a good starting point. After visibility into plant operations has been obtained, the initial segmentation can later be extended by further segmentation within Virtual Local Area Network (VLAN) segments. Another interesting aspect of segmentation raised in the document is the potential complexity. While discussing a particular security measure, they give an example of a situation where one must consider 400 area zones. We discuss this further in Sect. 6.

Combining zones and conduits within a Purdue reference architecture may be regarded as a defence in depth strategy. For example, DesRuisseaux [4] argues that proper defence in depth involves six steps. Step two involves separating networks by major function, and as example lists that a network can be divided into enterprise, plant, process, and field zones. Subsequently, all conduits between zones should be identified. In step three, named perimeter protection, these conduits should be protected. In step four, named network segmentation, existing zones can be divided into smaller zones. This division can be based on location or function.

Leander et al. [5] discuss the applicability of IEC 62443 in light of the expected transition to Industry 4.0 and increased use of Industrial Internet of Things (IIoT) devices. Specifically, regarding IEC 62443 zones, they argue that this concept may be challenged by IIoT devices relying on for instance cloud access and by the dynamic nature of Software Defined Networking.

Most requirements and discussions are made common for zones and conduits, without clarifying how the two types of segmentation are treated differently in practice. One exception is the European technical specification CLC/TS 50701 [6] that has an annex (A) dedicated to handling conduits. Here it is suggested that there are only three types of conduits:

1. Conduits implementing a transparent (i.e., logical) gateway connecting zones of the same SL.
2. Filtering conduits as firewall appliance, allowing a zone of lower or equal security to communicate with a zone of a higher SL.
3. Unidirectional conduit as a data diode or network tap, allowing output from a higher SL zone to others.

Here, the conduits may relate to logical functions implemented with the network devices and not necessarily the hardware involved. The technical specification comments that IEC 62443 does not clearly advice on whether network devices should be part of zones or conduits, and an interpretation is that a network device acting as a boundary protection of a zone can belong to the zone as well as the conduit.

Soderi et al. [7] have interpreted the practical application of CLC/TS 50701 when the conduit is a wireless communication within a train control system. Here, they assume that the connected zones have the same SL, meaning that a transparent gateway is the preferred type of conduit. In a wireless communication, additional measures are needed to achieve transparency when exposed to cyber attacks. Their solution is to use a host identify protocol (HIP) that provides transparency with end-to-end encryption, mutual authentication, and internet protocol security.

Kern et al. [8] have modelled and made an UML like visualization of an architecture of zones, data flows, and their related attributes and functions, following the basic requirements in IEC 62443. In their work, they consider conduits as the means of transporting data flows and they argue that the SL assigned to network components is determined by the highest SL assigned to the data flow that runs through them.

The target SL of a zone is determined by the highest SL of the network components within that zone.

Schlehuber et al. [9] propose an application layer gateway (ALG) to tackle the challenge when conduits are connecting zones with different SL requirements. An ALG can be configured to allow the desired type of connection (or layer) to connect specific zones. The configuration may include whitelist filtering for each of the implemented layers. The ALG is also envisioned to be able to forward traffic to a Security Operations Centre for analysis. This solution of including an ALG implies that the same conduit can be used to provide different levels of security, depending on the direction of the data flow.

3 IEC 62443 and Its Application

The IEC 62443 standard takes a holistic view on the cybersecurity of IACS, covering the technical, procedural and people aspects, for the different roles involved in the supply chain. These roles include the product supplier, service provider, and asset owner. The standard has been in development since 2009, and an overview of its different parts can be seen in Table 1. We now proceed to introduce a few selected concepts from the standard.

3.1 Application of Zones

IEC 62443 has introduced *zones* and *conduits* as means to group assets in a network architecture that share common security requirements. Unfortunately, IEC 62443 does not limit itself to only one definition of the term zone, and variants have been introduced with new parts of the standard being published, as shown in Table 2. A conduit is a special case of a zone, where the primary role is to group assets that play a role in connecting two or more zones.

Many industries apply a layered network architecture similar to the Purdue model, and IEC 62443 may have, for this reason, adopted this model as a reference architecture. The Purdue model splits a plant's network and systems hierarchically into 5-6 layers, with a clear separation of the operational technology (OT) side for plant operation, monitoring, data acquisition; and the IT side with the office networks and externally connected cloud solutions and applications. Each layer groups networks and assets that share common operational functions.

The cyber-risk assessment determines to what extent zones have similar boundaries as the Purdue model layers. For example, layer 2 with human machine interface may end up with the same security requirements, while layer 1 may decide on stricter security requirements for safety controllers compared to process controllers. Plants with a large outreach and many distributed systems, may group assets that belong to

Table 1 The IEC 62443 series

Number	Name	Year	Status
62443-1-1	Terminology, concepts, and models	2009	Published (newer draft exists)
62443-1-3	System security compliance metrics	2021	Draft
62443-2-1	Establishing an industrial automation and control systems security program	2010	Published (newer draft exists)
62443-2-2	IACS security program ratings	2022	Draft
62443-2-3	Patch management in the IACS environment	2015	Published (newer draft exists)
62443-2-4	Security program requirements for IACS service providers	2019	Published
62443-3-1	Security technologies for industrial automation and control systems	2009	Published
62443-3-2	Security risk assessment for system design	2020	Published
62443-3-3	System security requirements and security levels	2019	Published
62443-4-1	Secure product development lifecycle requirements	2018	Published
62443-4-2	Technical security requirements for IACS components	2019	Published

more than one layer into the same zone. A specific zone can therefore include some or all assets within a layer, or alternatively include assets in several layers.

3.2 *Application of Conduits*

A conduit is a "particular type of security zone" introduced in IEC 62443-1-1:2009 [2]. The term is defined in several parts of the IEC 62443 standards, as listed in Table 3. In summary, we can say that a conduit is a grouping of communication assets or channels. The purpose of a conduit is to secure the communication

Table 2 IEC 62443 definitions of a zone

Standard	Definition
62443-1-1 (2009)	Zone: grouping of logical or physical assets that share common security requirements NOTE: A zone has a clear border with other zones. The security policy of a zone is typically enforced by a combination of mechanisms both at the zone edge and within a zone. Zones can be hierarchical in the sense that they can be comprised or a collection of sub-zones
62443-3-2 (2020)	Zone: grouping of logical or physical assets based upon risk or other criteria, such as criticality of assets, operational function, physical or logical location, required access (for example, least privilege principles) or responsible organization NOTE: Collection of logical or physical assets that represents partitioning of a system under consideration on the basis of their common security requirements, criticality (for example, high financial, health, safety, or environmental impact), functionality, logical or physical (including location) relationship
62443-3-3 (2019) 62443-4-2 (2019)	Zone: grouping of logical or physical assets that share common security requirements NOTE: A zone has a clear border. The security policy of a zone is typically enforced by a combination of mechanisms both at the zone edge and within the zone
62443-4-1 (2018)	Zone: collection of entities that represents partitioning of a System under consideration on the basis [of] their functional, logical and physical (including location) relationship NOTE: Zones are often created on the basis of common security requirements, criticality (e.g., high financial, health, safety, or environmental impact), functionality, logical or physical (including location) relationship

exchange from one zone to another. However, apart from IEC 62443-1-1:2009 [2], the other parts of the standard do not to any large degree discuss the concept of conduits. Based on part 1-1, a conduit appears to include physical communication assets such as wires, routers, and network management devices. If we assume that the end points of the conduit are also a part of the conduit, the contents widen to also include the physical devices and applications using the channels in the conduit. This view can be supported by the example given of a conduit being the communication channels within a single computer.

Table 3 62443 definitions of a conduit

Standard	Definition
62443-1-1 (2009)	Conduit: logical grouping of communication assets that protects the security of the channels it contains NOTE: This is analogous to the way a physical conduit protects cables from physical damage
62443-3-2 (2020)	Conduit: a logical grouping of communication channels that share common security requirements connecting two or more zones [the standard defines a channel as *"specific logical or physical communication link between assets"*, with the associated note *"A channel facilitates the establishment of a connection"*]
62443-3-3 (2019) 62443-4-2 (2019)	Conduit: logical grouping of communication channels, connecting two or more zones, that share common security requirements NOTE: A conduit is allowed to traverse a zone as long as the security of the channels contained within the conduit is not impacted by the zone

Conduits do not appear to be as thoroughly discussed by the standard as is the case for zones. For zones IEC 62443-3-2:2020 [10] has clear requirements on what types of equipment and systems should be placed in separate zones, and IEC 62443-3-3:2019 [11] provides clear requirements for how a zone can achieve the different SLs. Adding to this, the established Purdue model provides a starting point for how the larger network can be segmented into different areas.

Given that a conduit is a special type of zone one can assume that many of the requirements applicable to zones are also applicable to conduits. However, since the standard does introduce the term conduit as a particular type of security zone, we believe more clarification of exactly what constitutes the "particular" aspect of it could be useful.

3.3 Application of SLs

In this paper we use the IEC 62443 definitions of SL, and specifically the definitions shown in Table 4. However, the definition of SLs can vary slightly between different sources, and we therefore start our discussion of SLs by referring to some of these definitions. The 62443-3-3:2013 [11] definition of SL is "measure of confidence that the IACS is free from vulnerabilities and functions in the intended manner". The 62443-3-2:2020 [10] uses almost the exact same definition, but replaces "IACS"

Table 4 SLs as defined in 62443-3-2:2020 [10] Annex A

SL	Description
1	Protection against casual or coincidental violation
2	Protection against intentional violation using simple means with low resources, generic skills and low motivation
3	Protection against intentional violation using sophisticated means with moderate resources, IACS specific skills and moderate motivation
4	Protection against intentional violation using sophisticated memes with extended resources, IACS specific skills and high motivation

with "SuC". The Global Cybersecurity Alliance [12] uses the same definition, but also include zones and conduits, in addition to SuC.

What further complicates the matter is that Annex A of part 3-2 defines the four SLs as protection against violations from increasingly sophisticated attackers, as shown in Table 4. These definitions, focusing on level of protection, are also in line with Gordon [13] and Kobes [14], who states that "security level values express the security capabilities of the automation solution".

The IEC 62443 furthermore operates with three types of SL, as shown below. However, for the discussion in this paper, SL-T is the most relevant.

SL-C: Capability Security Level (in product or component)
SL-T: Target Security Level (defined by requirements and specifications)
SL-A: Achieved Security Level (the actual security level achieved "as built")

Determining SL-T is covered by zone and conduit requirement (ZCR) 5.6 [10]. This requirement states that an SL target shall be established for zones and conduits. A corporate risk matrix with tolerable risk is used as input for the requirement. The standard does not detail how the SL shall be established and instead states that there is no "prescribed method for establishing SL-T". It does however describe some potential approaches for determining SL-T, including basing it on the definition of SL, and basing it on unmitigated cyber risk and the organization's tolerable level of risk.

4 Allocation of Assets to Zones and Conduits

In this section, we present architectural principles for establishing conduits and associated zones. The process of creating zones and conduits is defined in ZCR 3 in 62443-3-2:2020 [10]. The overarching principle for dividing an IACS into zones and conduits is the risk assessment. However, we believe it could be beneficial with a more detailed discussion of how to best manage the connection between zones and conduits. As a contribution towards this, we suggest principles for the detailed architecture of determining the structure of zones and conduits. The guiding idea is that:

- A conduit does not implement any connections of its own, it merely transports connections implemented by others.
- All functionality for configuring the conduit is placed in one or more network device configuration zones. However, the conduit itself is not a zone, as it has no configuration interface.

Following this, we assume that a conduit and the associated network device configuration zone(s) have two main responsibilities:

- Protection and containment of zones from each other, by dropping all unintended connections between zones.
- Protection of the availability, integrity and confidentiality of allowed connections between zones.

Based on these principles, we believe that communication assets connecting zones should be split into a data plane, which we consider to be the conduit, and a control plane, which is managed through the different network device configuration zones. As an example, a firewall typically consists of a set of rules applied to incoming traffic, and an interface/application for managing and defining such rules. The rules will be applied to the communication in the data plane, while the changing, adding, and deleting rules will happen through the interface placed in the network device configuration zone.

Another example can be that of a Virtual Private Network (VPN) set up between two routers in different zones. In this example, we believe that the conduit should be the data pipe set up by routing rules along the way between the two zones, and the VPN endpoints providing encryption. The network device configuration zones will include the software used to establish and configure the VPN, along with the configuration interfaces for any network devices between the two zones. In the case of a physical level data diode this will not be part of any network device configuration zone, assuming that it cannot be configured in any way.

We illustrate these concepts in Fig. 1. Zone A and B are both connected to a router, and communicate through a conduit set up by this router (indicated by the green arrow). The router can for instance create this conduit by placing the connections to Zone A and Zone B on a dedicated Virtual Local Area Network (VLAN). The security of the conduit and the connected zones can further be enhanced by applying firewall rules or intrusion detection analysis to the data flowing through the conduit.

The router can implement or be part of several different conduits. For instance if Zone B requires a remote connection, another conduit (indicated by the orange arrow) can be established by creating a new VLAN to facilitate this connection. While the green conduit between zone A and B is primarily managed from the router in Fig. 1, the orange conduit may pass through several routers and switches, each with a network device configuration zone.

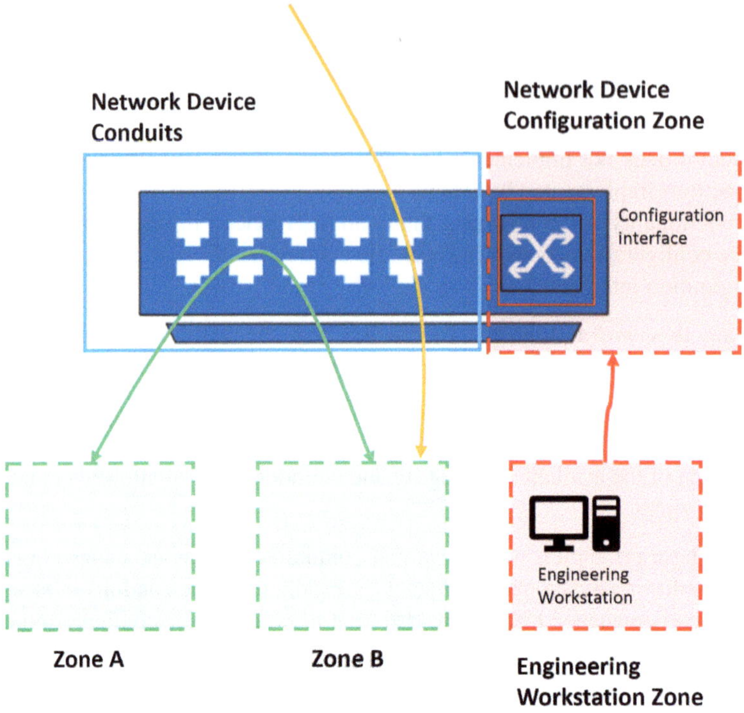

Fig. 1 In the case of a router, the data being transmitted is placed in the conduit, while configuration of the router is placed in the Network Device Configuration Zone

In Fig. 1, we have placed the router configuration in its own zone, although it can also be part of a larger zone including other devices. The same applies to the engineering workstation used to configure the router. This can either be placed in its own zone or in a larger zone. Furthermore, we argue that two aspects related to this are of particular importance. The first is to harden the network device interface, so that only the minimum required number of engineering workstations can reach and configure it. The second is to ensure that the engineering workstation has minimum the same SL as the network device configuration zone.

Following our suggestion to place all configuration interfaces into network device configuration zones, we believe it only makes sense to assign an SL to these configuration zones. Using these principles the conduit will be reduced to a data pipe which does not have any interface or functionality to secure by itself, since this is moved to dedicated configuration zones.

5 Securing Conduits Connecting Zones with Different SLs

As stated in ZCR 5.6 in IEC 62443-3-2:2020 [10], an SL shall be allocated to zones and conduits. In this section we propose a process for securing a conduit, and the associated network device configuration zones, when the conduit is connecting zones with potentially different SLs. We express the process in the form of a flow chart, and the steps involved in the process are as follows:

1. In the first step, we evaluate if all zones connected by the conduit have the same SL.
2. If all involved zones have the same SL, we assign the same SL to the network device configuration zones, provided that the existing SL of the network device configuration zone is not higher. As many conduits may go through a network device, the SL of the network device configuration zone should be the highest of any zone connected by any of the conduits. Once this step is completed, the process terminates.
3. In the case that the involved zones have different SLs, we assign the highest SL to the network device configuration zones. Similarly as for step 2, we account for the scenario where a previously analysed conduit has resulted in a higher SL for the network device configuration zones.
4. In step 4, we enumerate all data flows or communication channels that are part of the conduit.
5. In step 5, we evaluate whether any of the data flows perform write operations from zones with lower levels of criticality or lower integrity requirement to zones with higher levels of criticality or integrity requirements.
6. If this is the case, one should assess the potential risk if this data flow is compromised by an attacker and, if necessary, implement suitable countermeasures. Examples may include proxy servers limiting what interfaces the zone exposes, or intrusion detection systems flagging suspicious traffic.
7. In the step 7, we check if there are remaining data flows to be evaluated, and if so return to step 5. Otherwise, the process terminates.

An interesting situation arises when a conduit connects zones with different SLs, particularly because we assign the responsibility of protecting access to a zone to the conduits (or more precisely to the zones configuring the conduits). We reason that these network device configuration zones will need to have an SL equal to the highest SL of the zones they are tasked with protecting.

This approach may in some ways be different from the IEC 62443 standard [2], which seemed to cater for scenarios where the conduit has a lower SL than the zone. In this case the conduit is labelled as untrusted, according to part 1-1 (2009) [2]. Following this scenario where the conduit is untrusted, the communication security becomes the responsibility of the individual channel. However, we questioned the usefulness of implementing a conduit in the first place if the channels passing through it will have to provide their own security (Fig. 2).

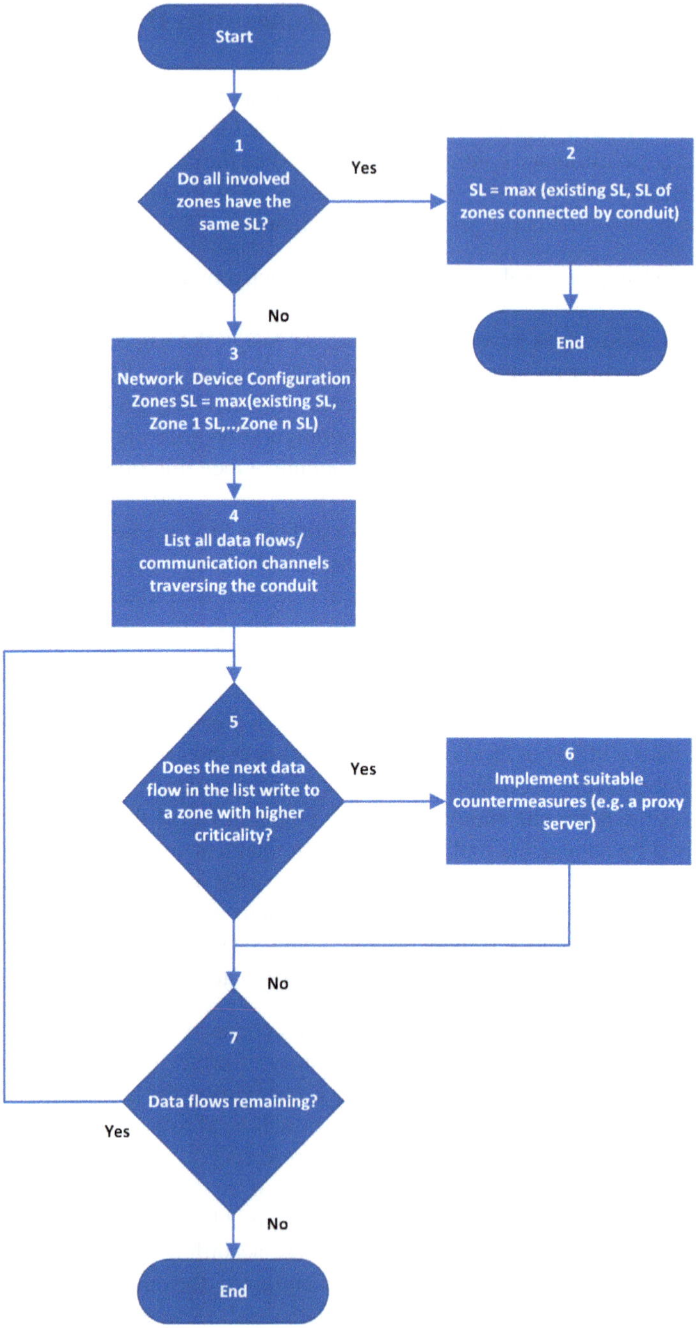

Fig. 2 The seven elements of the proposed process for securing conduits between zones with different SLs

With our approach, and in line with the commonly accepted philosophy for ICS security where availability and integrity is regarded as more important than confidentiality, we believe that the network device configuration zones should ensure that the availability and integrity of higher SL zones cannot be compromised from lower SL zones. We note that this may imply further security controls in addition to segmentation and boundary protection. When discussing zones and conduits, much of the security discussion tends to revolve around such security controls. While segmentation and boundary protection are efficient security controls, a discussion of the security of zones and conduits must not overlook the aspect of what trust is placed in the data flowing through conduits and between zones. As an example, we can assume to have a server in an SL 1 zone, and a client in an SL 4 zone. While segmentation and boundary protection are suitable for protecting against unintended connections, it may not protect against misuse of legitimate connections.

Similarly, while a VPN connection will protect the confidentiality and integrity of data transmitted across the network, it will not protect against an attacker who has obtained a foothold on the host where the VPN connection originates.

To also defend against these scenarios where an attacker is able to exploit a legitimate connection, it may be necessary to set up intrusion detection functionality on the network devices implementing the conduits, or other methods for detecting or preventing malicious activity.

6 Practical Challenges of Using Zones and Conduits

In this section we highlight what we perceive to be practical challenges with using zones and conduits in practice.

In addition to being a grouping of assets, a zone shall have a set of attributes, according to 62443-1-1 (2009) [2]. These attributes are: security policies, asset inventory, access requirements and controls, threats and vulnerabilities, consequences of a security breach, authorised technology (i.e., a dynamic list of technologies allowed and not allowed in the zone), and a change management process. An aspect of this which should not be overlooked is to what extent this is scalable. As an example, an oil and gas installation may implement 30–50 zones; for cases where a company has many installations, the number of zones quickly grows large. Another example is the Industrial Automation Security Design Guide 2.0 from Cisco, introduced in Sect. 2, which speaks of scenarios with up to 400 area zones.

We would furthermore argue that the approach of listing threats and vulnerabilities for each zone is problematic for three reasons: it is economically infeasible, it has previously shown to offer limited value for the effort invested, and by the time an installation is in operation the list developed during the design phase may already be outdated.

Regarding the allocation of SLs to zones and conduits, we believe it would be beneficial with more guidance regarding how the overall architecture should influence the selection of the SL for a specific zone or conduit. If one fails to take advantage of the fact that zones in lower parts of the Purdue model are typically protected by zones higher up in the Purdue model, the result may be overly expensive and inefficient security. As an example, the most critical components are typically at the bottom of the Purdue model, and consist of embedded devices with limited features. In cases where these are of high criticality, there have been attempts in the industry to give these high SL values (SL-3 or SL-4), with the result of a lot of non-compliance due to missing features, or due to undesirable connectivity with other components. Provided that the Purdue model is properly implemented, the embedded devices on the bottom of the model are significantly protected by the surrounding networks and components, and should be left with limited responsibilities to protect themselves. Enforcing large numbers of requirements on embedded devices will push the suppliers to replace these devices with more standardized computer architectures and operating systems, to fulfil the requirements. This is not desirable as it leads to devices with increased need of anti-malware solutions, regular / frequent patching etc., which is undesirable for devices with high availability and high integrity demands.

A discussion of zones and particularly conduits should also include a discussion of communication channels, as a conduit is defined as a logical grouping of communication channels. The channel is in turn a specific communication link established between assets. We are of the opinion that concepts such as conduits, channels, communication links, sub-conduits, and sub-zones could benefit from clearer definitions, demarcation, and more examples of implementation. There should also be a clear justification in terms of added value for introducing additional concepts and definition.

Lastly, we believe that ensuring consistent definitions of such terms as zone, conduit and SL across the different part of the IEC 62443 standards would be beneficial for the users of the standard.

7 Further Work

The ideas and proposed principles in this paper are intended as a first proposal to the IACS community, and will have to be further tested and validated together with the industry in real world use cases. This paper should therefore be seen as a first contribution towards establishing detailed and commonly accepted practices for conduit and zone architecture, and subsequent SL allocation.

One particular aspect which we believe can benefit from further study is how to best secure connections between zones with different SLs. This can range from outlining specific recommendations for firewalls and intrusion detection deployment, to developing methods for detecting and preventing misuse of compromised legitimate connections, potentially by relating communication traffic to observed anomalies in the underlying process being controlled.

8 Conclusion

In this article we have investigated the details of conduit architecture and practical considerations of connecting zones with different SLs. Motivated by the widespread use of the zones and conduits paradigm, we propose principles for conduit and zone architecture and a pragmatic process for connecting zones with potentially different SLs. This process is expressed in the form of a flow chart. Lastly we highlight some key challenges with using the zones and conduit paradigm in practice.

Acknowledgements The research in this paper was supported by the Research Council of Norway through the Cybersecurity Barrier Management project (grant no. 326717). We are grateful to the CBM advisory committee for their insightful comments and discussion.

References

1. Øien, K., Hauge, S., Jaatun, M.G., Flå, L., Bodsberg, L.: A survey on cybersecurity barrier management in process control environments. In: Proceedings of 2022 IEEE International Conference on Cloud Computing Technology and Science. IEEE, Bangkok (2022)
2. IEC: IEC/TS 62443-1-1:2009 Security for industrial automation and control systems—Part 1-1: Terminology, concepts and models (2009)
3. Industrial Automation Security Design Guide 2.0—Chapter: Segment the Network into Smaller Trust Zones (2023). https://www.cisco.com/c/en/us/td/docs/Technology/industrial-automation-security-design-guide/m-segment-the-network-into-smaller-trust-zones.html
4. DesRuisseaux, D.: Practical Overview of Implementing IEC 62443 Security Levels in Industrial Control Applications. Schneider Electric Whitepaper (2018). https://technology-signals.com/wp-content/uploads/download-manager-files/78151_PracticalOverviewofImplementingIEC62443SecurityLevelsWhitePaper.pdf
5. Leander, B., Čaušević, A., Hansson, H.: Applicability of the IEC 62443 standard in Industry 4.0/IIoT. In: Proceedings of the 14th International Conference on Availability, Reliability and Security (2019)
6. CENELEC: Railway applications: Cybersecurity. ICLC TS 50701:2023 (2023)
7. Soderi, S., Masti, D.: Cybersecurity considerations for communication based train control: Iinatti. J. IEEE Access **1**, 92312–92321 (2023)
8. Kern, M., Taspolatoglu, E., Scheytt, F., Glock, T., Liu, B., Betancourt, V.P., Sax, E.: An architecture-based modeling approach using data flows for zone concepts in industry 4.0. In: 2020 IEEE International Symposium on Systems Engineering (ISSE), pp. 1–8 (2020). https://doi.org/10.1109/ISSE49799.2020.9272013
9. Schlehuber, C., Heinrich, M., Vateva-Gurova, T., Katzenbeisser, S., Suri, N. (eds.): Computer Safety, Reliability, and Security. SAFECOMP 2017. Lecture Notes in Computer Science (2017). https://doi.org/10.1007/978-3-319-66266-4_21
10. IEC: IEC 62443-3-2:2020 Security for industrial automation and control systems—Part 3-2: Security Risk assessment for system design (2020). https://webstore.iec.ch/publication/30727
11. IEC: IEC 62443-3-3:2013 Industrial communication networks—network and system security—Part 3-3: System security requirements and security levels (2013). https://webstore.iec.ch/preview/info_iec62443-3-3%7Bed1.0%7Den.pdf. ISBN 978-2-8322-1036-9
12. ISA-GCA: Security Lifecycles in the ISA/IEC 62443 Series—Security of Industrial Automation and Control Systems (2020). https://21577316.fs1.hubspotusercontent-na1.net/hubfs/21577316/2022%20ISA%20Website%20Redesigns/ISASecure/PDFs/Miscellaneous

%20PDFs/Documents-Articles-and-Technical-Papers/ISAGCA-Security-Lifecycles-whitepaper.pdf
13. Gordon, J.: The Essential Guide to the IEC 62443 industrial cybersecurity standards (2021). https://industrialcyber.co/features/the-essential-guide-to-the-iec-62443-industrial-cybersecurity-standards/
14. Kobes, P.: Guideline Industrial Security—IEC 62443 is easy. VDE VERLAG GMBH (2023)

Author Index

A
Abidoye, Ademola Philip, 129
Abubakar, Mwrwan, 61, 169
Agyepong, Enoch, 109
Ahmad, Jawad, 209
Ajibade-Ajibosin, Boluwatife, 27
Ali, Hisham, 209
Ani, Uchenna Daniel, 27
Anjali Kumari, 77
Azeez, Nureni Ayofe, 129

B
Bernsmed, Karin, 321
Bleiman, Rachel, 149
Buchanan, William J., 61, 209, 349

C
Christie, Jack, 225
Connell, Jacob, 349

D
Duin, Joe, 169

E
Egbu, Ibeabuchi, 303
Elnour, Se, 61
Elnour, Sirag, 61

F
Flå, Lars Halvdan, 287, 367

G
Gatram Sravan Kumar, 45
Gratte, Fredrik, 367

H
Hanssen, Geir Kjetil, 97

J
Jaatun, Martin Gilje, 97, 287, 321, 367
Josang, Audun, 209

K
Kamalakanta Sethi, 45, 77
Kasimatis, Dimitrios, 349
Keating, Paul, 61
Khan, Zaheer, 225
Kyriacou, Theocharis, 27

L
Legg, Phil, 225
Leitold, Ferenc, 195
Lemoudden, Mouad, 209
Lo, Carol, 225
Lundteigen, Mary Ann, 367

M
Mavroeidis, Vasileios, 209
McKay, Annelize, 303
McKeown, Sean, 169

M. G. Jaatun et al. (eds.), *Proceedings of the International Conference on Cybersecurity, Situational Awareness and Social Media*, Springer Proceedings in Complexity,
https://doi.org/10.1007/978-981-96-0401-2

N
Narasimhan Kovalai, 3
Nzeribe, Ogechukwu Juliet, 129

O
Onwubiko, Cyril, 109
Ophoff, Jacques, 303

P
Padmalochan Bera, 3
Papadopoulos, Pavlos, 349
Pasumarthy, Lakshmi Rama Kiran, 209
Pitropakis, Nikolaos, 349
Piyush Joshi, 45

R
Rakesh Kumar Sanodiya, 45
Ramsundar Tanikella, 3
Rege, Aunshul, 149
Rezaeifar, Zeinab, 225
Rojalini Tripathy, 3

S
Sarigiannidis, Panagiotis, 349
Skjelvik, Alvhild, 265
Skytterholm, Andrea, 287
Solnør, Åse Marie, 321
Sørlien, Silje Marie, 321
Sreevallabh Karanam, 77
Swejan Annabathini, 77
Swinney, Carolyn J., 241

T
Turner, Mark, 27

V
Vyver Van der, Charles, 129

W
Win, Thu Yein, 225
Woods, John C., 241

Y
Yang, Bian, 265

MIX
Papier aus verantwortungsvollen Quellen
Paper from responsible sources
FSC® C105338

If you have any concerns about our products,
you can contact us on
ProductSafety@springernature.com

In case Publisher is established outside the EU,
the EU authorized representative is:
Springer Nature Customer Service Center GmbH
Europaplatz 3, 69115 Heidelberg, Germany

Printed by Libri Plureos GmbH
in Hamburg, Germany